Sustainability Sciences in Asia and Africa

Agroecosystem Dynamics and Sustainable Practices

Sustainability as a scientific endeavor requires broad understanding of the interconnections in our global environment. It is therefore positioned at the crossroads of research in natural sciences, technology and human sciences. This book series supports the global efforts by providing timely coverage of the progress, opportunities, and challenges towards achieving sustainable development goals (SDGs) in Asia and Africa. It fosters the research in transdisciplinary academic fields spanning across agriculture and global food security, health and diseases, access to clean water and affordable energy, climate resilience and effects of global climate change on biodiversity, water and land. This series is planned to narrate the stories of scientific development and progression towards SDGs from the regions of Asia and Africa. The overall aim of the series is to provide a comprehensive publication platform and act as a knowledge engine in the growth of sustainability sciences with a special focus on developing nations. The series constitute of four sub-series each focusing on a specific theme closely linked to a set of SDGs. Each sub-series is leaders and experts in agricultural sciences, environmental sciences, biomedicine and medical sciences respectively. Books in this series are useful for environmentalists, biotechnologists, plant scientists, medical practitioners, conservation biologists, policy makers, NGOs and scientific bodies working in the areas of sustainability sciences in the developing regions.

This book series publishes research associated with the prevailing opportunities and challenges faced by agroecosystems in Asia and Africa. It also explores the future scope of agroecological techniques and offers guidance on sustainable management of these systems. The series discusses important areas such as sustainable agricultural production, soil health, water resource management, biodiversity conservation, animal health and productivity, insect management strategies, genome editing and climate change mitigation. Further, it includes the challenges faced by African agriculture in arid and semi-arid regions, particularly in response to climate change. It highlights issues such as water shortages, soil degradation, desertification, and the need for sustainable farming techniques.

Sustainable agroecosystem maintenance entails strategies that improve overall environmental health and economic viability while limiting negative environmental impacts. The series takes a multidisciplinary approach to investigate creative ways for boosting sustainable agriculture, increasing food security, and improving lives for people in global south. This book series acts as a global publishing platform to collate and disseminate knowledge on fostering resilient and productive agroecosystems, thus enhancing accessibility, and fostering informed decision-making.

Jen-Tsung Chen
Editor

Nanobiosensors for Crop Monitoring and Precision Agriculture

Springer

Editor
Jen-Tsung Chen
Department of Life Sciences
National University of Kaohsiung
Kaohsiung, Taiwan

ISSN 2730-6771 ISSN 2730-678X (electronic)
Sustainability Sciences in Asia and Africa
ISSN 3059-247X ISSN 3059-2488 (electronic)
Agroecosystem Dynamics and Sustainable Practices
ISBN 978-981-96-8334-5 ISBN 978-981-96-8335-2 (eBook)
https://doi.org/10.1007/978-981-96-8335-2

This Springer imprint is published by the registered company Springer Nature Singapore Pte Ltd.
The registered company address is: 152 Beach Road, #21-01/04 Gateway East, Singapore 189721, Singapore

Preface

Nanobiotechnology has been intensively investigated for decades and is thus well recognized as a promising approach for advancing sustainable agriculture, particularly providing precision and smart tools based on engineered nanomaterials/nanoparticles (NPs). Among NPs, nanosensors and nanobiosensors have the potential to revolutionize agricultural practices by sensing, detecting, and monitoring cellular processes such as signal transduction and plant hormone dynamics, crop performances, and soil/environmental conditions for precision and even real-time management of fertilizers, pesticides, plant growth/yield/quality, stress responses, and post-harvest processes.

This book offers comprehensive and updated knowledge for readers interested in materials science and nanobiotechnology, particularly focusing on the detailed aspects of nanobiosensors for agri-food applications. It provides a complete guide to nanobiosensors, from their fundamentals to current agricultural applications, through a series of review chapters organized by various experts in the field. The book includes subtopics presented in individual chapters that cover optical nanobiosensors, calorimetric nanobiosensors, immuno-nanobiosensors, enzymatic nanobiosensors, and apta-nanobiosensors, summarizing their synthesis, functions, and current agricultural applications. Additionally, an updated overview of ethical considerations and regulations regarding nanotechnology in agri-food applications is included.

The book editor, Dr. Jen-Tsung Chen, thanks all contributors for their innovative and insightful chapters. The support and guidance from the publisher are greatly appreciated.

Kaohsiung, Taiwan — Jen-Tsung Chen

Contents

Editor and Contributors

About the Editor

Jen-Tsung Chen is Professor of Cell Biology in the Department of Life Sciences at the National University of Kaohsiung, Taiwan, where he teaches courses in cell biology, genomics, proteomics, plant physiology, and plant biotechnology. His research focuses on bioactive compounds, plant molecular biology, bioinformatics, and systems pharmacology. He serves on the editorial boards of several prestigious journals, including Plant Methods and GM Crops & Food. He has authored multiple books on drug discovery, herbal medicine, nanotechnology, plant biotechnology, functional genomics, and CRISPR-based plant genome editing. In 2023 and 2024, he was recognized as one of the "World's Top 2% Scientists" by Elsevier and Stanford University.

Contributors

Banzeer Ahsan Abbasi Department of Botany, Rawalpindi Women University, Rawalpindi, Pakistan

Maryam Afzaal Department of Biochemistry, Government College University, Faisalabad, Punjab, Pakistan

Shivangi Agarwal Department of Microbiology, Kanya Gurukul, Gurukula Kangri (Deemed to be University), Haridwar, Uttarakhand, India

Ayesha Akbar Department of Chemistry, Government College University, Faisalabad, Punjab, Pakistan

Jameel M. Al-Khayri Department of Agricultural Biotechnology, College of Agriculture and Food Sciences, King Faisal University, Al-Ahsa, Saudi Arabia

Julio C. Anchondo Páez Centro de Investigación en Alimentación y Desarrollo, Chihuahua, México

Prodipto Bishnu Angon Faculty of Agriculture, Bangladesh Agricultural University, Mymensingh, Bangladesh

Institute of Agronomy, Hungarian University of Agriculture and Life Sciences (MATE), Gödöllő, Hungary

Anshu Department of Medical Oncology, All India Institute of Medical Sciences, Delhi, India

Shamima Nasrin Asha Breeding Division, Bangladesh Sugarcrop Research Institute, Pabna, Bangladesh

Sanam Ashraf Department of Plant Sciences, Faculty of Biological Sciences, Quaid-i-Azam University, Islamabad, Pakistan

Jyoti Bala Department of Pharmaceutical Sciences and Technology, Maharaja Ranjit Singh Punjab Technical University, Bathinda, Punjab, India

Madhumita Barooah Department of Agricultural Biotechnology, Assam Agricultural University, Jorhat, Assam, India

Beda Saurav Behera P. G. Department of Biosciences and Biotechnology, Fakir Mohan University, Balasore, Odisha, India

Amit Bhatia Department of Pharmaceutical Sciences and Technology, Maharaja Ranjit Singh Punjab Technical University, Bathinda, Punjab, India

Rohit Bhatia Chitkara College of Pharmacy, Chitkara University, Rajpura, Punjab, India

Samina Bibi Department of Plant Sciences, Quaid-i-Azam University, Islamabad, Pakistan

Sudipta Sankar Bora Silchar Regional Forensic Science Laboratory, Ghungur, Silchar, Assam under the Directorate of Forensic Science, Kahilipara, Guwahati, Assam, India

Wadie Chalgham Department of Mechanical and Aerospace Engineering, University of California, Los Angeles, CA, USA

Fung Swrangshee Daimari Department of Botany, Boroland University, Deborgaon, Assam, India

Brijnandan Singh Dehiya Department of Chemical Engineering, Deenbandhu Chhotu Ram University of Science and Technology, Murthal, Haryana, India

Mohsen Mohamed Elsharkawy Agricultural Botany Department, Faculty of Agriculture, Kafrelsheikh University, Kafr Elsheikh, Egypt

Samreen Fatima Department of Biological Sciences, University of Baltistan, Gilgit Baltistan, Pakistan

Cristina L. Franco Lagos Centro de Invesstigación en Alimentación y Desarrollo, Chihuahua, México

Kuldeep Giri Department of Computer Science and Engineering, Gurukula Kangri (Deemed to be University), Haridwar, Uttarakhand, India

Sadhana Giri Department of Microbiology, Kanya Gurukul, Gurukula Kangri (Deemed to be University), Haridwar, Uttarakhand, India

Rahul Gogoi Department of Agricultural Biotechnology, Assam Agricultural University, Jorhat, Assam, India

Vinod Goyal Department of Botany and Plant Physiology, CCS Haryana Agricultural University, Hisar, Haryana, India

Zubaria Haakim Department of Plant Sciences, Faculty of Biological Sciences, Quaid-i-Azam University, Islamabad, Pakistan

Javed Iqbal Department of Botany, Bacha Khan University, Charsadda, Khyber Pakhtunkhwa, Pakistan

Israt Jahan Department of Genetics and Plant Breeding, Bangladesh Agricultural University, Mymensingh, Bangladesh

Ambreen Kalsoom Department of Physics, The Government Sadiq College Women University, Bahawalpur, Punjab, Pakistan

Jasmine Kaur Department of Pharmaceutical Sciences and Technology, Maharaja Ranjit Singh Punjab Technical University, Bathinda, Punjab, India

Punam Kumari P. G. Department of Biosciences and Biotechnology, Fakir Mohan University, Balasore, Odisha, India

Pawan Kumar Department of Pharmaceutical Sciences and Technology, Maharaja Ranjit Singh Punjab Technical University, Bathinda, Punjab, India

Ram Kumar Chitkara College of Pharmacy, Chitkara University, Rajpura, Punjab, India

Department of Pharmaceutical Sciences and Technology, Maharaja Ranjit Singh Punjab Technical University, Bathinda, Punjab, India

Tariq Mahmood Department of Plant Sciences, Faculty of Biological Sciences, Quaid-i-Azam University, Islamabad, Pakistan

Muhammad Majeed Department of Botany, University of Gujrat, Gujrat, Punjab, Pakistan

Shweta Mehrotra Department of Biotechnology, Guru Jambheshwar University of Science & Technology, Hisar, India

Maida Mobeen Department of Botany, University of Gujrat, Gujrat, Punjab, Pakistan

Sujit Mondal Faculty of Agriculture, Patuakhali Science and Technology University, Patuakhali, Bangladesh

Alan Álvarez Monge Centro de Invesstigación en Alimentación y Desarrollo, Chihuahua, México

Shiza Mukhtar Department of Plant Sciences, Quaid-i-Azam University, Islamabad, Pakistan

Jamila Naz Department of Forestry and Range Management, Bahauddin Zakariya University, Multan, Punjab, Pakistan

Erick H. Ochoa Chaparro Centro de Invesstigación en Alimentación y Desarrollo, Chihuahua, México

Akim Oladokoun Department of Geomatics and Ecosystem Modeling, Laboratory of Botany and Plant Ecology (LBEV), University of Lome (UL), Lome, Togo

Juan J. Patiño Cruz Centro de Invesstigación en Alimentación y Desarrollo, Chihuahua, México

Pratik Priyadarsi Puhan P. G. Department of Biosciences and Biotechnology, Fakir Mohan University, Balasore, Odisha, India

Umar Masood Quraishi Department of Plant Sciences, Quaid-i-Azam University, Islamabad, Pakistan

Waseem Akhtar Quresh Cholistan Institute of Desert, The Islamia University of Bahawalpur, Bahawalpur, Punjab, Pakistan

Carlos A. Ramírez Estrada Centro de Invesstigación en Alimentación y Desarrollo, Chihuahua, México

Muhammad Ramzan Department of Botany, Government Graduate College, Layyah, Punjab, Pakistan

Abhisek Rath Department of Plant Pathology, Assam Agricultural University, Jorhat, Assam, India

Zonaira Saeed Institute of Botany, University of the Punjab, Lahore, Punjab, Pakistan

Abhaya Kumar Sahu P. G. Department of Biosciences and Biotechnology, Fakir Mohan University, Balasore, Odisha, India

Md. Arif Sakil Department of Biochemistry and Molecular Biology, Bangladesh Agricultural University, Mymensingh, Bangladesh

Esteban Sánchez Centro de Invesstigación en Alimentación y Desarrollo, Chihuahua, México

Sarita Department of Botany and Plant Physiology, CCS Haryana Agricultural University, Hisar, Haryana, India

Hridesh Harsha Sarma Department of Agronomy, Assam Agricultural University, Jorhat, Assam, India

Madhurjya Ranjan Sharma Department of Agricultural Biotechnology, Assam Agricultural University, Jorhat, Assam, India

Rishita Shrivastva Department of Microbiology, Kanya Gurukul, Gurukula Kangri (Deemed to be University), Haridwar, Uttarakhand, India

Haramritpal Singh Sidhu Department of Mechanical Engineering, Giani Zail Singh Campus College of Engineering & Technology, Maharaja Ranjit Singh Punjab Technical University, Bathinda, Punjab, India

Anil Sindhu Department of Biotechnology, Deenbandhu Chhotu Ram University of Science and Technology, Murthal, Haryana, India

Mahakdeep Singh Department of Pharmaceutical Sciences and Technology, Maharaja Ranjit Singh Punjab Technical University, Bathinda, Punjab, India

Siraj Uddin Department of Plant Sciences, Quaid-i-Azam University, Islamabad, Pakistan

Aftab Umar Department of Botany, University of Agriculture, Faisalabad, Punjab, Pakistan

Verinder Virk Department of Microbiology, Kanya Gurukul, Gurukula Kangri (Deemed to be University), Haridwar, Uttarakhand, India

Muhammad Waheed Department of Botany, Govt. Islamia Graduate College, Civil Lines, Lahore, Punjab, Pakistan

Virender Yadav Department of Chemical Engineering, Maharishi Kashyap Government Polytechnic Jattal, Panipat, Haryana, India

Meskul Zannat Department of Genetics and Plant Breeding, Bangladesh Agricultural University, Mymensingh, Bangladesh

Chapter 1
Nanobiosensors for Precision Agrobiotechnology

Pratik Priyadarsi Puhan, Abhaya Kumar Sahu, Beda Saurav Behera, and Punam Kumari

Abstract The growing demands for sustainable agricultural practices have intensified the focus on precision agrobiotechnology, a field that leverages cutting-edge technologies to enhance crop productivity. Nanotechnology and its applications in agriculture have appeared as a boon with enormous potential in transforming conventional farming practices into redefined farming activities. Nanobiosensors have become a valuable tool in this field, providing real-time monitoring and precise interventions in plant systems. This chapter provides an in-depth exploration of the role of nanobiosensors in precision agrobiotechnology, from their fundamental principles to their diverse applications in increasing crop protection and yield, phytoremediation, and soil health management. Here, we delve into the integration of nanobiosensors with plant genetic engineering, highlighting how nanoparticle (NP)-mediated targeted genome editing offers new frontiers in crop improvement.

Keywords Agrobiotechnology · Nanobiosensors · Nanoparticles · Stress

1.1 Introduction

The world's population is expected to increase to 8.5 billion by 2030 and 9.8 billion by 2050, which would create significant challenges for agriculture and food production systems (Alvarado et al. 2019; Marchiol 2018). The National Academies Press (2018) report indicates that global food demand will surge by 50–80% over the next 50 years because of the mounting population and rising income rates. Environmental stresses cause increasing difficulties in meeting the rising global food demand (Gupta et al. 2020). Biological stresses such as pests and pathogens create significant yield losses of 11–30% in major crops, including wheat, rice, maize, and

P. P. Puhan · A. K. Sahu · B. S. Behera · P. Kumari (✉)
P. G. Department of Biosciences and Biotechnology, Fakir Mohan University, Balasore, Odisha, India

J.-T. Chen (ed.), *Nanobiosensors for Crop Monitoring and Precision Agriculture*, Sustainability Sciences in Asia and Africa,
https://doi.org/10.1007/978-981-96-8335-2_1

soybean, especially affecting food-insecure areas (Savary et al. 2019; Van Esse et al. 2020).

Current agricultural systems demand fresh, sustainable solutions for continuing crop cultivation. Nanotechnology currently presents itself as an effective method to tackle agricultural issues through the synthesis of nanomaterials using physical, chemical, and biological approaches that enhance conventional technologies (Purohit et al. 2019; Mazhar et al. 2017). Among these advancements, nanobiosensors, sensors developed by integrating nanomaterials with highly specific biological materials, have emerged as transformative tools in precision agrobiotechnology. These sensitive, noninvasive devices are made with innovative nanobiotechnology techniques and offer readily gathered and analyzed real-time responses (Kaphle et al. 2018; Chamundeeswari et al. 2019; Kalyani et al. 2021).

The nanobiosensors perform signal measurement through transducers, which convert biological and chemical interactions into detectable outputs. Based on the transducer mechanisms, nanobiosensors can be further categorized into electrochemical, optical (fluorescence and surface plasmon resonance), piezoelectric, and calorimetric biosensors. Electrochemical biosensors perform via potentiometric, impedimetric, amperometric, and capacitive transducers to transform biological or chemical signals into quantifiable means. A potentiometric biosensor measures the potential created by the interaction of electroactive molecules, and an amperometric biosensor detects changes in current (Malhotra and Ali 2018). The binding preference of target particles with the sensor surface causes the impedimetric biosensor to record changes in both capacitance and charge conductance (Kim et al. 2019). Optical biosensors detect biological interactions through changes measured with both label-based and label-free detectors employing techniques of optical characteristics such as wavelength and refractive index, as well as intensity and polarization (Huertas et al. 2019). Optical labels enable the detection of color changes and photon particles at particular wavelengths to confirm sample analyte presence (Huertas et al. 2019). Calorimetric biosensors record the heat production (or dissipation) driven by the precise interaction of the analytes with receptors, correlating it to the analyte concentration (Mondal et al. 2022). Two different kinds of technologies are commonly employed: (1) thermopile-based devices, which rely on the Seebeck effect, and (2) thermistor-based devices, which concentrate on how resistance changes in response to temperature fluctuations.

Based on the biological recognition element, biosensors can be categorized into immunosensors, enzymatic biosensors, and aptamers. Target molecules are detected by immunosensors, which use the antigen–antibody interactions where antibodies serve as the key biological recognition elements. Immunosensors are of two different types: non-labeled sensors recognize physical alterations due to the formation of immunocomplex, and at the same time, the labeled sensors perform signal measurement through detectable markers. Enzymatic biosensors implement enzymes as biorecognition components to initiate specific reactions that produce electrochemical signals. Aptasensors become optimal detection methods for target analytes because aptamers ensure high stability, combined with specificity and cost-effectiveness. The sensors use aptamers to identify target molecules through the formation of

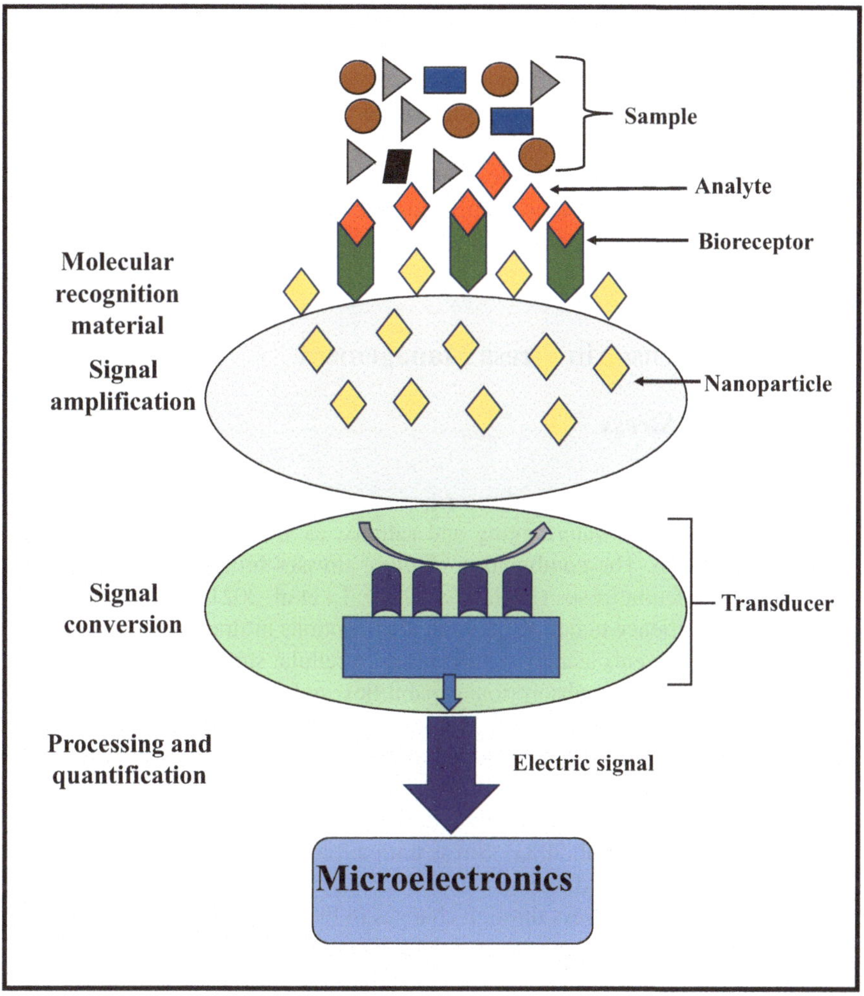

Fig. 1.1 Components of nanobiosensors

particular three-dimensional structures by implementing hydrogen bonds and van der Waals forces after Systematic Evolution of Ligands by Exponential Enrichment (SELEX) selection for better binding performance.

The three main parts of these sensors include biological recognition elements (aptamers, antibodies, peptides, and enzymes), transducers (gold and silver nanoparticles (NPs), carbon nanotubes, graphene oxide, and quantum dots), and detectors for signal conversion to readable data (Naresh and Lee 2021; Singh and Kumar 2021) (Fig. 1.1). The devices help farmers achieve real-time soil condition assessment and control abilities as they monitor soil fertility, as well as pH measurements, moisture content, and mineral concentration. These devices possess the capability to identify pests and evaluate soil productivity, together with nutrient deficiencies, before plant diseases emerge (Rai et al. 2015; Sekhon 2014). Single-walled carbon nanotubes (SWCNTs) together with nanowires have been used to enhance soil health, thus improving agricultural production rates (Ramnani et al. 2016). The integration of nanobiosensors with information technology systems allows distant farmers to receive monitoring updates and important data through remote networks. These networks enable real-time crop yield assessment and disease identification (Duhan et al. 2017; Srivastava et al. 2018).

1.2 Nanobiosensor in Stress Management

1.2.1 Abiotic Stress

Agricultural productivity and food security face major threats from environmental stresses, which include waterlogging and salinity, as well as drought and heavy metal contamination. The combination of these stresses brings about 51–82% of annual crop production losses (Kajla et al. 2015; Li et al. 2021; Oshunsanya et al. 2019). Plants experience major physiological disruptions in stressful environmental conditions as these events lead to modifications in cellular structure besides altering phytohormonal and light absorption capabilities and carbohydrate storage and affecting stomatal function and reactive oxygen species (ROS) production. The changes in biochemical and physiological processes reduce agricultural yields (Ansari et al. 2020; Husen 2022; Husen 2021a, b; Pérez-Llorca et al. 2023).

Different types of stress impact plant development, physiology, and production capabilities (Al-Khayri et al. 2023). Stress mitigation solutions remain essential for preserving crops and defending against food scarcity. Plants adopt multiple mechanisms to manage abiotic stress through changes in their physiology and production of osmoprotectants and the activation of detoxification pathways as well as antioxidant defense systems while generating stress-related proteins and expressing stress-responsive genes to achieve cellular homeostasis (Kocabay et al. 2013; Mashabela et al. 2023; Singh and Chauhan 2017).

Plants under abiotic stress produce damaging chemical substances, which include reactive oxygen species (ROS), including hydrogen peroxide (H_2O_2), superoxide anion (O_2•−), and hydroxyl radicals (OH•) (Choudhury et al. 2013; Hasanuzzaman et al. 2020). Plant stress symptoms include leaf shape modifications as well as changes in how the plant absorbs light, which serve as visible stress indicators. The recording of acoustic signals produced by stressed plants now enables researchers to collect data about their health status, according to Khait et al. (2023). Plant health assessment under different stress conditions benefits from detecting phenotypes combined with chemical compositions and phytohormones (Table 1.1) (Ansari et al. 2020).

Table 1.1 List of recent nanobiosensors developed for stress detection and monitoring

Nanosensors	Nanomaterial utilized	Sensing purpose	Detection limit	Sensing process	Reference
Heavy metal detection					
Single plasmonic NSs	Gold nano-rods (GNRs)	Hg^{2+}; Pb^{2+}	5 nm; 1 nM	GNR-sensing device with position-encoded aptamer functionality	Ye et al. (2022)
Triboelectric NSs	Ultrathin MXene	Zn^{2+}; Cr^{3+}; Cu^{2+}	10–300 μM	Simple vacuum filtration method	Wang et al. (2022b)
Electrochemical	Fc-NH_2-UiO-66	Pb(II); cd(II)	8.5 nm; 0.6 nM	Electrochemical impedance spectroscopy (EIS) and cyclic voltammetry (CV)	Wang et al. (2020)
	Ln-MOF/ZJU-27	Cd(II)	1.66 nm	Square wave anodic stripping voltammetry (SWASV)	Ye et al. (2022)
	3DGO-Py10	Cd^{2+}	3.6 μg L^{-1}	SWASV	Guo et al. (2020a, b)
	SWCNTs	Mercury	3 f M	Cyclic voltametric test based on vertical SWCNTs	Shi et al. (2017)
	$MnCo_2O_4$ nanoparticles	Pb^{2+}	8.06 nmol dm^{-3}	Linear sweep anodic stripping voltammetry (LSASV)	Antunović et al. (2019)

(continued)

Table 1.1 (continued)

Nanosensors	Nanomaterial utilized	Sensing purpose	Detection limit	Sensing process	Reference
Colorimetric NSs	Mesoporous silica NPs	Hg	60 pM	High affinity of Hg (II) for sulfur allows it to interact with solid S_2 and create new $Hg(S\text{-}R)_2$, which can break the linear dithioacetal bonds and cause cargo to be released	Jimenez-Falcao et al. (2020)
	Silver-coated Gold nanobipyramids (AuNBPs)	Hg	0.8 μM	A color shift caused by etching silver-coated Gold nanobipyramids (AuNBPs) makes it possible to detect Hg^{2+}	Qi et al. (2018)
	Au	Pd(II)	4.23 μM	Pd (II) caused APP-Gold nanoparticles (AuNPs) to aggregate more selectively than other metals, which led to the total and substantial elimination of surface plasmon resonance	Anwar et al. (2018)

SERS	Gold nanoparticles (AuNPs)	Cd	8 $\mu g\ kg^{-1}$	Radiometric sensing based on Gold nanoparticles (AuNP)-trimercapto triazine	Zuo et al. (2018)
Lateral flow assay	Gold nanoparticles (AuNPs)	Hg	2.53 nM	The DNA probe used in the portable sensor is biotinylated using Gold nanoparticles (AuNPs) and streptavidin	Guo et al. (2020a, b)
	AuNSs	Cd	0.18ngmL^{-1}	Using AuNSs' ability to quench the QD signal, competitive immune-chromatographic strips are constructed	Xiao et al. (2018)
	Core–shell Upconverting Nanoparticles (UCNPs)	Hg	5 ppb	Smartphone-based Upconverting Nanoparticles (UCNP)-aptamer integrated mobile sensing system	Jin et al. (2018)
Multimodal NS	Superparamagnetic Fe_2O_3 NPs	Hg	0.49 nM	Fluorescence quenching as Hg^{2+} concentrations increase	Satapathi et al. (2018)

(continued)

Table 1.1 (continued)

Nanosensors	Nanomaterial utilized	Sensing purpose	Detection limit	Sensing process	Reference
Pathogen detection					
Electrochemical biosensor	Colloidal Gold Nanoparticles (GNPs)	*Pseudomonas syringae*	214 pM	Electrochemical changes are evaluated using differential pulse voltammetry	Lau et al. (2017)
	Multi-walled carbon nanotube (MWCNTs)	*Ganoderma boninense*	0.041 mg L^{-1}	Modified electrode's electrocatalytic effects on oil palm leaves infected with *G. boninense* and healthy leaves	Isha et al. (2019)
	Gold nanoparticles (AuNP) microelectrode	*Avian influenza virus* (H_5N_1)	0.367 HAU ml^{-1}	CV	Zhang et al. (2017)
Cu (I) tris(pyrazolyl) borate-SWCNT	Cu	Necrosis	0.5 ppm	Chemiresistor	Cellini et al. (2017)
Optical NS	Selective single-walled carbon nanotube (SWCNT)	Wound signaling		Superior fluorescent probes called selective SWCNTs are able to track H_2O_2 produced in plants as a result of mechanical injury in real time	Lew et al. (2020a, b)
Pd (II)-pH dye-SiO_2 microspheres	Pd	*Fusarium* wilt	0.17 ppm	Colorimetry	Cellini et al. (2017)
Fluorescent NS	Carbon dots (CDs) and terbium ions (Tb^{3+})	*Citrus tristeza*	220 ng	Energy transfer via fluorescence resonance	Chen et al. (2018)

Anti-PPV IgG polyclonal antibody	Au	*Pox virus* (Plum)	10 pg mL^{-1}	Detection using electrochemical impedance spectroscopy	Dyussembayev et al. (2021)
Surface plasmon resonance nanosensor (SPRNS)	Au	Afatoxin B1 (*Aspergillus flavus*)	1.04 pg mL^{-1}	The reflectivity index of the gold electrode surface was impacted by the aflatoxin concentration	Akgönüllü et al. (2020)
CdSe-PEI-QD	Cd	Bacterial wilt	10 μg mL^{-1}	Fluorescence	Gao et al. (2018)
Chemiresistive biosensors	Polypyrrole (PPY) nanoribbons on Au microelectrode array (Au- MEAs)	Mosaic virus (Cucumber)	10 ng ml^{-1}	Amperometry	Chartuprayoon et al. (2015)
AuNP-ssDNA	Au	Bacterial wilt	7.5 ng	Colorimetry	Farber et al. (2019)
AgNP-ssDNA	Ag	*Phytophthora ramorum*	0.2 μM	DNA hybridization and nanoparticle functionalization	Li et al. (2020)
PtNP-IgG	Pt	Rot of potato	10_2 CFU	MALDI-TOF MS	Li et al. (2021)
TiO_2NP-dsDNA	TiO_2	*Phytophthora* spp.	35 nM	Electrochemistry	Li (2018)

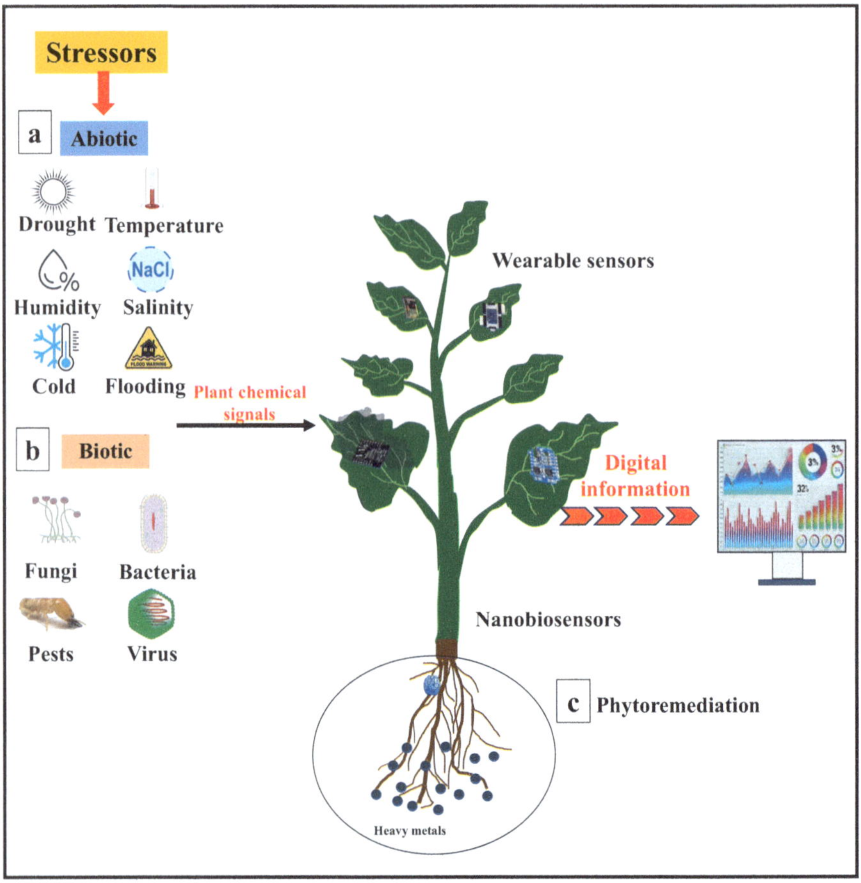

Fig. 1.2 Schematic representation of nanobiosensor application to combat various plant stress. (**a**) Abiotic stress (**b**) Biotic Stress (**c**) Phytoremediation

Nanosensors (NSs) developed by scientists enable real-time detection of stress-related plant chemicals to help reduce environmental stress effects (Fig. 1.2a). Son et al. (2023) developed a surface-enhanced Raman scattering (SERS)–based nanosensor to detect multiple stress-related chemicals inside living plants in real time. Similarly, Alova et al. (2020) developed O_2 nanosensors based on quartz micropipettes that easily detect oxygen reduction in *Chara corallina* plants exposed to waterlogging conditions by enhancing their tips with platinum and carbon layers. Furthermore, Lew et al. (2020a, b) developed an optical nanosensor probe that works without damaging the plants to detect H_2O_2-signaling molecules under stress conditions in plant species including *Arabidopsis thaliana*, *Spinacia oleracea*, and *Eruca sativa*. H_2O_2-specific detection under stress conditions becomes possible through H_2O_2-binding (GT)15 single-stranded carbon nanotubes (CNTs).

He et al. (2020) developed an electrochemical biosensor for stress hormone detection in *Glycine max* seedlings under salt stress, with salicylic acid detection sensitivity at 1.99 μM, while the auxin detection sensitivity reached 3.30 μM. Carbon nanopowder, along with multi-walled carbon nanotubes (MWCNTs), enabled this detection system to become more sensitive. Near-infrared fluorescent nanosensors were used for the detection of gibberellins (GAs) in lettuce plants under salt stress conditions, and bioactive GA compounds activate fluorescence intensity changes in sensors that incorporate polymer-wrapped engineered single-walled carbon nanotubes (SWCNTs) (Ang et al. 2021).

Sun et al. (2016) devised silicon nanoparticle (SiNP) sensors to detect drought effectively with greater accuracy and improve the ability of soils to retain water. Zhu et al. (2017) reported cytosolic glucose monitoring in rice (*Oryza sativa*) plants under different stress conditions using Fluorescence Resonance Energy Transfer (FRET)-based nanosensors, which detected glucose elevations between 0.2 and 20 μM.

The accumulation of heavy metals like cadmium (Cd) and lead (Pb) causes toxicity in the plant. Wang Xiang et al. (2019) created an electrochemical immunobiosensor system to screen vitronectin along with other glycoproteins that serve as biological indicators of heavy metal stress. The biosensor showed its ability to detect Cd and Pb through its detection thresholds, with 18.5 nmol/L and 25.6 nmol/L being the respective values. Quantum dot (QD)–based nanosensors serve to detect heavy metal ions present in plants (Wang et al. 2022a). SERS technology enabled Gold nanoparticles (AuNP)-trimercaptotriazine-based ratiometric sensing models to detect Cd with a threshold of 8 μg kg^{-1} (Zuo et al. 2018). The detection of heavy metal stress involves the use of two types of nanosensors, known as lateral flow assays and colorimetric sensors, with successful applications (Anwar et al. 2018; Gao et al. 2018, and Xiao et al. 2018).

1.2.2 Biotic Stress

The existence of several pathogens with pests and insects together creates severe challenges for sustainable agricultural productivity and global food security. Modern biological stress management relies on insecticides as well as pesticides and fungicides for control. Using these chemicals beyond recommended levels causes damage to ecosystems and enables pathogen resistance development. Nanobiosensors function as contemporary devices providing nature-friendly solutions for the early detection and sustainable control methods of plant biotic stress (Fig. 1.2b).

Precision agriculture advances through the use of nanomaterial-based biosensors because they speed up the process of detecting and tracking stress indicators. HeAptDNA-SWCNT sensors developed by Wu et al. (2020) serve as a detection method for hydrogen peroxide (H_2O_2), which functions as an important plant stress–signaling molecule to monitor plant health in different environmental and pathogen-induced stressful conditions in real time (Fig. 1.3). Lau et al. (2017) created an electrochemical sensor with outstanding detection sensitivity (214 pM) to detect *Pseudomonas syringae*, which causes bacterial speck in tomatoes.

Panferov et al. (2018) and Zhan et al. (2018) established lateral flow biosensors for *Phytophthora infestans* detection to identify late blight in potatoes and potato leafroll virus (PLRV). Khaledian et al. (2017) developed a gold nanoparticle

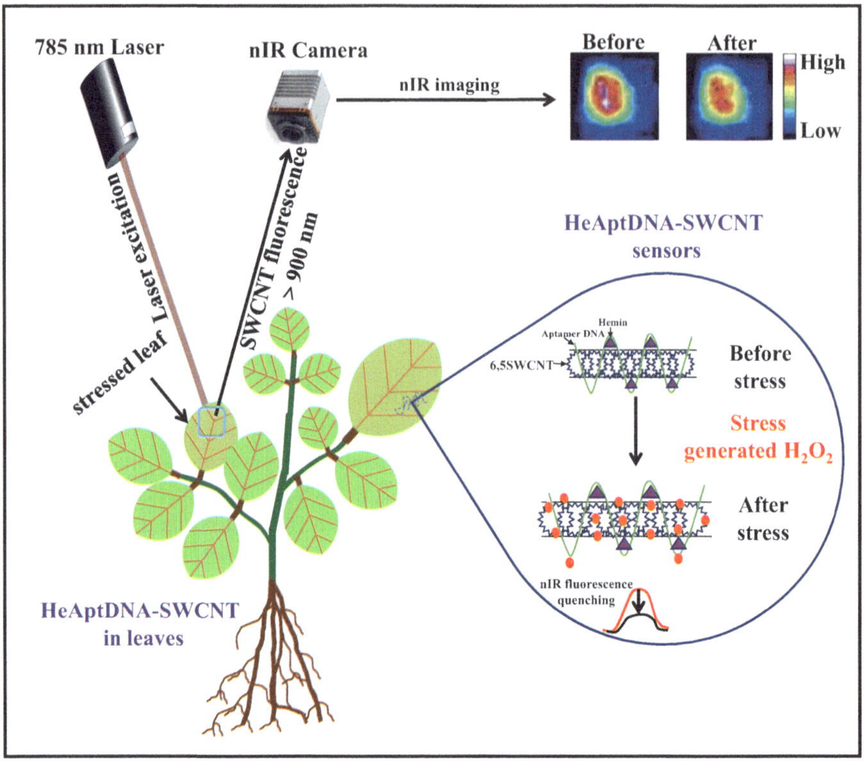

Fig. 1.3 Plant health monitoring by H2O2 HeAptDNA-SWCNT nanobiosensor

(GNP)-based biosensor detection method for *Ralstonia solanacearum*, which causes bacterial wilt throughout potato crops. These biosensors are validated as dependable equipment for real-time pathogen detection because of their high precision potential.

Biosensors equipped with quantum dots (QDs) and gold nanoparticles as advanced materials have demonstrated their ability for virus detection of *Citrus tristeza* virus and Grapevine virus A (Khater et al. 2019; Shojaei et al. 2016). Urease-functionalized nanosensors have been developed that detect bacteria at 10^2 CFU/LL levels using magnetic nanoparticles within 30 min (Sharma et al. 2021). Early detection through these systems proves indispensable for stopping diseases from advancing while reducing agricultural crop damage.

Nanobiosensors function as key instruments in the management of pest control operations. Micro-electromechanical system (MEMS) gas sensors have proven effective for detecting insect sex pheromones at highly sensitive levels. Brezolin et al. (2020) developed a sensor system that detected stink bugs' (*Euschistus heros*) pheromone even at minimal concentrations between 0.005 and 0.03 μg/mL. Moitra et al. (2016) designed silicon dioxide–based Micro-electromechanical system

(MEMS) devices that perform the detection of lepidopteran pests' female-specific pheromones in rice and chickpea, along with other crops. Nanobiosensors establish an adaptable system that provides accurate methods to manage biological threats in agriculture. Nanobiosensor implementations in agriculture deliver essential benefits for enduring agricultural production systems and safe food systems worldwide.

1.3 Nanoparticles in Phytoremediation

Nanotechnology has emerged as a revolutionary tool for enhancing the efficiency of phytoremediation, an eco-friendly approach to detoxifying contaminated soils and waters. Phytoremediation functions through the use of plants together with their microbial partners to eliminate pollutants that consist of heavy metals, along with organic substances and industrial chemicals that remain in the environment over time (Song et al. 2019). Traditional phytoremediation encounters barriers because remediation moves at a slow pace while pollutants maintain limited accessibility, which leads plants to experience strain from excessive contaminants. Nanophytoremediation systems achieve creative solutions for these obstacles by utilizing nanoparticles (NPs) along with their specific physicochemical properties (Aseri et al. 2024).

The application of nanoparticles in phytoremediation produces two beneficial effects: they directly clean up contaminants, and they help plants acquire and endure contaminants better. Nanoscale zero-valent iron (nZVI) functions as a key remedy to treat heavy metals and organic pollutants in numerous applications throughout the world. The chemical properties of nanomaterials combined with their large surface area deliver effective pollution removal capabilities for chromium stabilization and chlorinated hydrocarbon degradation to reduce their environmental persistence (Liang et al. 2017). Nano-hydroxyapatite successfully immobilizes lead (Pb) in polluted soils so that plants can extract it through bioavailability without environmental side effects (Rajput et al. 2022).

Studies have highlighted that specific nanomaterials enhance pollutant bioavailability, which improves their phytoextraction capabilities. Fullerene nanoparticles increase poplar tree uptake of trichloroethylene (TCE) by 82% above normal levels (Ma and Wang 2010). Research has revealed that the application of salicylic acid nanoparticles (SANPs) improved arsenic accumulation in *Isatis cappadocica* plants by raising root arsenic levels to 705 ppm and shoot arsenic accumulation to 1188 ppm (Ghani et al. 2022).

The way nanoparticles interact with plants plays an essential role in how effective they will be for phytoremediation purposes. Plant roots absorb nanoparticles, which move to aerial tissues by means of symplastic and apoplastic pathways. The absorption rate of nanomaterials depends on several influencing characteristics, including particle dimensions, surface composition, and nanomaterial chemical structure (Pradhan et al. 2014). The study of manganese nanoparticles (MnNPs) demonstrates that they activate root and shoot development in *Vigna radiata* plants alongside improving nitrogen intake and photosynthetic operation, which strengthens pollutant removal potential and plant resistance (Doria-Manzur et al. 2023).

Different phytoremediation research studies have displayed the beneficial effects of nanoparticles as an intervention. Bimetallic nanoparticles composed of Pd/Fe and Ni/Fe exhibit outstanding capabilities to destroy organic pollutants, including Polychlorinated Biphenyls and Polybrominated Diphenyl Ethers (PCBs and PBDEs), through their catalytic dechlorination and hydrogenation reactions (Yang et al. 2022). Nano-titanium dioxide (TiO_2) demonstrates successful application as a photocatalyst, which shortens the persistence times of harmful organics such as pyrene and phenanthrene in the soil (Romeh 2022).

Recent advances in nanotechnology have converted phytoremediation into a versatile, effective method to tackle environmental contaminants (Fig. 1.2c). Plants that implement nanoparticles in their systems become capable of remediating challenging contaminated sites and achieving enhanced pollutant degradation and plant tolerance as well as increased phytoavailability. Nanotechnology uses nanoparticles as essential tools to deal with tough-to-manage heavy metals, along with persistent organic pollutants and industrial toxins, by interacting with these contaminants at their molecular level.

1.4 Nano-enabled Plant Genetic Engineering

Plant genetic engineering receives transformative changes through the application of nanotechnology, which has revolutionized multiple scientific fields. Agrobacterium-mediated transfer and biolistic particle delivery methods currently encounter numerous obstacles when used for genetic transformation where host specificity and tissue damage, as well as low transformation efficiencies present major challenges (Wang et al. 2015). Nanoparticles represent a multipurpose system that enables genetic material transport into plant cells by crossing the rigid plant cell wall barrier (Demirer et al. 2019a). Nucleic acids, including DNA, RNA, and proteins, enter plant cells effectively through nanoparticles because of their nanometer-scale dimensions combined with their adaptable physicochemical properties (Zhang et al. 2019). Nanoparticles provide maximum genetic material accommodation because of their high surface-area-to-volume ratio, and they retain biocompatibility properties, which reduce cellular toxicity (Kwak et al. 2019). SWCNTs and MWCNTs act as penetrating agents that pass through plant cell walls to transport DNA efficiently into organelles without physical aids (Hendler-Neumark and Bisker 2019) (Fig. 1.4). The medical field now uses SWCNTs in gene delivery experiments because of their enhanced biocompatibility and operational effectiveness. Genetic materials remain stable during transfection because SWCNTs function as protective barriers against cellular enzyme damage. Demirer et al. 2019b, have found that SWCNTs deliver Small Interfering RNA (siRNA) as well as plasmid DNA for precise gene editing through biological mechanisms. SWCNTs showed successful DNA infiltration into plant cells in experimental research since they avoided nuclear genome integration, which minimized genetic migration risks (Kwak et al. 2019).

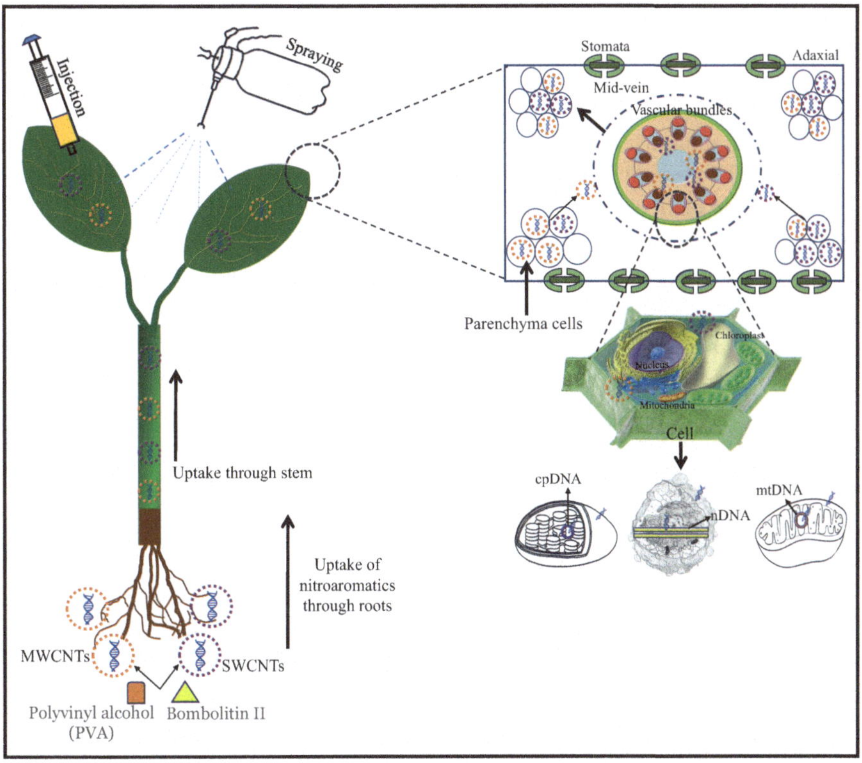

Fig. 1.4 Process of gene delivery via MWCNTs and SWCNTs nanobiosensors

The delivery of plasmid DNA with Fe_3O_4 nanoparticles under magnetic field influence represents an innovative technique known as magnetofection. According to Zhao et al. (2017a, b), their technique produces successful incorporation of genetic material into cotton pollen grains capable of transferring inherited traits to naturally pollinated offspring. The efficient modular strategy of genetic engineering relies on DNA nanostructures made from genetic material. Plants benefit from nanocubes and tetrahedron structures because these nanostructures effectively avoid enzymatic attacks while providing efficient delivery of nucleic acid materials (Li et al. 2019). By applying nanostructures, Zhang et al. (2019) delivered Small Interfering RNA (siRNA) that led to prominent gene suppression in plants *Nicotiana tabacum* and *Spinacia oleracea*. These delivery platforms feature an adjustable design that enables their application for different purposes. An innovative alternative comes from peptide-based nanoparticles. Plants accept genetic material delivery because these short chains penetrate cells and membranes effectively (Chuah and Numata 2018). Thagun et al. (2019) developed peptide-based chloroplast-targeting sequences that led DNA to plastids so that transient gene expression could happen without altering the nuclear genome. The technique demonstrates opportunities for precise genetic engineering of plant organelles, which creates new possibilities in plant biotechnology research. One eco-friendly option to transport genetic material through biological systems is layered double hydroxide (LDH) clay nanosheets. Research has demonstrated the excellent potential of this approach to activate RNAi pathways, which leads to increased pest and pathogen resistance (Mitter et al. 2017). The technology presents benefits that avoid traditional approach problems, but it still requires further optimization to enhance process efficiency and minimize environmental consequences. Plant genetic engineering will benefit from future applications when nanotechnology combines with CRISPR/Cas9 genome editing platforms and de novo regeneration methods. The combined usage of these technologies enables scientists to skip the conventional tissue culture section, which shortens the transgenic plant development process and minimizes required resources substantially. Nanotechnology continues advancing toward becoming integral for resolving global food security problems while building sustainable agricultural systems.

1.5 Nanoparticle-Mediated Targeted Genome Editing in Plants

Nanoparticle-based targeted genome editing systems in plants show promise as revolutionary methods in precision plant biology because they link nanotechnology with CRISPR/Cas-dependent genetic editing strategies. The introduction of CRISPR/Cas systems led to significant changes in plant breeding practices by allowing scientists to perform exact gene alterations at each nucleotide position. The challenge of effective genome-editing reagent delivery has persisted because of the rigid plant cell wall structure and the limitations associated with conventional transformation approaches such as *Agrobacterium*-mediated transfer and biolistic particle bombardment (Chen et al. 2019).

Nanotechnology introduces a modern approach that efficiently provides a universal non-integrative delivery system for CRISPR components and addresses many deficiencies of conventional gene delivery methods (Sharma and Lew 2022). Ribonucleoproteins (RNPs), single-guide RNAs (sgRNAs), and donor DNA templates receive optimum transport through nanoparticles because of their flexible physicochemical properties. Plant cell walls and membranes remain intact when nanotubes pass through and avoid time-intensive tissue culture regeneration protocols (Demirer et al. 2019a, b; Kwak et al. 2019). The brief nature of nanoparticle delivery systems lowers the risk of harmful effects and regulatory problems related to stable gene integration, which allows fast development of genetically modified plants that do not have transgenic inheritance problems.

Current advances in nanoparticle engineering have increased genome-editing reagent delivery effectively. CRISPR components reach their targets by using surface functionalization methods that add targeting ligands to precisely deliver contents to chloroplasts or mitochondria, thus broadening the capabilities of plant genome engineering (Thagun et al. 2019; Saminathan et al. 2022). DNA origami–based nanostructures serve as a programmable platform for multiple CRISPR component delivery, which enables controlled component release to enhance editing effectiveness (Deng et al. 2020; Lew et al. 2020a, b). These advancements reduced the cytotoxicity, which contributes to better survival rates of genetically modified crops.

The delivery systems utilizing nanoparticles for CRISPR have shown promise as they transition from laboratory applications to industrial agricultural purposes. Aerosolized nanoparticle formulations enable plant genome-editing reagents to enter plants directly through foliar sprays without relying on complicated tissue culture protocols. The editing process becomes simpler through this method, which allows increased scalability and better field-level applications. Improved formulation designs that enhance surface adhesion and water retention capabilities enable more efficient plant surface absorption under exterior conditions (Mitter et al. 2017; Su et al. 2019). The developments create opportunities for massive crop improvement focused on challenging issues like crop resistance to pests and climate changes and higher yield production.

1.6 Nanobiosensors in Omics

The omics-based method for studying engineered nanomaterial's (ENM) effects on plants utilizes significant nanobiosensors as essential analytical instruments. Plant research involving engineered nanomaterials (ENMs) depends on omics technologies through genomics, along with transcriptomics, proteomics, and metabolomics, for the rapid detection of biological changes and molecular-level mechanisms studying plant–ENM interactions (Majumdar and Keller 2021). The implementation of nanobiosensors for omics purposes allows direct monitoring of ENM-induced biomolecular modifications, which enhances knowledge about both enzymatic reactions and metabolic changes in plants. Nanobiosensors applied to proteomics and metabolomics amplify the detection power to identify plant stress biomarkers, which leads to a better understanding of metabolic pathways resulting from nanomaterial exposure. Omics-assisted nanobiosensors promote safer and more efficient designs of nano-enabled agrochemicals because they link molecular responses to phenotypic modifications.

1.7 Challenges and Future Directions

Nanobiosensors hold exceptional potential for transforming agricultural practices through precision agrobiotechnology applications. Implementation on a large scale poses various obstacles when dealing with emerging technologies. These range from nanoparticle toxicity and environmental concerns to technological limitations in field deployment, as well as regulatory hurdles and socioeconomic barriers. Nanobiosensors require proper solutions to maximize their potential for sustainable agricultural practice.

1.8 Environmental and Ecological Risks of Nanoparticles

The use of nanobiosensor technology poses serious threats regarding nanoparticle toxicity, which could significantly affect soil ecosystems along with plant systems in the long term. The uncontrolled implementation of nanoparticles that enhance plant growth stands as a potential environmental safety hazard because they demonstrate benefits for plant nutrient absorption and stress reduction. Immoderate usage of silver nanoparticles (AgNPs) with their known antimicrobial properties creates disruptions in soil microbial communities, which affect soil fertility alongside nutrient cycling processes (Heisnam et al. 2022; Bhanse et al. 2022). Metal-based nanoparticles, including zinc oxide (ZnO) and titanium dioxide (TiO_2), perform well when used properly, but environmental accumulation can modify water and soil biochemical characteristics (Song et al. 2019).

Researchers need to develop naturally degrading, eco-friendly nanoparticles that serve their purpose before dissolving after completion. Efficient biocompatible nanoparticles with protein-based structures or polysaccharide coatings reduce toxicity and ensure effective functionality (Aseri et al. 2024). Phytoremediation assisted by nanoparticles presents a modern technique for cleaning heavy metal pollutants out of soil and water (Bhanse et al. 2022). The integration of these materials into wider adoption requires additional assessment of their extended stability and environmental effects (Duan et al. 2021).

1.9 Scalability and Field Deployment Challenges

The successful laboratory demonstrations of nanobiosensors face a major obstacle when it comes to their transformation for use in open-field applications. The changing environmental conditions of agricultural fields, including temperature swings, variations in humidity, and microbial population, affect sensor reliability and product lifespan (Demirer et al. 2019a).

The practice of using nanosensors faces economic implementation barriers and distribution challenges. The high manufacturing expenses of nanosensors, together with restrictions on fabrication facilities, present financial barriers that affect low-resourced farmers, especially in developing nations (Lv et al. 2020). Scientists need to develop budget-friendly nanosensors, which can be manufactured efficiently while maintaining their operational excellence. Long-term field applications of sensor networks require self-tailing capabilities together with minimal maintenance needs (Soutter 2021).

Live monitoring of soil moisture and nutrient concentrations alongside disease detection relies on wireless nanosensor networks connected to IoT systems (Zhang et al. 2020). The technology enables automatic data-based farming decisions, which help farmers deliver precise fertilizers and pesticides at optimal times, thus minimizing waste and environmental degradation (Menezes-Silva et al. 2017). Reliable data transmission and smooth connectivity in rural agricultural areas need additional innovative approaches (Sharma et al. 2023).

1.10 Regulatory and Ethical Considerations

The development of regulatory systems for agricultural nanobiosensors continues to progress together with other biotechnological advances. Nanosensors operate through molecular interactions compared to traditional fertilizers and pesticides, which raise safety doubts, combined with environmental stability and multipartite organism interaction concerns (Fahad et al. 2017). A lack of international consensus regarding nanobiosensor regulation in precision agriculture results in inconsistent global adoption (Goulson 2013).

The international regulatory bodies, which include FAO and WHO alongside national agricultural agencies, need to implement organized safety-evaluating mechanisms for nanosensor-based technology platforms (Ahmad et al. 2020). The successful implementation of nanotechnology in agriculture depends on complete transparency in risk evaluation research and thorough assessments regarding nanoparticle safety, together with defined labels for nano-enhanced agricultural items (Rueda et al. 2017). To achieve responsible nanotechnology development and food safety, governments, together with research institutions, need to establish policies that ensure environmental safety (Pedersen et al. 2021).

1.11 Integration with AI and Smart Agriculture

The future success of nanobiosensor technology depends on its successful integration with artificial intelligence systems, which also serve smart agriculture. AI-driven predictive analytics combined with nanosensors leads to enhanced precision farming since these systems enable immediate detection of crop stress and health monitoring (Xu et al. 2020). Device-generated data processed through machine learning algorithms helps farmers detect upcoming infestations of pests, water-related problems, and nutrient shortages, thus allowing them to enact preventive strategies before major crop losses happen (Bailey-Serres et al. 2019). Automated irrigation, together with fertilization systems based on nanosensor data, represents a major transformative advance. Dynamic changes in water and nutrient delivery occur through automated systems that monitor current soil conditions and boost resource efficiency and reduce environmental effects (Thangthong et al. 2019). The continuous development of AI models presents an ongoing challenge to attain precise interpretations of nanosensor data across a wide range of crop species combined with different soil conditions and climate environments (Nakabayashi and Saito 2015). Hosting additional interdisciplinary studies between agronomists, data scientists, and nanotechnologists will lead to essential improvements in making the systems ready for extensive agricultural applications (Nelissen et al. 2018).

1.12 Farmer Adoption and Training

Nuclear biosensors face both technical and socioeconomic obstacles when farmers must implement them in agricultural systems. Farmers, mainly from developing nations, require information about nanosensors alongside their advantages since they lack sufficient technical understanding (Diaz and Rosenberg 2008). The widespread adoption of nanobiosensors requires farmers to gain knowledge through training programs, government-led initiatives, and industry collaborations as described in Wiebe et al. (2019).

The advancement of low-cost nanobiosensor kits needs public–private collaboration to fit traditional farming practices so that farmers will participate (Saudy et al. 2021). Demonstration farms combined with workshops and extension services must function as key tools to display nanobiosensor practical advantages for increased crop yield, cheaper inputs, and better soil conditions (Etesami and Jeong 2018).

1.13 Conclusion

Nanobiosensors transform precision agrobiotechnology through their ability to deliver on-the-spot information about plants' health and soil properties and environmental condition monitoring. The use of nanobiosensors in modern agriculture represents a substantial opportunity for raising crop production levels, improving resource distribution, and creating a defense against climate change in farming systems. Nanobiosensors serve as a sustainability tool for agriculture through their ability to detect diseases early and improve nutrient precision while decreasing chemical substances in farming operations. Various obstacles need resolution for nanobiosensors to achieve their maximum capabilities. Future research in nanobiosensor development will advance by addressing operational challenges along with regulatory and environmental issues and cost-effectiveness, as well as practical use in fields.

The success of nanobiosensors in agricultural applications depends on developing biodegradable nanomaterials and implementing sensors with AI-controlled analytics alongside affordable monitoring networks to facilitate their transition from academic labs into agricultural settings. Nanobiosensors in agrobiotechnology will reach their future potential through effective partnerships between scientists and policymakers as well as farmers. An environment dedicated to knowledge sharing and responsible innovation, combined with sustainable deployment practices, permits nanotechnology to create vital changes within agricultural development. The ongoing progress in nanobiosensor research will produce techniques that boost both food production security and the long-term resistance of global farming operations despite environmental stressors.

References

Ahmad M, Anjum NA, Asif A, Ahmad A (2020) Real-time monitoring of glutathione in living cells using genetically encoded Fluorescence Resonance Energy Transfer (FRET)-based ratiometricnanosensor. Sci Rep 10(1):1–9

Akgönüllü S, Yavuz H, Denizli A (2020) SPR nanosensor based on molecularly imprinted polymer film with gold nanoparticles for sensitive detection of aflatoxin B1. Talanta 219:1–8

Al-Khayri JM, Rashmi R, Surya Ulhas R, Sudheer WN, Banadka A, Nagella P, Almaghasla MI (2023) The role of nanoparticles in response of plants to abiotic stress at physiological, biochemical, and molecular levels. Plan Theory 12(2):1–26

Alova A, Erofeev A, Gorelkin P, Bibikova T, Korchev Y, Majouga A, Bulychev A (2020) Prolonged oxygen depletion in microwounded cells of Chara corallina detected with novel oxygen nanosensors. J Exp Bot 71(1):386–398

Alvarado K, Bolanos M, Camacho C, Quesada E, Baudrit JV (2019) Nanobiotechnology in agricultural sector: overview and novel applications. J Biomater Nanobiotechnol 10:120–141

Ang MCY, Dhar N, Khong DT, Lew TTS, Park M, Sarangapani S, Strano M (2021) Nanosensor detection of synthetic auxins in planta using corona phase molecular recognition. ACS Sensors 6(8):3032–3046

Ansari MH, Lavhale S, Kalunke RM, Srivastava PL, Pandit V, Gade S, Singh AV (2020) Recent advances in plant nanobionics and nanobiosensors for toxicology applications. Curr Nanosci 16(1):27–41

Antunović V, Ilić M, Baošić R, Jelić D, Lolić A (2019) Synthesis of MnCo2O4 nanoparticles as modifiers for simultaneous determination of Pb (II) and Cd (II*)*. PLoS One 14(2):1–19

Anwar A, Minhaz A, Khan NA, Kalantari K, Afifi ABM, Shah MR (2018) Synthesis of gold nanoparticles stabilized by a pyrazinium thioacetate ligand: a new colorimetric nanosensor for detection of heavy metal Pd (II). Sensors Actuators B Chem 257:875–881

Aseri V, Chopade RL, Godara V, Lohar S, Kumari P, Mavry B, Sankhla MS (2024) Advancing phytoremediation of contaminated soil through nanoparticle-based approaches: current progress and future perspectives. Lett Appl Nanobiosci 13(2):1–15

Bailey-Serres J, Parker JE, Ainsworth EA, Oldroyd GE, Schroeder JI (2019) Genetic strategies for improving crop yields. Nature 575(7781):109–118

Bhanse P, Kumar M, Singh L, Awasthi MK, Qureshi A (2022) Role of plant growth-promoting rhizobacteria in boosting the phytoremediation of stressed soils: opportunities, challenges, and prospects. Chemosphere 303:1–24

Brezolin AN, Martinazzo J, Steffens J, Steffens C (2020) Nanostructured cantilever sensor using with Pani/MWCNT-COOH nanocomposites applied in the detection of pheromone. J Mater Sci Mater Electron 31(8):6008–6016

Cellini A, Blasioli S, Biondi E, Bertaccini A, Braschi I, Spinelli F (2017) Potential applications and limitations of electronic nose devices for plant disease diagnosis. Sensors 17(11):1–13

Chamundeeswari M, Jeslin J, Verma ML (2019) Nanocarriers for drug delivery applications. Environ Chem Lett 17:849–865

Chartuprayoon N, Zhang M, Bosze W, Choa YH, Myung NV (2015) One-dimensional nanostructures based bio-detection. Biosens Bioelectron 63:432–443

Chen BB, Liu ML, Zhan L, Li CM, Huang CZ (2018) Terbium (III) modified fluorescent carbon dots for highly selective and sensitive ratiometry of stringent. Anal Chem 90(6):4003–4009

Chen K, Wang Y, Zhang R, Zhang H, Gao C (2019) CRISPR/Cas genome editing and precision plant breeding in agriculture. Annu Rev Plant Biol 70(1):667–697

Choudhury S, Panda P, Sahoo L, Panda SK (2013) Reactive oxygen species signaling in plants under abiotic stress. Plant Signal Behav 8(4):1-6

Chuah JA, Numata K (2018) Stimulus-responsive peptide for effective delivery and release of DNA in plants. Biomacromolecules 19(4):1154–1163

Demirer GS, Zhang H, Goh NS, Chang R, Landry MP (2019a) Nanotubes effectively deliver Small Interfering RNA (siRNA) to intact plant cells and protect Small Interfering RNA (siRNA) against nuclease degradation. BioRxiv 564427:1–29

Demirer GS, Zhang H, Goh NS, González-Grandío E, Landry MP (2019b) Carbon nanotube–mediated DNA delivery without transgene integration in intact plants. Nat Protoc 14(10):2954–2971
Deng S, Li X, Liu S, Chen J, Li M, Chew SY et al (2020) Codelivery of CRISPR-Cas9 and chlorin e6 for spatially controlled tumor-specific gene editing with synergistic drug effects. Sci Adv 6(29):1–14
Diaz RJ, Rosenberg R (2008) Spreading dead zones and consequences for marine ecosystems. Science 321(5891):926–929
Doria-Manzur A, Sharifan H, Tejeda-Benitez L (2023) Application of zinc oxide nanoparticles to promote remediation of nickel by Sorghum bicolor: metal ecotoxic potency and plant response. Int J Phytoremediation 25(1):98–105
Duan L, Ouyang K, Xu X, Xu L, Wen C, Zhou X et al (2021) Nanoparticle delivery of CRISPR/Cas9 for genome editing. Front Genet 12:1–10
Duhan JS, Kumar R, Kumar N, Kaur P, Nehra K, Duhan S (2017) Nanotechnology: the new perspective in precision agriculture. Biotechnol Rep 15:11–23
Dyussembayev K, Sambasivam P, Bar I, Brownlie JC, Shiddiky MJ, Ford R (2021) Biosensor technologies for early detection and quantification of plant pathogens. Front Chem 9:1–13
Etesami H, Jeong BR (2018) Silicon (Si): review and future prospects on the action mechanisms in alleviating biotic and abiotic stresses in plants. Ecotoxicol Environ Saf 147:881–896
Fahad S, Bajwa AA, Nazir U, Anjum SA, Farooq A, Zohaib A et al (2017) Crop production under drought and heat stress: plant responses and management options. Front Plant Sci 8:1–16
Farber C, Mahnke M, Sanchez L, Kurouski D (2019) Advanced spectroscopic techniques for plant disease diagnostics. A review. TrAC Trends Anal Chem 118:43–49
Gao G, Jiang YW, Sun W, Wu FG (2018) Fluorescent quantum dots for microbial imaging. Chin Chem Lett 29(10):1475–1485
Ghani MI, Saleem S, Rather SA, Rehmani MS, Alamri S, Rajput VD et al (2022) Foliar application of zinc oxide nanoparticles: an effective strategy to mitigate drought stress in cucumber seedling by modulating antioxidant defense system and osmolytes accumulation. Chemosphere 289:1-8
Goulson D (2013) An overview of the environmental risks posed by neonicotinoid insecticides. J Appl Ecol 50(4):977–987
Guo X, Cui R, Huang H, Li Y, Liu B, Wang J et al (2020a) Insights into the role of pyrrole doped in three-dimensional graphene aerogels for electrochemical sensing Cd (II). J Electroanal Chem 871:1–8
Guo Z, Kang Y, Liang S, Zhang J (2020b) Detection of Hg (II) in adsorption experiment by a lateral flow biosensor based on streptavidin-biotinylated DNA probes modified gold nanoparticles and smartphone reader. Environ Pollut 266:1–8
Gupta A, Rico-Medina A, Caño-Delgado AI (2020) The physiology of plant responses to drought. Science 368(6488):266–269
Hasanuzzaman M, Bhuyan MB, Zulfiqar F, Raza A, Mohsin SM, Mahmud JA et al (2020) Reactive oxygen species and antioxidant defense in plants under abiotic stress: revisiting the crucial role of a universal defense regulator. Antioxidants 9(8):1–52
He KC, Wang HR, Yang H, Sun LJ, Liu W, Bao N (2020) Digitalized pencil trace modified electrodes for real time evaluation of salicylic acid in detached Arabidopsis thaliana leaves during regeneration. Anal Chim Acta 1120:59–66
Heisnam P, Moirangthem A, Singh YD, Dutta P, Devi CV, Hazarika BN (2022) Exploitation of silver nanoparticles in bioremediation. In: Bioremediation. CRC Press, pp 143–155
Hendler-Neumark A, Bisker G (2019) Fluorescent single-walled carbon nanotubes for protein detection. Sensors 19(24):1–16
Huertas CS, Calvo-Lozano O, Mitchell A, Lechuga LM (2019) Advanced evanescent-wave optical biosensors for the detection of nucleic acids: an analytic perspective. Front Chem 7:1–25
Husen A (2021a) Morpho-anatomical, physiological, biochemical and molecular responses of plants to air pollution. In: Harsh environment and plant resilience: molecular and functional aspects. Springer, Cham, pp 203–234
Husen H (2021b) Plant performance under environmental stress. Springer

Husen A (ed) (2022) Environmental pollution and medicinal plants. CRC Press
Isha A, Akanbi FS, Yusof NA, Osman R, Mui-Yun W, Abdullah SNA (2019) An NMR metabolomics approach and detection of Ganoderma boninense-infected oil palm leaves using MWCNT-based electrochemical sensor. J Nanomater 2019(1):15–20
Jimenez-Falcao S, Villalonga A, Parra-Nieto J, Llopis-Lorente A, Martinez-Ruiz P, Martinez-Manez R, Villalonga R (2020) Dithioacetal-mechanized mesoporous nanosensor for Hg (II) determination. Microporous Mesoporous Mater 297:110054
Jin B, Yang Y, He R, Park YI, Lee A, Bai D et al (2018) Lateral flow aptamer assay integrated smartphone-based portable device for simultaneous detection of multiple targets using upconversion nanoparticles. Sensors Actuators B Chem 276:48–56
Kajla M, Yadav VK, Khokhar J, Singh S, Chhokar RS, Meena RP, Sharma RK (2015) Increase in wheat production through management of abiotic stresses: a review. J Appl Nat Sci 7(2):1070–1080
Kalyani N, Goel S, Jaiswal S (2021) On-site sensing of pesticides using point-of-care biosensors: a review. Environ Chem Lett 19(1):345–354
Kaphle A, Navya PN, Umapathi A, Daima HK (2018) Nanomaterials for agriculture, food and environment: applications, toxicity and regulation. Environ Chem Lett 16:43–58
Khait I, Lewin-Epstein O, Sharon R, Saban K, Goldstein R, Anikster Y et al (2023) Sounds emitted by plants under stress are airborne and informative. Cell 186(7):1328–1336
Khaledian S, Nikkhah M, Shams-bakhsh M, Hoseinzadeh S (2017) A sensitive biosensor based on gold nanoparticles to detect Ralstonia solanacearum in soil. J Gen Plant Pathol 83:231–239
Khater M, De La Escosura-Muñiz A, Quesada-González D, Merkoçi A (2019) Electrochemical detection of plant virus using gold nanoparticle-modified electrodes. Anal Chim Acta 1046:123–131
Kim M, Iezzi R Jr, Shim BS, Martin DC (2019) Impedimetric biosensors for detecting vascular endothelial growth factor (VEGF) based on poly (3, 4-ethylene dioxythiophene)(PEDOT)/gold nanoparticle (Au NP) composites. Front Chem 7:1–11
Kocabay O, Emregul E, Aydın SS, Aras S (2013) Detection of superoxide radicals in tomato plants exposed to salinity, drought, cold and heavy metal stress using CMC-G-SOD biosensor. Artif Cell Nanomed Biotechnol 41(5):352–358
Kwak SY, Lew TTS, Sweeney CJ, Koman VB, Wong MH, Bohmert-Tatarev K, Strano MS (2019) Chloroplast-selective gene delivery and expression in planta using chitosan-complexed single-walled carbon nanotube carriers. Nat Nanotechnol 14(5):447–455
Lau HY, Wu H, Wee EJ, Trau M, Wang Y, Botella JR (2017) Specific and sensitive isothermal electrochemical biosensor for plant pathogen DNA detection with colloidal gold nanoparticles as probes. Sci Rep 7(1):1–7
Lew TTS, Koman VB, Silmore KS, Seo JS, Gordiichuk P, Kwak SY, Strano MS (2020a) Real-time detection of wound-induced H2O2 signalling waves in plants with optical nanosensors. Nat Plants 6(4):404–415
Lew TTS, Park M, Wang Y, Gordiichuk P, Yeap WC, Mohd Rais SK, Strano MS (2020b) Nanocarriers for transgene expression in pollen as a plant biotechnology tool. ACS Mater Lett 2(9):1057–1066
Li LL (2018) Self-assembled nanomaterials for bacterial infection diagnosis and therapy. In: In vivo self-assembly nanotechnology for biomedical applications. Springer, Singapore, pp 57–88
Li M, Liu J, Deng M, Ge Z, Afshan N, Zuo X, Li Q (2019) Rapid transmembrane transport of DNA nanostructures by chemically anchoring artificial receptors on cell membranes. ChemPlusChem 84(4):323–327
Li Z, Yu T, Paul R, Fan J, Yang Y, Wei Q (2020) Agricultural nanodiagnostics for plant diseases: recent advances and challenges. Nanosc Adv 2(8):3083–3094
Li Z, Liu Y, Hossain O, Paul R, Yao S, Wu S, Wei Q (2021) Real-time monitoring of plant stresses via chemiresistive profiling of leaf volatiles by a wearable sensor. Matter 4(7):2553–2570

Liang J, Yang Z, Tang L, Zeng G, Yu M, Li X, Luo Y (2017) Changes in heavy metal mobility and availability from contaminated wetland soil remediated with combined biochar-compost. Chemosphere 181:281–288
Lv Z, Jiang R, Chen J, Chen W (2020) Nanoparticle-mediated gene transformation strategies for plant genetic engineering. Plant J 104(4):880–891
Ma X, Wang C (2010) Fullerene nanoparticles affect the fate and uptake of trichloroethylene in phytoremediation systems. Environ Eng Sci 27(11):989–992
Majumdar S, Keller AA (2021) Omics to address the opportunities and challenges of nanotechnology in agriculture. Crit Rev Environ Sci Technol 51(22):2595–2636
Malhotra BD, Ali MA (2018) Nanomaterials in biosensors: fundamentals and applications. In: Nanomaterials for biosensors. Elsevier, pp 1–74
Marchiol L (2018) Nanotechnology in agriculture: new opportunities and perspectives. New Visions Plant Sci 9(4):pp.121-141
Mashabela MD, Masamba P, Kappo AP (2023) Applications of metabolomics for the elucidation of abiotic stress tolerance in plants: a special focus on osmotic stress and heavy metal toxicity. Plan Theory 12(2):1–18
Mazhar T, Shrivastava V, Tomar RS (2017) Green synthesis of bimetallic nanoparticles and its applications: a review. J Pharm Sci Res 9(2):102–109
Menezes-Silva PE, Sanglard LM, Ávila RT, Morais LE, Martins SC, Nobres P, DaMatta FM (2017) Photosynthetic and metabolic acclimation to repeated drought events play key roles in drought tolerance in coffee. J Exp Bot 68(15):4309–4322
Mitter N, Worrall EA, Robinson KE, Li P, Jain RG, Taochy C, Xu ZP (2017) Clay nanosheets for topical delivery of RNAi for sustained protection against plant viruses. Nat Plants 3(2):1–10
Moitra P, Bhagat D, Pratap R, Bhattacharya S (2016) A novel bio-engineering approach to generate an eminent surface-functionalized template for selective detection of female sex pheromone of Helicoverpaarmigera. Sci Rep 6(1):1–12
Mondal R, Dam P, Chakraborty J, Paret ML, Katı A, Altuntas S, Husen A (2022) Potential of nanobiosensor in sustainable agriculture: the state-of-art. Heliyon 8(12):1–20
Nakabayashi R, Saito K (2015) Integrated metabolomics for abiotic stress responses in plants. Curr Opin Plant Biol 24:10–16
Nelissen H, Sun XH, Rymen B, Jikumaru Y, Kojima M, Takebayashi Y, Inzé D (2018) The reduction in maize leaf growth under mild drought affects the transition between cell division and cell expansion and cannot be restored by elevated gibberellic acid levels. Plant Biotechnol J 16(2):615–627
Naresh V, Lee N (2021) A review on biosensors and recent development of nanostructured materials-enabled biosensors, Sensors, 21(4):1109
Oshunsanya SO, Nwosu NJ, Li Y (2019) Abiotic stress in agricultural crops under climatic conditions. In: Sustainable agriculture, forest and environmental management. Springer, Singapore, pp 71–100
Panferov VG, Safenkova IV, Byzova NA, Varitsev YA, Zherdev AV, Dzantiev BB (2018) Silver-enhanced lateral flow immunoassay for highly-sensitive detection of potato leafroll virus. Food Agric Immunol 29(1):445–457
Pedersen O, Nakayama Y, Yasue H, Kurokawa Y, Takahashi H, Heidi Floytrup A, Nakazono M (2021) Lateral roots, in addition to adventitious roots, form a barrier to radial oxygen loss in Zeanicaraguensis and a chromosome segment introgression line in maize. New Phytol 229(1):94–105
Pérez-Llorca M, Pollmann S, Müller M (2023) Ethylene and jasmonatessignaling network mediating secondary metabolites under abiotic stress. Int J Mol Sci 24(6):1–29
Pradhan S, Patra P, Mitra S, Dey KK, Jain S, Sarkar S, Goswami A (2014) Manganese nanoparticles: impact on non-nodulated plant as a potent enhancer in nitrogen metabolism and toxicity study both in vivo and in vitro. J Agric Food Chem 62(35):8777–8785

Purohit J, Chattopadhyay A, Singh NK (2019) Green synthesis of microbial nanoparticle: approaches to application. In: Microbial nanobionics: volume 2, basic research and applications. Springer, Cham, pp 35–60
Qi Y, Zhao J, Weng G, Li J, Zhu J, Zhao J (2018) Modification-free colorimetric and visual detection of Hg2+ based on the etching from core-shell structural Au-Ag nanorods to nanorices. Sensors Actuators B Chem 267:181–190
Rai M, Ribeiro C, Mattoso L, Duran N (eds) (2015) Nanotechnologies in food and agriculture, vol 33. Springer, Cham
Rajput VD, Minkina T, Upadhyay SK, Kumari A, Ranjan A, Mandzhieva S, Verma KK (2022) Nanotechnology in the restoration of polluted soil. Nano 12(5):1–16
Ramnani P, Saucedo NM, Mulchandani A (2016) Carbon nanomaterial-based electrochemical biosensors for label-free sensing of environmental pollutants. Chemosphere 143:85–98
Romeh AA (2022) Nano-phytoremediation: using plants and nanomaterials to environmental pollution remediation. In: Nanotechnology for environmental pollution decontamination. Apple Academic Press, pp 445–468
Rueda M, Godoy O, Hawkins BA (2017) Spatial and evolutionary parallelism between shade and drought tolerance explains the distributions of conifers in the conterminous United States. Glob Ecol Biogeogr 26(1):31–42
Saminathan A, Zajac M, Anees P, Krishnan Y (2022) Organelle-level precision with next-generation targeting technologies. Nat Rev Mater 7(5):355–371
Satapathi S, Kumar V, Chini MK, Bera R, Halder KK, Patra A (2018) Highly sensitive detection and removal of mercury ion using a multimodal nanosensor. Nano Struct Nano Obj 16:120–126
Saudy HS, El-Bially M, Ramadan KA, Abo El-Nasr EK, Abd El-Samad GA (2021) Potentiality of soil mulch and sorghum extract to reduce the biotic stress of weeds with enhancing yield and nutrient uptake of maize crop. GesundePflanzen 73(4):555–564
Savary S, Willocquet L, Pethybridge SJ, Esker P, McRoberts N, Nelson A (2019) The global burden of pathogens and pests on major food crops. Nat Ecol Evol 3(3):430–439
Sekhon BS (2014) Nanotechnology in agri-food production: an overview. Nanotechnol Sci Appl:31–53
Sharma P, Lew TTS (2022) Principles of nanoparticle design for genome editing in plants. Front Genome Edit 4:1–8
Sharma P, Singh J, Rani A, Sharma P (2021) A review on biosensors and nanosensors application in agroecosystems. Nanoscale Res Lett 16(1):1–24
Sharma K, Tyagi S, Vikal S, Devi A, Gautam YK, Singh BP (2023) Sustainable nanomaterials for environmental remediation. In: Handbook of green and sustainable nanotechnology: fundamentals, developments and applications. Springer, Cham, pp 933–971
Shi L, Wang Y, Chu Z, Yin Y, Jiang D, Luo J, Jin W (2017) A highly sensitive and reusable electrochemical mercury biosensor based on tunable vertical single-walled carbon nanotubes and a target recycling strategy. J Mater Chem B 5(5):1073–1080
Shojaei TR, Salleh MAM, Sijam K, Rahim RA, Mohsenifar A, Safarnejad R, Tabatabaei M (2016) Fluorometric immunoassay for detecting the plant virus Citrus tristeza using carbon nanoparticles acting as quenchers and antibodies labeled with CdTe quantum dots. Microchim Acta 183:2277–2287
Singh A, Chauhan PS (2017) Ecological significance of soil-associated plant growth–promoting biofilm-forming microbes for stress management. In: Biofilms in plant and soil health. Wiley, pp 291–326
Singh AK, Das A, Kumar P (2021) Nanobiosensors and their applications. In Nanotechnology (pp. 249-288). Jenny Stanford Publishing
Son WK, Choi YS, Han YW, Shin DW, Min K, Shin J, Kwak SY (2023) In vivo surface-enhanced Raman scattering nanosensor for the real-time monitoring of multiple stress signalling molecules in plants. Nat Nanotechnol 18(2):205–216
Song B, Xu P, Chen M, Tang W, Zeng G, Gong J, Ye S (2019) Using nanomaterials to facilitate the phytoremediation of contaminated soil. Crit Rev Environ Sci Technol 49(9):791–824

Soutter W (2021) Nanotechnology in agriculture. AZoNano https://www.azonano.com/article.aspx?ArticleID=3141z
Srivastava AK, Dev A, Karmakar S (2018) Nanosensors and nanobiosensors in food and agriculture. Environ Chem Lett 16:161–182
Su Y, Ashworth V, Kim C, Adeleye AS, Rolshausen P, Roper C, Jassby D (2019) Delivery, uptake, fate, and transport of engineered nanoparticles in plants: a critical review and data analysis. Environ Sci Nano 6(8):2311–2331
Sun D, Hussain HI, Yi Z, Rookes JE, Kong L, Cahill DM (2016) Mesoporous silica nanoparticles enhance seedling growth and photosynthesis in wheat and lupin. Chemosphere 152:81–91
Thagun C, Chuah JA, Numata K (2019) Targeted gene delivery into various plastids mediated by clustered cell-penetrating and chloroplast-targeting peptides. Adv Sci 6(23):1–14
Thangthong N, Jogloy S, Punjansing T, Kvien CK, Kesmala T, Vorasoot N (2019) Changes in root anatomy of peanut (Arachis hypogaea L.) under different durations of early season drought. Agronomy 9(5):1–18
The National Academies Press (2018) Science breakthroughs to advance food and agricultural research by 2030. The National Academies Press
Van Esse HP, Reuber TL, van der Does D (2020) Genetic modification to improve disease resistance in crops. New Phytol 225(1):70–86
Wang Xiang WX, Cheng MengZhu CM, Yang Qing YQ, Wei HaiYan WH, Xia Ao XA, Wang LiHong WL, Huang XiaoHua HX (2019) A living plant cell-based biosensor for real-time monitoring invisible damage of plant cells under heavy metal stress. Sci Total Environ 697:1–10
Wang S, Li S, Liu Q, Wu K, Zhang J, Wang S, Fu X (2015) The OsSPL16-GW7 regulatory module determines grain shape and simultaneously improves rice yield and grain quality. Nat Genet 47(8):949–954
Wang X, Qi Y, Shen Y, Yuan Y, Zhang L, Zhang C, Sun Y (2020) A ratiometric electrochemical sensor for simultaneous detection of multiple heavy metal ions based on ferrocene-functionalized metal-organic framework. Sensors Actuators B Chem 310:1–13
Wang X, Kong L, Zhou S, Ma C, Lin W, Sun X, Wang P (2022a) Development of QDs-based nanosensors for heavy metal detection: a review on transducer principles and in-situ detection. Talanta 239:1–13
Wang E, Cao Y, Bai Y, Gai Y, Shan Y, Li Q, Li Z (2022b) A triboelectric nanosensor based on ultrathin MXene composite paper for heavy metal ion detection. J Micromech Microeng 32(4):1–8
Wiebe K, Robinson S, Cattaneo A (2019) Climate change, agriculture and food security: impacts and the potential for adaptation and mitigation. In: Sustainable food and agriculture. Food and Agricultural Organization of the United Nations (FAO) and Elsevier, pp 55–74
Wu H, Nißler R, Morris V, Herrmann N, Hu P, Jeon SJ, Giraldo JP (2020) Monitoring plant health with near-infrared fluorescent H2O2 nanosensors. Nano Lett 20(4):2432–2442
Xiao M, Fu Q, Shen H, Chen Y, Xiao W, Yan D, Tang Y (2018) A turn-on competitive immunochromatographic strips integrated with quantum dots and gold nano-stars for cadmium ion detection. Talanta 178:644–649
Xu F, Fu C, Yang Y (2020) Water affects morphogenesis of growing aquatic plant leaves. Phys Rev Lett 124(3):1–6
Yang Z, Yang F, Liu JL, Wu HT, Yang H, Shi Y, Chen KM (2022) Heavy metal transporters: functional mechanisms, regulation, and application in phytoremediation. Sci Total Environ 809:1-21
Ye W, Yu M, Wang F, Li Y, Wang C (2022) Multiplexed detection of heavy metal ions by single plasmonic nanosensors. Biosens Bioelectron 196:1–6
Zhang Q, Li L, Qiao Z, Lei C, Fu Y, Xie Q, Ying Y (2017) Electrochemical conversion of Fe3O4 magnetic nanoparticles to electroactive Prussian blue analogues for self-sacrificial label biosensing of avian influenza virus H5N1. Anal Chem 89(22):12145–12151
Zhang H, Demirer GS, Zhang H, Ye T, Goh NS, Aditham AJ, Landry MP (2019) DNA nanostructures coordinate gene silencing in mature plants. Proc Natl Acad Sci 116(15):7543–7548
Zhang Y, Pribil M, Palmgren M, Gao C (2020) A CRISPR way for accelerating improvement of food crops. Nat Food 1(4):200–205

Zhao WW, Xu JJ, Chen HY (2017a) Photoelectrochemical enzymatic biosensors. Biosens Bioelectron 92:294–304

Zhao X, Meng Z, Wang Y, Chen W, Sun C, Cui B, Cui H (2017b) Pollen magnetofection for genetic modification with magnetic nanoparticles as gene carriers. Nat Plants 3(12):956–964

Zhu Q, Wang L, Dong Q, Chang S, Wen K, Jia S, Wang Y (2017) Fluorescence Resonance Energy Transfer (FRET)-based glucose imaging identifies glucose signalling in response to biotic and abiotic stresses in rice roots. J Plant Physiol 215:65–72

Zuo Q, Chen Y, Chen ZP, Yu RQ (2018) Quantification of cadmium in rice by surface-enhanced Raman spectroscopy based on a ratiometric indicator and conical holed enhancing substrates. Anal Sci 34(12):1405–1410

Chapter 2
Optical Nanobiosensors in Precision Agriculture

Ram Kumar, Pawan Kumar, Mahakdeep Singh, Rohit Bhatia, Jyoti Bala, Haramritpal Singh Sidhu, and Anil Sindhu

Abstract Precision agriculture (PA) integrates advanced technologies to optimize agricultural practices, ensuring efficient resource use and sustainability. By employing tools such as remote sensing, artificial intelligence, geographic information systems (GIS), and GPS, PA addresses the challenges of food security amid growing global populations and limited natural resources. Optical nanobiosensors represent a cutting-edge innovation in PA, offering high sensitivity, specificity, and real-time monitoring capabilities. These sensors facilitate applications including soil health monitoring, nutrient management, disease detection, and water quality optimization. They utilize nanotechnology to enhance agricultural decision-making, reduce agrochemical use, and improve crop yields while mitigating environmental impacts. Despite challenges like high implementation costs and technical barriers, integrating nanobiosensors with the Internet of Things (IoT) and artificial intelligence (AI) holds significant potential for transforming both smallholder and commercial farming. This paradigm shift promotes sustainable agricultural practices, ensuring productivity and ecological balance in the face of global food demands. Despite

R. Kumar
Chitkara College of Pharmacy, Chitkara University, Rajpura, Punjab, 140401, India

Department of Pharmaceutical Sciences and Technology, Maharaja Ranjit Singh Punjab Technical University, Bathinda, Punjab, India

P. Kumar · M. Singh · J. Bala
Department of Pharmaceutical Sciences and Technology, Maharaja Ranjit Singh Punjab Technical University, Bathinda, Punjab, India

R. Bhatia (✉)
Chitkara College of Pharmacy, Chitkara University, Rajpura, Punjab 140401, India

H. S. Sidhu
Department of Mechanical Engineering, Giani Zail Singh Campus College of Engineering & Technology, Maharaja Ranjit Singh Punjab Technical University, Bathinda, Punjab, India

A. Sindhu
Department of Biotechnology, Deenbandhu Chhotu Ram University of Science and Technology, Murthal, India

J.-T. Chen (ed.), *Nanobiosensors for Crop Monitoring and Precision Agriculture*, Sustainability Sciences in Asia and Africa,
https://doi.org/10.1007/978-981-96-8335-2_2

significant benefits, the widespread adoption of optical nanobiosensors in developing regions faces barriers such as high costs, limited access to the Internet and electricity, and the need for technical training. Innovative solutions such as Internet of Things (IoT) integration and AI-driven analytics are enhancing the accessibility and scalability of these technologies, enabling predictive insights and automated decision-making processes. Furthermore, advancements in nanomaterial synthesis, miniaturization, and field-ready sensor designs are making these tools increasingly practical for both smallholder and commercial farmers. By addressing implementation challenges and fostering global collaboration, optical nanobiosensors can revolutionize agricultural systems, offering economic benefits through reduced input costs and increased yield quality while contributing to environmental sustainability and long-term food security. This comprehensive application of nanotechnology in PA exemplifies a balanced approach to productivity and ecological stewardship.

Keywords Agricultural innovation · Sustainable farming · Crop disease · Pest monitoring

2.1 Introduction

2.1.1 Precision Agriculture

Food is fundamental for human survival, making its consistent production and supply essential to meet the growing needs of the global population. However, the increasing demand for food and limited natural resources like water pose significant challenges. In this context, sustainable agriculture becomes critical, offering solutions to ensure food security while conserving resources. One promising approach is precision agriculture (PA), particularly in developing countries where resources are often scarce. Precision agriculture uses technology to boost food production efficiency, helping farmers achieve higher yields while using fewer resources. This approach is centered on data and employs tools like remote sensing, artificial intelligence (AI), geographic information systems (GIS), and GPS. By gathering and analyzing data, precision agriculture provides farmers with specific insights tailored to the unique conditions of their fields. These insights help them make better decisions, addressing the variability in agricultural environments and optimizing farming practices (Rozin et al. 1999; Cox 2002).

PA includes several critical technologies like remote sensing, GIS, GPS, and data analysis that work together to help farmers monitor crops, map fields, and use insights to improve productivity efficiently (Godfray et al. 2010). Precision agriculture provides several key benefits, including improved efficiency by minimizing waste and optimizing the use of water, fertilizers, and pesticides, which helps farmers save costs. It also supports environmental sustainability by reducing the overuse of chemicals and conserving water resources. Additionally, precision agriculture enables farmers to use accurate crop data to enhance productivity and achieve higher yields (Gomes and Leta 2012; Bongiovanni and Lowenberg-DeBoer 2004). Precision agriculture holds significant promise, but its adoption in developing countries comes with challenges. The high cost of technology often makes it unaffordable for small-scale farmers with limited budgets. Many farmers also face restricted access to reliable Internet and electricity, which are essential for using advanced tools. Additionally, a lack of awareness and understanding about these technologies further hinders their effective use (McBratney et al. 2005).

Organizations such as Precision Agriculture for Development (PAD) work to support farmers by offering essential information and services. They use mobile technology to provide tailored agricultural advice suited to local conditions, helping improve productivity and promote sustainable practices. By focusing on collaborative approaches and local knowledge, these organizations aim to make precision agriculture more accessible and relevant to the unique challenges faced by farmers in developing regions. Precision agriculture represents a transformative approach that can significantly improve food production efficiency while addressing sustainability concerns. However, for its benefits to be fully realized in developing countries, concerted efforts are needed to overcome financial, technological, and educational barriers (Khorram et al. 2016; Abdullahi and Sheriff 2022).

2.1.2 Nanobiosensors

The global population is expected to reach 8.5 billion by 2030 and 9.8 billion by 2050, putting immense pressure on natural resources and food production systems. This growth brings challenges like shrinking farmland, food shortages, and competition for resources, making it essential to adopt advanced agricultural practices to boost crop yields in tough conditions. Biosensors, which combine biological components with physical transducers, play a vital role in this effort by detecting and measuring substances such as pathogens, metabolites, and pollutants. They work by turning biological responses into electrical signals, enabling real-time monitoring. Their main parts include a bioreceptor, which identifies the target substance, and a transducer, which converts the interaction into a readable signal. Nanotechnology, which operates at the scale of atoms and molecules, has revolutionized many fields by improving material properties. In biosensing, it helps create nanobiosensors that use the special traits of nanomaterials to enhance sensitivity, speed, and accuracy, making them powerful tools for detecting biological agents (Marchiol 2018).

Nanotechnology is proving to be a promising solution to challenges in the agri-food sector by improving agricultural practices and boosting food production efficiency. Through advanced materials, it offers innovative ways to enhance traditional methods. Various approaches, including physical, chemical, and biological methods, have been developed to synthesize nanomaterials tailored for agricultural applications, providing effective solutions for modern farming needs (Purohit et al. 2019; Mazhar et al. 2017).

2.1.3 Nanobiosensors: Enhancing Agricultural Monitoring

A key innovation in this field is the development of nanobiosensors, which combine nanomaterials with biological components to improve sensing capabilities for environmental monitoring and crop management. These sensors are characterized by their noninvasive nature, high sensitivity, and real-time responsiveness, allowing for the early detection of plant diseases and monitoring of soil conditions. Nanobiosensors utilize various nanomaterials, such as nanotubes, nanowires, and nanoparticles, which enhance their performance through unique properties that improve detection limits and response times. Nanobiosensors are specialized biosensors that utilize nanomaterials to enhance their detection capabilities. These devices are designed to detect biological agents, such as antibodies, proteins, nucleic acids, pathogens, and metabolites, with high sensitivity. The core structure of a nanobiosensor consists of a biorecognition element that binds to the target analyte and a transducer that converts the binding event into an electrical signal. This technology integration allows farmers to make informed decisions based on real-time data, ultimately leading to better crop management practices (Anubhav Kaphle et al. 2018; Verma et al. 2019; Behuria et al. 2021; Omara et al. 2019).

2.2 Types of Nanobiosensors

Biosensors can be classified into different groups depending on the method of signal transduction, as described in the following sections (Fig. 2.1).

2.2.1 Electrochemical Nanobiosensor

Electrochemical nanobiosensors use an electrode transducer to detect electrons produced from the interaction between a bioreceptor and an analyte, offering a measurable way to assess contaminants. These biosensors are highly practical, featuring portability, the ability to conduct on-site measurements, minimal sample preparation, and reduced use of reagents. They also eliminate the need for organic solvents, making them efficient tools for in-process control (Thévenot et al. 2001; Bunney et al. 2017).

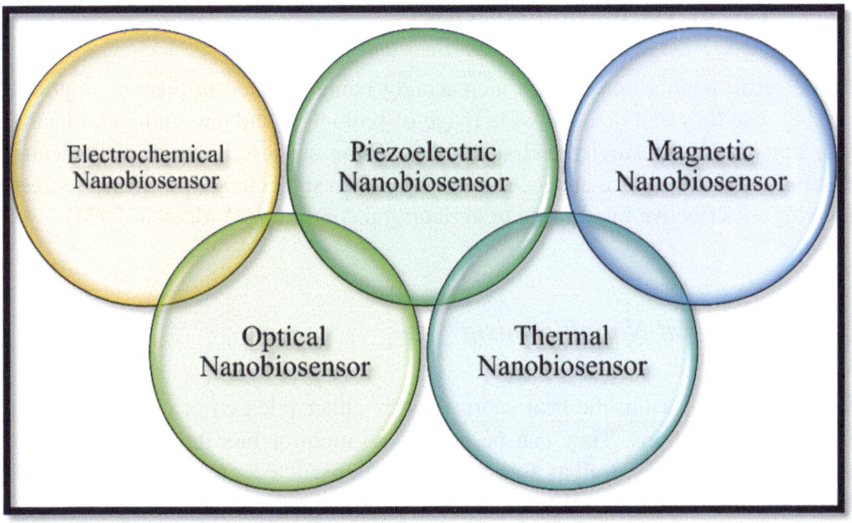

Fig. 2.1 Types of nanobiosensors in precision agriculture

2.2.2 *Optical Nanobiosensor*

An optical nanobiosensor is a device designed to detect and analyze biological molecules or chemical substances by utilizing the interaction of light with the target analyte. These sensors provide real-time, highly sensitive, and specific measurements, making them reliable tools for various applications. Over the years, optical biosensors have become increasingly prominent due to their precision and versatility (Uniyal et al. 2023; Shrikrishna et al. 2024).

2.2.3 *Magnetic Nanobiosensor*

Magnetic nanobiosensors are an innovative technology that uses magnetic nanoparticles (MNPs) for detecting biological molecules. They leverage the unique properties of MNPs, such as a large surface area, superparamagnetic behavior, and compatibility with biological systems, to enhance the sensitivity and specificity of detection. This makes them highly effective for various biological applications (Wang et al. 2024a, 2024b).

2.2.4 Piezoelectric Nanobiosensor

Piezoelectric nanobiosensors are increasingly being adopted in precision agriculture because they can detect a wide range of biological and environmental factors with remarkable sensitivity and specificity. These sensors leverage piezoelectric materials, which produce an electrical charge when subjected to mechanical stress, allowing for effective monitoring of agricultural conditions (Virk et al. 2024).

2.2.5 Thermal Nanobiosensor

These devices measure the heat energy that is either released or absorbed during biochemical reactions. They can be utilized to monitor biocatalysts in food and clinical applications as well as assess soil conditions, detect crop diseases and pests during storage, evaluate soil toxicity, and determine microbial load, among other uses (dos Santos et al. 2024).

2.2.6 Optical Nanobiosensor

Optical biosensors use light properties such as reflection, refraction, diffraction, absorption, fluorescence, and resonance to detect and measure biological analytes. These analytes include antigens, antibodies, and various biomarkers, making optical biosensors valuable for precise and reliable detection in biological applications (Fig. 2.2). Optical biosensors are based on the principle of using an optical transducer to detect interactions between biological analytes and sensing elements, which may include enzymes, antibodies, antigens, receptors, nucleic acids, whole cells, or tissues. The transducers are classified based on their mechanisms, such as surface plasmon resonance (SPR), interferometers, refractometers, absorption, fluorescence, phosphorescence, and refractive index (RI). These transducers generate a physical signal that is then processed by an electronic system for analysis (Gandhi et al. 2022).

Major categories of optical biosensors and their fundamental operating principles are outlined in the following section.

2.2.6.1 Resonance-Based Biosensor

The surface plasmon resonance (SPR) biosensor is widely used for monitoring interactions such as DNA–protein, antigen–antibody, and drug–protein. It plays a key role in label-free biosensing due to its high sensitivity to changes in the refractive index (RI) and biomolecular behavior near the sensing surface. Traditional SPR

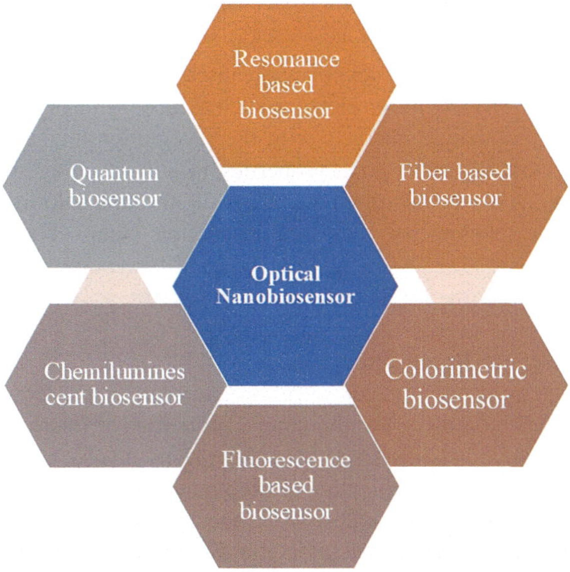

Fig. 2.2 Types of optical nanobiosensors

sensors use prismatic configurations like the Otto and Kretschmann setups, with the Kretschmann configuration being favored for its simpler fabrication, where a metal layer is applied to the prism surface. In biosensing, bioreceptors on the metal surface interact with target analytes like proteins or cells in a liquid, causing changes in the refractive index based on the analyte's concentration and properties. When light hits the prism-metal boundary, the resulting evanescent wave excites surface plasmons at the metal-dielectric interface, enhancing light intensity at specific resonance angles or wavelengths, which creates distinct SPR resonance peaks in the reflection spectrum (Lv et al. 2024; Xiao et al. 2024).

2.2.6.2 Fiber-Based Biosensor

Fiber-based biosensors use a fiber-optic component as a transducer to convert biochemical interactions into measurable optical signals. The core mechanism involves the interaction of light with a biological recognition element fixed on the fiber surface. When target molecules bind to these recognition elements, changes in light properties, such as intensity, phase, or wavelength, occur, allowing for the detection of specific analytes (Bosch et al. 2007).

2.2.6.3 Colorimetric Biosensor

Colorimetric biosensors detect color changes caused by chemical reactions between the target analyte and a recognition element, such as an enzyme or a specific binding molecule. When the analyte interacts with the recognition element, it triggers a reaction that alters the sensor's optical properties, usually resulting in a visible color change. This change can be measured quantitatively through colorimetric methods, often using spectrophotometry or visual inspection (Jeon et al. 2022; Piriya et al. 2017).

2.2.6.4 Fluorescence-Based Biosensor

Fluorescence-based biosensors use the unique properties of fluorescent molecules to detect biological interactions. These sensors typically involve a fluorescent probe that emits light when excited, allowing for the quantification of specific analytes through changes in fluorescence intensity. The mechanism relies on binding an analyte to a recognition element, which alters the fluorophore's environment, causing measurable changes in fluorescence. This technology is widely used in medical diagnostics and environmental monitoring due to its high sensitivity and specificity. Fluorescence biosensors include single fluorophore sensors and Förster resonance energy transfer (FRET) sensors, which involve energy transfer between donor and acceptor fluorophores. The addition of nanomaterials has further improved their performance, enabling detection at lower concentrations and in more complex samples (Ligler 2008; Strianese et al. 2012).

2.2.6.5 Terahertz-Based Biosensor

Terahertz (THz)-based biosensors operate in the terahertz frequency range (0.1–10 THz) and take advantage of the unique interaction between terahertz radiation and biological materials. These sensors use the spectral signatures of biomolecules in the THz range to enable label-free detection and characterization of various biological substances. The principle behind THz biosensing involves measuring changes in the amplitude and phase of terahertz waves as they interact with a sample, providing insights into molecular vibrations, rotations, and other dynamic processes within biomolecules. This makes THz biosensors highly effective for detecting proteins, nucleic acids, and complex biological structures. They are becoming valuable tools in medical diagnostics, drug development, and food safety due to their nondestructive nature and high sensitivity (Shur and Liu 2022; Konoplev et al. 2022).

2.2.6.6 Chemiluminescent Biosensor

This type of biosensor uses chemical reactions that produce light to detect specific analytes. In these sensors, a chemiluminescent substrate reacts with an enzyme or other catalyst attached to a biological recognition element. The light emitted during this reaction is directly related to the concentration of the target analyte. These biosensors are very sensitive and are useful in a wide range of applications, including environmental monitoring, safety checks, and food testing (Calabretta et al. 2021).

2.2.6.7 Quantum Biosensors

Quantum biosensors use quantum phenomena like superposition and entanglement to improve sensitivity and resolution beyond what traditional sensors can achieve. These sensors often include quantum dots or other nanostructures that have special optical properties when they interact with biomolecules. The basic principle involves measuring changes in quantum states or properties that occur when these sensors interact with target analytes. This allows for highly sensitive detection of biomolecules, even at very low concentrations. As a result, quantum biosensors hold great potential for use in medical diagnostics, environmental monitoring, and biosecurity (Bass and Doser 2024).

2.3 Fabrication of Optical Nanobiosensors

The advancement in nanotechnology has propelled the development of optical nanobiosensors tailored for agricultural applications. Fabrication methods for optical nanobiosensors encompass both traditional and advanced nanotechnological approaches aimed at enhancing sensitivity, specificity, and application range. These sensors leverage optical properties, such as fluorescence, plasmon resonance, and luminescence, to detect biochemical changes and provide precise, real-time information on various agricultural parameters. Techniques such as nanosphere lithography and surface plasmon resonance–based methodologies are central to the development of nanostructures on optical fibers for biosensing applications. For instance, gold nanostructures fabricated using nanosphere lithography have been shown to effectively detect biomarkers like carbohydrate antigen 19–9, with high sensitivity and low detection limits. Other advancements include the green synthesis of nanoparticles and the integration of aptamers, enzymes, and antibodies to improve biorecognition components in environmental monitoring and clinical diagnostics (Reddy et al. 2024). Furthermore, single-walled carbon nanotubes (SWCNTs) have emerged as pivotal in the development of nanosensors due to their resistance to photobleaching and high biological sensitivity, as demonstrated in gynecological cancer detection using implantable devices combined with AI algorithms (Yaari 2024). These fabrication methods are complemented by novel

Table 2.1 Key methods for the fabrication of optical nanobiosensors based on research findings

Methods	Description	References
Femtosecond laser direct writing	This method is used to create protein single nanowires with excellent optical properties, such as waveguiding and biocompatible transmission. It enables the customization of nanoscale structures for label-free detection applications	Sun et al. (2015)
Microemulsion-based polymerization	Fabrication of spherical nanosensors, such as PEBBLEs, involves a generalized polymerization approach. These optical sensors encapsulate fluorescent indicators in a polyacrylamide matrix for robust in vivo calibrations	Kopelman et al. (1999)
Localized surface plasmon resonance (LSPR) techniques	Metal nanostructures, including gold nanorings, are fabricated using lithography and optimized for biosensing through real-time, label-free detection of biomolecules	Lagae (2012)
Organic-inorganic hybrid microgels	Hybrid microgels embedded with fluorescent quantum dots are used for glucose nanobiosensors. Methods like "breathing in" integrate nanoparticles for high sensitivity in physiological conditions	Zhou et al. (2010)
Fiber-optic techniques	Optical fibers are drawn to nanoscale tips and functionalized with bioreceptor probes for intracellular analysis. This technique uses evanescent fields to excite target molecules	BA et al. (2018)
Sol-gel techniques	Nanobiosensors are constructed by integrating nanosystems and organic molecules into gels. This improves enzyme immobilization and detection sensitivity for various applications	Saeed et al. (2020)
Nanograting structures	Fabrication via photolithography creates nanograting devices that monitor changes in refractive index, enabling highly sensitive optical sensing	Loiseau et al. (2019)

approaches like microfluidic integration and the incorporation of label-free detection systems to enhance precision and portability, crucial for pathogen detection and point-of-care diagnostics (Roy 2022). Overall, the convergence of nanotechnology, advanced fabrication techniques, and bio-photonics is transforming optical biosensor design, promising robust and scalable applications in healthcare and environmental management. Table 2.1 shows the key methods used for the fabrication of optical nanobiosensors based on research findings.

2.4 Application of Optical Nanobiosensors in Precision Agriculture

Optical nanobiosensors transform precision agriculture by providing real-time, high-sensitivity tools for monitoring and optimizing agricultural practices. These advancements in soil health monitoring drive more sustainable farming practices, reduce input costs, and enhance crop yield quality. These sensors integrate nanotechnology and optical detection mechanisms to improve decision-making in farming. Below are the key applications.

2.4.1 Soil Health Monitoring

Soil health monitoring is a cornerstone of precision agriculture, where maintaining an optimal soil environment is crucial for sustainable crop production. Optical nanobiosensors play a vital role in assessing various soil parameters with high sensitivity, speed, and accuracy. These sensors utilize nanostructures like quantum dots, plasmonic nanoparticles, or fluorescent probes to detect and measure soil components such as nutrients, pH levels, moisture content, and contaminants.

2.4.2 Nutrient Detection and Management

Optical nanobiosensors can detect essential nutrients like nitrogen, phosphorus, and potassium in the soil. Farmers can apply fertilizers more precisely by monitoring these levels, reducing overuse and environmental runoff (Lee et al. 2019).

2.4.3 pH and Salinity Measurement

Soil pH is a critical factor influencing nutrient availability. Nanobiosensors equipped with fluorescent or colorimetric probes can rapidly measure pH changes. Similarly, salinity levels affecting plant water uptake can be monitored using plasmonic nanostructures, enabling better water and fertilizer management (Love et al. 2005).

2.4.4 Pathogen and Pest Detection

Crop diseases and pest infestations are significant challenges in agriculture. Optical nanobiosensors can detect pathogens and pests at an early stage by identifying specific biomarkers or volatile organic compounds emitted by plants under attack. These sensors utilize advanced recognition elements like DNA probes or antibodies to provide rapid and accurate identification of threats. Early detection facilitates timely interventions, reducing the need for broad-spectrum pesticide use and lowering crop losses. This targeted approach enhances both environmental and economic sustainability (Zhianmanesh et al. 2023).

2.4.5 Early Detection of Plant Stress

Plant stress, caused by factors such as drought, nutrient deficiencies, or extreme temperatures, can severely impact yield quality and quantity. Optical nanobiosensors help monitor stress markers like reactive oxygen species (ROS) or plant hormone levels (e.g. abscisic acid) in real time. By providing early warnings of stress, these sensors enable farmers to take corrective actions, such as adjusting irrigation schedules or applying specific nutrients, thereby preventing significant yield losses. This proactive approach ensures resilient crop production under varying climatic conditions (Luu et al. 2024).

2.4.6 Precision Water Management

Water is one of the most critical resources in agriculture. Optical nanobiosensors play a vital role in precision water management by measuring soil moisture levels and plant water uptake efficiently. Integrated with Internet of Things (IoT) devices, these sensors can provide real-time data on water requirements, allowing automated irrigation systems to supply water only when and where needed. This targeted irrigation reduces water wastage, lowers energy costs, and improves water use efficiency in crops. Such advancements are crucial for sustainable farming, particularly in water-scarce regions (Shanbhag et al. 2023).

2.4.7 Contaminant Detection

Soil contaminants such as heavy metals, pesticides, or microbial pathogens can degrade soil health and harm crop yields. Nanobiosensors functionalized with specific recognition elements (e.g., enzymes and DNA probes) can identify these pollutants at trace levels, ensuring safe and productive farmland (Faustino et al. 2016).

2.4.8 Real-Time Monitoring

Optical nanobiosensors are often integrated with IoT devices, allowing real-time monitoring and remote data collection. This feature enables dynamic responses to soil condition changes, optimizing resource use and improving sustainability (Taghizadeh et al. 2020).

2.4.9 Environmental Monitoring

By providing precise soil health data, these sensors minimize the overuse of agrochemicals, reduce environmental pollution, and help maintain soil biodiversity. This contributes to the overall ecological balance and long-term agricultural productivity.

2.5 Advantages and Challenges in Using Optical Nanobiosensors

Optical nanobiosensors are increasingly recognized for their potential to enhance precision agriculture by providing rapid, sensitive, and specific detection of various agricultural parameters. However, while they offer numerous benefits, there are also significant challenges associated with their deployment and scalability. Below is a detailed overview of the advantages and challenges of using optical nanobiosensors in precision agriculture.

2.5.1 Advantages

(a) *High Sensitivity and Specificity*: Optical nanobiosensors can detect analytes at very low concentrations, often down to the picomolar range. This sensitivity is crucial for monitoring nutrients, pathogens, and environmental conditions in real time. The use of nanomaterials enhances the interaction between the sensor and the target analyte, leading to improved specificity and reduced background noise during detection (Malik et al. 2024; Lamba et al. 2024; Bramley et al. 2007).
(b) *Rapid Detection*: These sensors provide quick results, allowing farmers to make timely decisions regarding crop management, such as nutrient application or disease control. This rapid response can significantly improve crop yield and health (Kumar 2020).
(c) *Cost-Effectiveness*: Although initial development costs may be high, the long-term benefits include reduced resource waste (like fertilizers and water) and increased efficiency in agricultural practices. This can lead to overall cost savings for farmers.
(d) *Versatility*: Optical nanobiosensors can be designed to detect a wide range of analytes, including nutrients, pathogens, pesticides, and soil moisture levels. This versatility makes them suitable for various applications in precision agriculture (Srivastava et al. 2018).
(e) *Sustainability*: By optimizing inputs like water and fertilizers based on real-time data, these sensors contribute to more sustainable agricultural practices that minimize environmental impact (Alengebawy et al. 2021).

2.5.2 Challenges in Deployment and Scaling

(a) *Technical Complexity*: The integration of nanotechnology into existing agricultural systems requires sophisticated technical knowledge and expertise. Farmers may need training to use these advanced technologies effectively (Civciristov and Halls 2019).
(b) *Calibration and Reliability*: Ensuring that sensors remain calibrated over time is critical for maintaining accuracy. Regular recalibration can be resource-intensive and may not always be feasible in field conditions (Tovar-Lopez 2023; Dey et al. 2012).
(c) *Environmental Factors*: Optical nanobiosensors can be sensitive to environmental conditions such as temperature, humidity, and light interference. These factors can affect sensor performance and reliability in diverse agricultural settings (Pongnumkul et al. 2015; Bhalla et al. 2020).

2.5.3 Issues Related to Cost, Sensitivity, and Environmental Conditions

(a) *Cost of Development*: While the operational costs may decrease over time, the initial investment required for developing high-quality optical nanobiosensors can be significant. This may limit accessibility for small-scale farmers (Han et al. 2023).
(b) *Sensitivity Limitations*: Despite their high sensitivity, some optical nanobiosensors may still struggle with detecting certain analytes due to interference from other substances present in complex agricultural matrices (e.g., soil or plant extracts) (Mondal and Jain 2021; Yadav et al. 2022).
(c) *Durability Concerns*: The longevity of optical nanobiosensors in harsh agricultural environments is a concern. Protective coatings may be necessary to shield sensors from corrosive elements or physical wear, but these coatings must not interfere with sensor function (Gupta et al. 2022; Eivazzadeh-Keihan et al. 2022).

2.6 Innovation and Recent Advances in Optical Biosensors in Precision Agriculture

Optical nanobiosensors are becoming increasingly vital in precision agriculture, enabling enhanced monitoring and management of crops through innovative technologies. Recent advancements focus on the integration of nanomaterials, trends toward miniaturization, and the application of artificial intelligence (AI) for data analysis.

2.6.1 Advances in Nanomaterials for Enhanced Detection

Nanomaterials significantly improve the sensitivity and specificity of optical nanobiosensors used in agricultural applications. Key developments include:

(a) *Metallic Nanoparticles*: Gold and silver nanoparticles enhance optical signals through localized surface plasmon resonance (LSPR), allowing for the detection of specific plant pathogens or nutrient levels at very low concentrations. This capability is crucial for timely interventions in crop management (Arellano Vidal and Govan 2024).
(b) *Quantum Dots*: These semiconductor nanoparticles are employed for their unique fluorescence properties, enabling multiplex detection of various analytes simultaneously. In precision agriculture, they can be used to monitor plant health by detecting stress markers or nutrient deficiencies (Ji et al. 2017).
(c) *Carbon-Based Nanomaterials*: Graphene and carbon nanotubes are utilized for their high conductivity and surface area, which enhance the performance of biosensors. They can be functionalized to detect specific biomolecules related to plant health or soil conditions, providing real-time data for farmers (Sheikh et al. 2021).

2.6.2 AI and Data Integration with Optical Nanobiosensors

1. *Machine Learning Algorithms*: These algorithms analyze complex datasets generated by nanobiosensors, identifying patterns that indicate plant health or environmental stressors. This capability allows for predictive analytics, enabling proactive management strategies (Kulkarni et al. 2022).
2. *Real-Time Monitoring*: AI facilitates real-time processing of sensor data, allowing farmers to make informed decisions based on current conditions rather than relying solely on historical data. This immediacy is crucial for managing dynamic agricultural environments effectively (Lim et al. 2015).
3. *Data Fusion Techniques*: By combining data from multiple sources (e.g., satellite imagery, weather data, and sensor readings), AI enhances the accuracy of agricultural assessments. This integrated approach supports more precise recommendations for irrigation, fertilization, and pest control (Taha et al. 2024).

2.6.3 Future Perspectives

1. *Integration of Optical Nanobiosensors with IoT in Agriculture*

Integrating optical nanobiosensors with the Internet of Things (IoT) technology can facilitate real-time monitoring and data transfer, enhancing decision-making processes in agriculture. This connectivity enables remote monitoring of soil health, moisture, and nutrient levels, promoting efficient resource management and reducing operational costs. IoT integration allows continuous monitoring of crop health, enabling early intervention and minimizing losses due to pests or nutrient deficiencies.

2. *Miniaturization and Field-Ready Designs*

In the coming years, we can expect significant advancements in the miniaturization of optical nanobiosensors, making them even more portable and field-friendly. These sensors will be smaller, more affordable, and easier to deploy across large agricultural landscapes. With more robust and durable designs, they can withstand the harsh environmental conditions typical in agriculture, such as high humidity, extreme temperatures, and soil erosion. The reduced size and enhanced practicality will also make these sensors more accessible to smallholder farmers, who may have previously been deterred by the high costs or technical challenges associated with implementing nanotechnology.

3. *Advanced Nanomaterials for Better Sensitivity and Specificity*

Future research is likely to lead to the development of new nanomaterials that further improve the sensitivity, specificity, and versatility of optical nanobiosensors. For example, quantum dots, plasmonic nanoparticles, and biofunctionalized materials could be optimized to detect a broader range of soil nutrients, contaminants, and pathogens with higher accuracy. This would help farmers monitor more variables simultaneously, offering a comprehensive understanding of soil health, crop conditions, and environmental factors in real time. Furthermore, the development of multifunctional sensors that can simultaneously measure several parameters (e.g., pH, salinity, and nutrient levels) will streamline the data collection process and reduce costs.

4. *Widespread Adoption in Developing Regions*

While the adoption of optical nanobiosensors in developed countries is already making strides, challenges remain in developing regions. The high costs of technology, lack of technical expertise, and limited infrastructure (such as unreliable electricity and Internet access) have hindered broader implementation. However, future advancements in low-cost, solar-powered, and low-maintenance sensors, combined with initiatives for providing technical training and financial support to smallholder farmers, could overcome these barriers. Moreover, the global collaboration between governments, NGOs, and private-sector companies could help bring affordable, scalable solutions to these regions, enhancing food security and improving livelihoods (Sharma et al. 2021).

5. *Enhancing Environmental Sustainability*

As agricultural practices evolve, the focus on sustainability will become even more critical. Optical nanobiosensors hold great potential for minimizing the environmental footprint of farming by reducing the overuse of water, fertilizers, and pesticides. By enabling precision farming techniques, these sensors help farmers apply resources more efficiently, reducing waste and minimizing the ecological impacts of farming practices. As regulatory frameworks around environmental protection become more stringent, the role of nanobiosensors in ensuring compliance with sustainability standards will grow. The use of these sensors can also help maintain soil health by identifying early signs of contamination or degradation, allowing for timely interventions that prevent long-term environmental damage (Kumar et al. 2024).

References

Abdullahi HS, Sheriff RE (2022) Introduction to deep learning in precision agriculture: farm image feature detection using unmanned aerial vehicles through classification and optimization process of machine learning with convolution neural network. In: Deep learning for sustainable agriculture. Academic, pp 81–107

Alengebawy A, Abdelkhalek ST, Qureshi SR, Wang MQ (2021) Heavy metals and pesticides toxicity in agricultural soil and plants: ecological risks and human health implications. Toxics 9(3):42

Anuar AA, Sani MAA, Sani NS, Yusof MI, Kumalasari SD (2024) A review on plant diseases classification methods using machine learning algorithms. Analyzing education, sustainability, and innovation. Multidiscip Res Perspect 91–98.

Anubhav Kaphle AK, Navya PN, Akhela Umapathi AU, Daima HK (2018) Nanomaterials for agriculture, food and environment: applications, toxicity and regulation. Environ Chem Lett 16(1):43–58

Arellano Vidal CL, Govan JE (2024) Machine learning techniques for improving nanosensors in agroenvironmental applications. Agronomy 14(2):341

BA AEN, YM G (2018). An electrochemical investigation in the anticorrosive properties of silver nanoparticles for the acidic corrosion of aluminium. J Electrochem 24(1):10

Bass SD, Doser M (2024) Quantum sensing for particle physics. Nat Rev Phys 6:1–11

Behuria PR, Ojha S, Nayak L, Kumar J, Kar DG, Jaiswal N, Rajvir A The transformative potential of precision farming in India: A comprehensive review.

Bhalla N, Pan Y, Yang Z, Payam AF (2020) Opportunities and challenges for biosensors and nanoscale analytical tools for pandemics: COVID-19. ACS Nano 14(7):7783–7807

Bongiovanni R, Lowenberg-DeBoer J (2004) Precision agriculture and sustainability. Precision Agric 5:359–387

Bosch ME, Sánchez AJR, Rojas FS, Ojeda CB (2007) Recent development in optical fiber biosensors. Sensors 7(6):797–859

Bunney J, Williamson S, Atkin D, Jeanneret M, Cozzolino D, Chapman J (2017) The use of electrochemical biosensors in food analysis. Curr Res Nutr Food Sci J 5(3):183–195

Calabretta MM, Zangheri M, Calabria D, Lopreside A, Montali L, Marchegiani E, Trozzi I, Guardigli M, Mirasoli M, Michelini E (2021) Based immunosensors with bio-chemiluminescence detection. Sensors 21(13):4309

Civciristov S, Halls ML (2019) Signalling in response to sub-picomolar concentrations of active compounds: pushing the boundaries of GPCR sensitivity. Br J Pharmacol 176(14):2382–2401

Cox S (2002) Information technology: the global key to precision agriculture and sustainability. Comput Electron Agric 36(2–3):93–111

Dey D, Mummert L and Sukthankar R (2012) January. Classification of plant structures from uncalibrated image sequences. In 2012 IEEE workshop on the applications of computer vision (WACV), pp 329–336

dos Santos OAL, dos Santos MS, Antunes Filho S, Backx BP (2024) Nanotechnology for the control of plant pathogens and pests. Plant Nano Biol 8:100080

Duveiller G, Defourny P, Desclée B and Mayaux P (2008) Deforestation in Central Africa: Estimates at regional, national and landscape levels by advanced processing of systematically-distributed Landsat extracts. Remote Sens Environ 112(5):1969–1981

Eivazzadeh-Keihan R, Noruzi EB, Chidar E, Jafari M, Davoodi F, Kashtiaray A, Gorab MG, Hashemi SM, Javanshir S, Cohan RA, Maleki A (2022) Applications of carbon-based conductive nanomaterials in biosensors. Chem Eng J 442:136183

Faustino V, Catarino SO, Lima R, Minas G (2016) Biomedical microfluidic devices by using low-cost fabrication techniques: a review. J Biomech 49(11):2280–2292

Gandhi RR, Chellam JAI, Prabhu TN, Kathirvel C, Sivaramkrishnan, M, Ramkumar M.S (2022) March. Machine learning approaches for smart agriculture. In 2022 6th International Conference on Computing Methodologies and Communication (ICCMC), (pp 1054–1058). IEEE.

Godfray HCJ, Crute IR, Haddad L, Lawrence D, Muir JF, Nisbett N, Pretty J, Robinson S, Toulmin C, Whiteley R (2010) The future of the global food system. Philos Trans R Soc B Biol Sci 365(1554):2769–2777

Gomes JFS, Leta FR (2012) Applications of computer vision techniques in the agriculture and food industry: a review. Eur Food Res Technol 235:989–1000

Gupta A, Mehta SK, Kunal K, Mukhopadhyay K, Singh S (2022) Quantum dots as promising nanomaterials in agriculture. In: Agricultural nanobiotechnology. Woodhead Publishing, pp 243–296

Hadary O, Marshall L, Menache I, Pan A, Greeff EE, Dion D, Dorminey S, Joshi S, Chen Y, Russinovich M and Moscibroda T (2020) Protean:{VM} allocation service at scale. In 14th USENIX Symposium on Operating Systems Design and Implementation (OSDI 20) (pp. 845–861)

Han D, Hosamo H, Ying C, Nie R (2023) A comprehensive review and analysis of nanosensors for structural health monitoring in bridge maintenance: innovations, challenges, and future perspectives. Appl Sci 13(20):11149

Jeon HJ, Kim HS, Chung E, Lee DY (2022) Nanozyme-based colorimetric biosensor with a systemic quantification algorithm for noninvasive glucose monitoring. Theranostics 12(14):6308

Ji W, Adamchuk V, Chen S, Biswas A, Leclerc M, Rossel RV (2017) The use of proximal soil sensor data fusion and digital soil mapping for precision agriculture. In: Pedometrics, p 298

Kalyani Y, Collier R (2021) A systematic survey on the role of cloud, fog, and edge computing combination in smart agriculture. Sensors, 21(17) p: 5922

Khorram S, van der Wiele CF, Koch FH, Nelson SA, Potts MD, Khorram S, van der Wiele CF, Koch FH, Nelson SA, Potts MD (2016) Future trends in remote sensing. In: Principles of applied remote sensing. Springer, New York, pp 277–285

Konoplev G, Agafonova D, Bakhchova L, Mukhin N, Kurachkina M, Schmidt MP, Verlov N, Sidorov A, Oseev A, Stepanova O, Kozyrev A (2022) Label-free physical techniques and methodologies for proteins detection in microfluidic biosensor structures. Biomedicines 10(2):207

Kulkarni MB, Ayachit NH, Aminabhavi TM (2022) Recent advancements in nanobiosensors: current trends, challenges, applications, and future scope. Biosensors 12(10):892

Kumar S (2020) Abiotic stresses and their effects on plant growth, yield and nutritional quality of agricultural produce. Int J Food Sci Agric 4(4):367–378

Kumar P, Chugh P, Ali SS, Chawla W, Sushmita S, Kumar R, Raval AV, Shamim S, Bhatia A, Kumar R (2024) Trends of Nanobiosensors in modern agriculture systems. Appl Biochem Biotechnol:1–24

Koppelman SJ, Knulst AC, Koers WJ, Penninks AH, Peppelman H, Vlooswijk R, Pigmans I, van Duijn G, Hessing M (1999) Comparison of different immunochemical methods for the detection and quantification of hazelnut proteins in food products. J Immunol Methods 229(1–2): 107–120

Lagae BJ (2012) Community-based participatory action research: An emerging alternative (Doctoral dissertation, University of Miami)

Lamba R, Chauhan P, Rani P, Sachdeva RK, Jain A and Choudhury T (2024) November Precision Agriculture: A Machine Learning Approach to Crop Recommendation. In 2024

4th International Conference on Technological Advancements in Computational Sciences (ICTACS) (pp: 1281–1286)
Lee S, Yeom B, Kim Y, Cho J (2019) Layer-by-layer assembly for ultrathin energy-harvesting films: piezoelectric and triboelectric nanocomposite films. Nano Energy 56:1–15
Ligler FS (2008) Fluorescence-based optical biosensors. In: Biophotonics. Springer, Berlin/Heidelberg, pp 199–215
Lim EK, Kim T, Paik S, Haam S, Huh YM, Lee K (2015) Nanomaterials for theranostics: recent advances and future challenges. Chem Rev 115(1):327–394
Loiseau A, Asila V, Boitel-Aullen G, Lam M, Salmain M, Boujday S (2019) Silver-based plasmonic nanoparticles for and their use in biosensing. Biosensors 9(2):78
Love JC, Estroff LA, Kriebel JK, Nuzzo RG, Whitesides GM (2005) Self-assembled monolayers of thiolates on metals as a form of nanotechnology. Chem Rev 105(4):1103–1170
Luu T, Gristwood K, Knight JC, Jörg M (2024) Click chemistry: reaction rates and their suitability for biomedical applications. Bioconjug Chem 35(6):715–731
Lv J, Wang J, Yang L, Liu W, Haihao F, Chu PK, Liu C (2024) Recent advances of optical fiber biosensors based on surface plasmon resonance: sensing principles, structures, and prospects. Sens Diagn
Malik JA, Saleem M, Almutairi S, Qadri S, Raza MA, Alsanoosy T, Chaudhary UMD (2024) Implementing fog computing in precision agriculture for real-time soil health monitoring and data management. In: Computational intelligence in internet of agricultural things. Springer, Cham, pp 371–400
Marchiol L (2018) Nanotechnology in agriculture: new opportunities and perspectives. New Visions Plant Sci 9(4):161
Mazhar T, Shrivastava V, Tomar RS (2017) Green synthesis of bimetallic nanoparticles and its applications: a review. J Pharm Sci Res 9(2):102
McBratney A, Whelan B, Ancev T, Bouma J (2005) Future directions of precision agriculture. Precis Agric 6:7–23
Mondal K, Jain AG (2021) Precision agriculture with nano-based biosensors. Precis Agric 2:15
N, Bramley RGV, Davis R, Bartels R, Schmidt E (2007) Precision agriculture options for the Australian sugarcane industry.
Omara AED, Elsakhawy T, Alshaal T, El-Ramady H, Kovács Z, Fári M (2019) Nanoparticles: a novel approach for sustainable agro-productivity. Environ Biodivers Soil Secur 3(2019):29–62
Piriya VSA, Joseph P, Kiruba Daniel SCG, Lakshmanan S, Kinoshita T, Muthusamy S (2017) Colorimetric sensors for rapid detection of various analytes. Mater Sci Eng C 78:1231–1245
Pongnumkul S, Chaovalit P, Surasvadi N (2015) Applications of smartphone-based sensors in agriculture: a systematic review of research. J Sens 2015(1):195308
Purohit J, Chattopadhyay A, Singh NK (2019) Green synthesis of microbial nanoparticle: approaches to application. In: Microbial nanobionics: volume 2, basic research and applications. Springer, Cham, pp 35–60
Rai M and Ingle A (2012) Role of nanotechnology in agriculture with special reference to management of insect pests. Applied microbiology and biotechnology 94, pp:287–293
Reddy TV, Reddy RA, Prasanna KS, Shiva SS, Meghana S and Reddy TSS (2024) September. Design and Developing AI-Driven Agro-sage for Optimal Precision Agriculture. In 2024 5th International Conference on Smart Electronics and Communication (ICOSEC) (pp: 1538–1542)
Roy P (2022) Development of optical biosensors for the diagnosis of pathogens. In: Biosensors for emerging and re-emerging infectious diseases. Academic, pp 137–168
Rozin P, Fischler C, Imada S, Sarubin A, Wrzesniewski A (1999) Attitudes to food and the role of food in life in the USA, Japan, Flemish Belgium and France: possible implications for the diet–health debate. Appetite 33(2):163–180
Saeed M, Alshammari Y, Majeed SA, Al-Nasrallah E (2020) Chemical vapour deposition of graphene—synthesis, characterisation, and applications: a review. Molecules 25(17):3856

Shanbhag A, Abramovich RS, Kingsnorth J, Tjokrosetio D, Mikulskyte O and Rothenbuchner J (2023) A new approach for the search of biosignatures and assessment of habitability on Mars using a swarm of wind-driven mobile impactors. In Proceedings of the 74th International Astronautical Congress.

Sharma P, Pandey V, Sharma MMM, Patra A, Singh B, Mehta S, Husen A (2021) A review on biosensors and nanosensors application in agroecosystems. Nanoscale Res Lett 16:1–24

Sheikh JA, Cheema SM, Ali M, Amjad Z, Tariq JZ, Naz A (2021) IoT and AI in precision agriculture: Designing smart system to support illiterate farmers. In Advances in Artificial Intelligence, Software and Systems Engineering: Proceedings of the AHFE 2020 Virtual Conferences on Software and Systems Engineering, and Artificial Intelligence and Social Computing, July 16-20, 2020, USA (pp 490–496). Springer International Publishing

Shrikrishna NS, Sharma R, Sahoo J, Kaushik A, Gandhi S (2024) Navigating the landscape of optical biosensors. Chem Eng J 490:151661

Shur M, Liu X (2022, March) Biomedical applications of terahertz technology. In: Advances in terahertz biomedical imaging and spectroscopy, vol 11975. SPIE, p 1197502

Strianese M, Staiano M, Ruggiero G, Labella T, Pellecchia C, D'Auria S (2012) Fluorescence-based biosensors. In: Spectroscopic methods of analysis: methods and protocols. Humana Press, Totowa, pp 193–216

Sun YL, Sun SM, Wang P, Dong WF, Zhang L, Xu BB, Chen QD, Tong LM, Sun HB (2015) Customization of protein single nanowires for optical biosensing. Small (Weinheim an der Bergstrasse, Germany) 11(24):2869–2876

Taghizadeh S, Vazehan R, Beheshtian M, Sadeghinia F, Fattahi Z, Mohseni M, Arzhangi S, Nafissi S, Kariminejad A, Najmabadi H and Kahrizi K (2020) Molecular Diagnosis of Hereditary Neuropathies by Whole Exome Sequencing and Expanding the Phenotype Spectrum. Archives of Iranian Medicine (AIM) (7)

Taha BA, Kadhim AC, Addie AJ, Haider AJ, Azzahrani AS, Raizada P, Rustagi S, Chaudhary V, Arsad N (2024) Advancing cancer diagnostics through multifaceted optical biosensors supported by nanomaterials and artificial intelligence: A panoramic outlook. Microchem J:111307

Thévenot DR, Toth K, Durst RA, Wilson GS (2001) Electrochemical biosensors: recommended definitions and classification. Biosens Bioelectron 16(1–2):121–131

Tovar-Lopez FJ (2023) Recent progress in micro-and nanotechnology-enabled sensors for biomedical and environmental challenges. Sensors 23(12):5406

Uniyal A, Srivastava G, Pal A, Taya S, Muduli A (2023) Recent advances in optical biosensors for sensing applications: a review. Plasmonics 18(2):735–750

Verma S, Lu S, Kenis PJ (2019) Co-electrolysis of CO2 and glycerol as a pathway to carbon chemicals with improved technoeconomics due to low electricity consumption. Nat Energy 4(6):466–474

Virk V, Deepak H, Taneja K, Srivastava R, Giri S (2024) Amelioration in nanobiosensors for the control of plant diseases: current status and future challenges. Front Nanotechnol 6:1310165

Wang M, Jin L, Hang-Mei Leung P, Wang-Ngai Chow F, Zhao X, Chen H, Pan W, Liu H, Li S (2024a) Advancements in magnetic nanoparticle-based biosensors for point-of-care testing. Front Bioeng Biotechnol 12:1393789

Wang Q, Zhang D, Qian Y, Yin X, Wang L, Zhang S, Wang Y (2024b) Research on fiber optic surface Plasmon resonance biosensors: a review. Photon Sens 14(2):240201

Xiao L, Wang G, Wang E, Liu S, Chang J, Zhang P, Zhou H, Wei Y, Zhang H, Zhu Y, Shi Z (2024) Spatiotemporal co-optimization of agricultural management practices towards climate-smart crop production. Nat Food 5(1):59–71

Yaari Z (2024) Developing optical nanosensors for biomedical applications. In: Electrochemical society meeting abstracts 245, vol 8. The Electrochemical Society, Inc, pp 852–852

Yadav A, Pal S, Singh N (2022) Precision tools assisted nutrient management for enhancing nutrient use efficiency. Vigyan Varta SP1:75–79

Yao H, and Clark R.L, (1999) Development of topographic maps for precision farming with medium accuracy GPS receivers. Applied Engineering in Agriculture, 16(6), pp.629-636

Yao Y, Miao Y, Huang S, Gao L, Ma X, Zhao G, Jiang R, Chen X, Zhang F, Yu K and Gnyp M.L, (2012). Active canopy sensor-based precision N management strategy for rice. Agronomy for sustainable development 32, pp.925-933

Zhianmanesh M, Gilmour A, Bilek MM, Akhavan B (2023) Plasma surface functionalization: a comprehensive review of advances in the quest for bioinstructive materials and interfaces. Appl Phys Rev 10(2):021301

Zhou T, Lu Y, Wang B (2010) Integrating TTF and UTAUT to explain mobile banking user adoption. Comput Hum Behav 26(4):760–767

Chapter 3
Calorimetric Nanobiosensors: Revolutionizing Precision Agriculture for Sustainable Crop Management

Verinder Virk, Sadhana Giri, Kuldeep Giri, and Rishita Shrivastva

Abstract Calorimetric nanobiosensors have become an innovative tool in precision agriculture, offering a cutting-edge method for assessing vital agricultural parameters, such as soil nutrient levels, plant health, and environmental stress. These sensors detect and measure heat changes generated by biochemical reactions, enabling real-time, precise data collection that aids in optimizing crop management strategies. By leveraging nanomaterials with calorimetric techniques, these sensors can monitor nutrient absorption, detect plant pathogens, and assess environmental stress factors, thus reducing the overuse of fertilizers and pesticides. This chapter delves into the methodologies involved in designing and applying calorimetric nanobiosensors, showcasing their role in enhancing agricultural productivity and promoting sustainable practices. Moreover, it addresses challenges such as production costs, technical expertise, and integration within existing agrarian frameworks. Prospects for large-scale deployment are also discussed, emphasizing the potential of these sensors for revolutionizing modern agriculture.

Keywords Calorimetric biosensors · Precision agriculture · Nanomaterials · Nutrient monitoring · Sustainable practices

V. Virk · S. Giri (✉) · R. Shrivastva
Department of Microbiology, Kanya Gurukul, Gurukula Kangri (Deemed to be University), Haridwar, Uttarakhand, India

K. Giri (✉)
Department of Computer Science and Engineering, Gurukula Kangri (Deemed to be University), Haridwar, Uttarakhand, India
e-mail: Kuldeep.giri@gkv.ac.in

J.-T. Chen (ed.), *Nanobiosensors for Crop Monitoring and Precision Agriculture*, Sustainability Sciences in Asia and Africa,
https://doi.org/10.1007/978-981-96-8335-2_3

3.1 Introduction

The global population is projected to reach 9.8 billion by 2050, intensifying the need for innovative approaches to food production. As climate change and resource depletion threaten traditional farming methods, precision agriculture has emerged as a critical solution. This data-driven approach leverages technologies such as sensors, drones, and advanced analytics to optimize crop management and reduce environmental impacts (Zhao et al. 2021a, b). Among these technologies, calorimetric nanobiosensors have garnered significant attention due to their potential to provide real-time, accurate data for a range of agricultural applications (Fig. 3.1). Calorimetric biosensors work by measuring the heat released or absorbed during biochemical reactions. These sensors are highly sensitive and capable of detecting small fluctuations in temperature that occur as a result of processes such as nutrient uptake, pathogen invasion, or plant stress responses. When coupled with nanomaterials, these sensors become even more efficient, with their high surface area and reactivity enhancing their sensitivity. By detecting biochemical changes at the

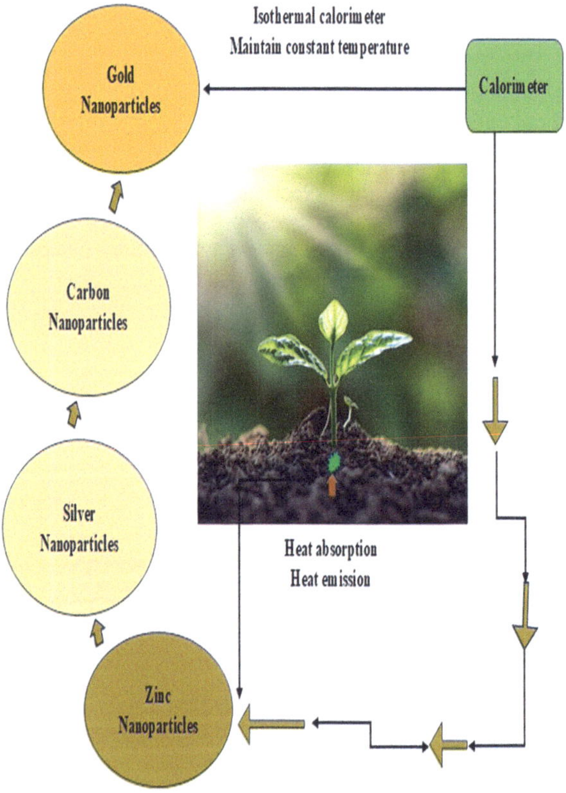

Fig. 3.1 Application of calorimetric sensors in agriculture

molecular level, calorimetric sensors provide insights that can significantly improve decision-making in crop management, ultimately leading to more efficient and sustainable farming practices. The principles behind calorimetric nanobiosensors are grounded in thermodynamic reactions, where changes in heat correspond to specific biochemical events. These sensors are designed to convert thermal energy variations into measurable signals, enabling the monitoring of complex biochemical processes in plants. For instance, during nutrient uptake, specific enzymes catalyze reactions that generate heat (Zhao et al. 2021a, b). Calorimetric sensors can detect these subtle changes, allowing for precise monitoring of nutrient dynamics in the soil–plant interface. This capability is particularly valuable in tailoring fertilizer applications to match the specific needs of crops, thereby reducing waste and minimizing environmental pollution.

One of the most promising applications of calorimetric nanobiosensors lies in disease management. Early detection of plant pathogens is crucial for mitigating crop losses, which amount to billions of dollars annually worldwide. Traditional diagnostic methods, such as polymerase chain reaction (PCR) or enzyme-linked immunosorbent assay (ELISA), while effective, are time-consuming and require specialized equipment (Jones and White 2021). In contrast, calorimetric sensors offer rapid, on-site detection of pathogens by identifying heat changes associated with pathogen–host interactions. For example, when a fungal pathogen invades a plant, specific metabolic reactions occur that produce distinct thermal signatures. These signals can be captured by calorimetric sensors, enabling farmers to take timely action to prevent the spread of disease.

The integration of nanotechnology further enhances the performance of calorimetric biosensors. Nanomaterials such as gold nanoparticles (AuNPs), carbon nanotubes (CNTs), and graphene have unique properties, including high surface area, excellent conductivity, and strong thermal sensitivity. These characteristics make them ideal for improving the detection capabilities of biosensors. For instance, gold nanoparticles can amplify the thermal signals generated during biochemical reactions, making it possible to detect minute changes that might otherwise go unnoticed. Additionally, the use of nanomaterials enables the miniaturization of biosensors, facilitating their deployment in remote and resource-limited settings. Stress monitoring is another critical area where calorimetric nanobiosensors are making a significant impact. Plants respond to various abiotic stresses, such as drought, salinity, and temperature extremes, through biochemical changes that involve heat generation or absorption. By detecting these changes, calorimetric sensors can provide real-time insights into the stress levels of plants, allowing farmers to implement targeted interventions. For example, calorimetric sensors can monitor the plant's water stress levels during a drought and guide irrigation schedules to optimize water use. Such precision in stress management is vital for enhancing crop resilience to climate change. Advancements in calorimetric nanobiosensors have also led to their application in monitoring soil health; however, there are challenges in this field (Fig. 3.2). Soil is a dynamic system where microbial activities and chemical processes generate heat. By analyzing these thermal signatures, biosensors can provide valuable information about soil fertility, microbial diversity, and nutrient

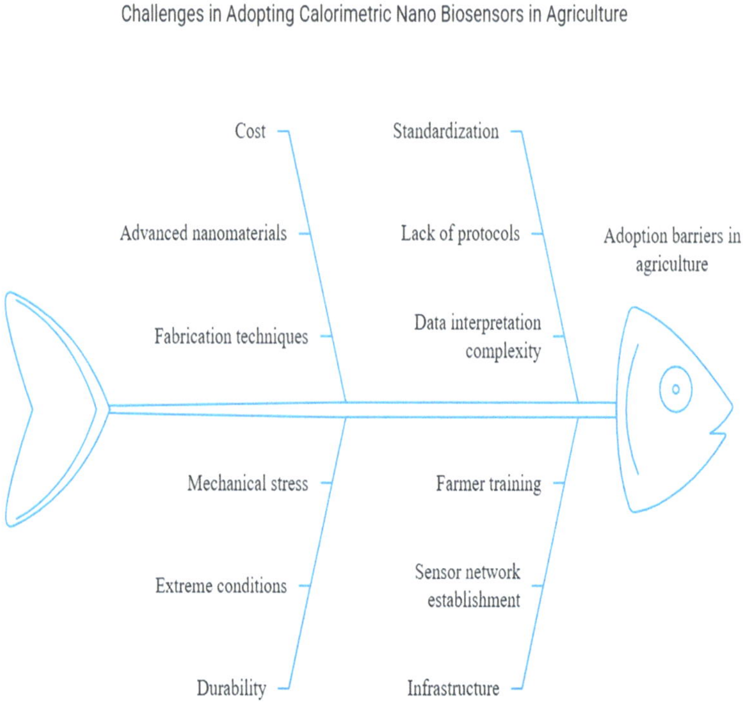

Fig. 3.2 Challenges in adopting colorimetric nanobiosensors in agriculture

availability. This data can inform sustainable soil management practices, ensuring that agricultural productivity is maintained without compromising environmental integrity. For instance, calorimetric sensors can detect imbalances in soil nutrient cycles, prompting timely interventions to restore soil health. Despite their immense potential, the adoption of calorimetric nanobiosensors in agriculture faces several challenges (Zhao and Liang 2022). One major hurdle is the cost associated with the development and deployment of these sensors. The integration of advanced nanomaterials and the need for precise fabrication techniques make these devices expensive, limiting their accessibility to smallholder farmers. Additionally, the durability and stability of calorimetric sensors under field conditions remain a concern. Factors such as extreme temperatures, humidity, and mechanical stress can affect the performance of these sensors, necessitating the development of robust and resilient designs (Ahmed and Yadav 2020). Another challenge is the lack of standardized protocols for the use of calorimetric biosensors in agricultural settings. The interpretation of thermal data requires a deep understanding of plant physiology and biochemistry, which may not be readily available to all users. To address this, efforts are being made to develop user-friendly interfaces and decision support systems that can simplify the interpretation of sensor data. For example, integrating calorimetric sensors with wireless communication technologies and artificial intelligence

(AI) can enable real-time data analysis and provide actionable recommendations to farmers. The scaling up of calorimetric nanobiosensors for large-scale agricultural use also requires significant investments in infrastructure and training. Establishing a network of sensors across vast agricultural landscapes involves logistical challenges and substantial financial resources. Furthermore, farmers need to be trained in using and maintaining these sensors to ensure their effective implementation. Collaborative efforts between researchers, policymakers, and industry stakeholders are essential to overcome these barriers and promote the widespread adoption of calorimetric biosensors.

Future research in the field of calorimetric nanobiosensors is focused on addressing these challenges and exploring new frontiers. One promising direction is the development of multifunctional sensors that can simultaneously monitor multiple parameters, such as temperature, pH, and nutrient levels (Fig. 3.3). Such sensors would provide a comprehensive picture of plant health and soil conditions, enabling holistic decision-making. Additionally, advances in nanotechnology are paving the way for the creation of self-powered sensors that can harvest energy from environmental sources, reducing the need for external power supplies. The integration of calorimetric biosensors with precision agriculture platforms is another exciting area

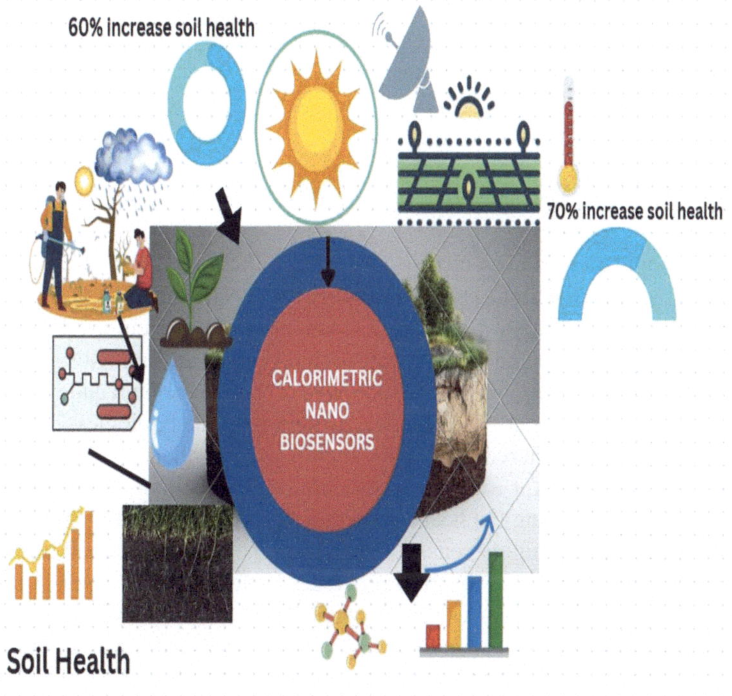

Fig. 3.3 Sensors that simultaneously monitor multiple parameters, such as temperature, pH, and nutrient levels, of soil and water samples

of innovation. By combining sensor data with satellite imagery, weather forecasts, and machine learning algorithms, it is possible to create predictive models that optimize crop management strategies. For instance, calorimetric sensors can detect early signs of stress in crops, which, when combined with predictive models, can guide farmers in adjusting their practices to mitigate potential yield losses. Such integrated approaches have the potential to revolutionize agriculture by making it more data-driven, adaptive, and sustainable.

Recent advancements in material science are also contributing to the evolution of calorimetric nanobiosensors. The use of bioinspired materials, such as silk fibroin and cellulose nanofibers, is being explored to enhance the biocompatibility and environmental sustainability of these sensors. These materials not only improve the performance of biosensors but also align with the principles of green chemistry, reducing the ecological footprint of sensor production. Furthermore, the incorporation of nanostructured polymers and metal-organic frameworks (MOFs) is opening new possibilities for designing sensors with unprecedented sensitivity and specificity. The potential of calorimetric nanobiosensors extends beyond agriculture to include applications in environmental monitoring, food safety, and healthcare. For example, these sensors can be used to detect contaminants in water or soil, ensuring the safety of agricultural inputs. In the context of food safety, calorimetric sensors can monitor the freshness and quality of produce, reducing food waste and ensuring consumer health. The versatility of these sensors makes them a valuable tool for addressing the interconnected challenges of food security, environmental sustainability, and public health.

3.2 Calorimetric Sensing

The global population is projected to reach 9.8 billion by 2050, intensifying the need for innovative approaches to food production. As climate change and resource depletion threaten traditional farming methods, precision agriculture has emerged as a critical solution. This data-driven approach leverages technologies such as sensors, drones, and advanced analytics to optimize crop management and reduce environmental impacts (Zhao et al. 2021a, b). Among these technologies, calorimetric nanobiosensors have garnered significant attention due to their potential to provide real-time, accurate data for a range of agricultural applications. Calorimetric biosensors work by measuring the heat released or absorbed during biochemical reactions. These sensors are highly sensitive and capable of detecting small fluctuations in temperature that occur as a result of processes such as nutrient uptake, pathogen invasion, or plant stress responses. When coupled with nanomaterials, these sensors become even more efficient, with their high surface area and reactivity enhancing their sensitivity (Lee and Patel 2023). By detecting biochemical changes at the molecular level, calorimetric sensors provide insights that can significantly improve decision-making in crop management, ultimately leading to more efficient and sustainable farming practices (Smith and Thompson 2020). At the heart of calorimetric sensors is the measurement of thermal changes in response to biochemical reactions. Every biochemical process, from nutrient absorption to pathogen interactions, results in heat exchange, either through the release or absorption of energy. Calorimetric sensors are designed to detect these temperature fluctuations with high precision. The core principles behind calorimetric sensing are based on thermodynamic changes during biochemical events. For example, when plants absorb nutrients from the soil, it involves enzymatic reactions that either absorb or release heat. This heat change can be detected by the sensor, providing insights into the plant's nutrient uptake and the overall health of the soil. Recent research highlights the versatility of calorimetric sensors in agricultural applications. For instance, Miller and Kumar (2022) demonstrated the use of calorimetric techniques to monitor plant health by detecting stress-related thermal signatures. These insights enable proactive interventions to mitigate yield losses caused by abiotic and biotic stresses. Furthermore, nanotechnology advancements have enhanced the performance of these sensors. The incorporation of materials like graphene and carbon nanotubes has improved their thermal sensitivity and durability, making them suitable for real-time field deployment (Lee and Patel 2023). One significant advantage of calorimetric biosensors is their ability to detect early-stage infections in plants. Brown and Green (2019) noted that these sensors could identify specific thermal patterns associated with pathogen–host interactions, enabling timely management strategies. This rapid detection is critical in addressing diseases that can devastate crops, such as fungal infections or bacterial blights. Additionally, Smith and Thompson (2020) emphasized the role of nanotechnology in enhancing sensor specificity, allowing for the differentiation between various stressors and pathogens.

Soil health monitoring is another critical application area. Calorimetric sensors provide insights into microbial activity and nutrient cycles by detecting heat changes during biochemical soil processes. Jones and White (2021) reported that integrating these sensors with AI platforms could transform soil management practices, enabling data-driven decisions to optimize fertility and reduce environmental impacts. Such integration also supports the precise application of fertilizers and other soil amendments, ensuring sustainable resource use. Despite their potential, the widespread adoption of calorimetric sensors faces challenges. Cost remains a barrier, as advanced nanomaterials and precise fabrication techniques make these devices expensive (Ahmed and Yadav 2020). Durability under varying field conditions is another concern, necessitating the development of robust designs that withstand mechanical stress and environmental fluctuations. Additionally, interpreting the complex data generated by these sensors requires specialized knowledge, which can be a limiting factor for practical implementation. Emerging research is addressing these challenges through innovation. For example, Zhao et al. (2021a, b) highlighted efforts to develop cost-effective fabrication methods using bioinspired materials like cellulose nanofibers. These materials not only reduce production costs but also improve sensor biocompatibility.

Furthermore, advancements in wireless communication and Internet of Things (IoT) integration are simplifying data interpretation, enabling farmers to access actionable insights through user-friendly interfaces (Jones and White 2021).

Prospects for calorimetric sensors include the development of multifunctional platforms capable of monitoring various parameters simultaneously, such as nutrient levels, temperature, and stress markers. This comprehensive approach promises to enhance crop management strategies by providing a holistic view of plant and soil health (Smith and Thompson 2020). Additionally, the integration of AI and machine learning algorithms with sensor data is paving the way for predictive models that optimize farming practices, mitigate risks, and enhance productivity (Zhao et al. 2021a, b).

3.3 Nanomaterials in Calorimetric Biosensors

Nanomaterials exhibit unique properties at the nanometer scale (1–100 nm). These properties include high surface area, high reactivity, quantum effects, and enhanced optical, thermal, and electrical behaviors. When incorporated into biosensors, these features significantly enhance the sensitivity and selectivity of the sensors (Kumar and Roy 2021). Calorimetric biosensors, which measure changes in heat as a result of biochemical reactions, benefit immensely from the integration of nanomaterials. This paper explores the role of various nanomaterials, such as nanoparticles (NPs), carbon nanotubes, gold nanorods (AuNRs), and quantum dots (QDs), in enhancing the performance of calorimetric biosensors, with a particular focus on their applications in agriculture (Hossain and Hassan 2021).

3.4 The Role of Nanoparticles in Calorimetric Biosensors

Nanoparticles (NPs) are among the most commonly used nanomaterials in biosensor development due to their unique physical and chemical properties. Gold nanoparticles (AuNPs), for instance, have been widely applied in calorimetric biosensors due to their excellent heat transfer efficiency, stability, and biocompatibility. Their high surface-area-to-volume ratio enables the immobilization of a significant number of biomolecules, thereby enhancing the detection sensitivity of the sensor (Wang and Zhang 2020).

AuNPs are particularly effective in detecting plant stress or pathogen presence. When conjugated with specific antibodies or ligands, they can target molecules indicative of plant health, such as stress hormones or pathogen-specific antigens (Lee and Park 2021). The calorimetric response, resulting from heat changes during these biochemical interactions, can provide accurate and real-time data. Sharma et al. (2020a, b) demonstrated that AuNP-based biosensors could detect plant pathogens at extremely low concentrations, significantly improving early detection and management strategies in agriculture.

3.5 Carbon Nanotubes in Calorimetric Biosensors

Carbon nanotubes (CNTs) have unique electrical and thermal properties, making them valuable nanomaterials in calorimetric biosensors. Their high aspect ratio and ability to conduct heat efficiently allow them to amplify the thermal signals generated during biochemical reactions. Moreover, CNTs can be functionalized with various biomolecules, such as enzymes, antibodies, or nucleic acids, to enhance their specificity and versatility in detecting target analytes. In agriculture, functionalized CNTs have been used to monitor nutrient uptake and detect disease biomarkers. For example, CNT-based biosensors can identify molecules like ethylene, a stress-related hormone in plants (Gupta and Singh 2023). When ethylene interacts with functionalized CNTs, the reaction generates a heat change detectable by the calorimetric biosensor. This capability enables real-time monitoring of plant stress levels, providing valuable insights for optimizing agricultural practices (Zhang and Wei 2022).

3.6 Gold Nanorods and Their Applications

Gold nanorod (AuNR) is another class of nanomaterials used in calorimetric biosensors. Unlike spherical nanoparticles, nanorods have anisotropic shapes, which provide unique optical and thermal properties. Their ability to absorb and convert light into heat efficiently makes them ideal for applications requiring precise temperature measurements. In agricultural biosensing, AuNRs have been employed to detect plant stress markers, soil nutrient levels, and the presence of harmful pathogens (Ali and Rahman 2021). Their strong plasmonic resonance enables enhanced thermal sensitivity, allowing the detection of minute temperature changes associated with biochemical interactions. For instance, Liu et al. (2021) reported the use of AuNR-based biosensors for detecting nitrate levels in soil, facilitating better nutrient management and crop yield optimization.

3.7 Quantum Dots in Calorimetric Biosensors

Quantum dots (QDs) are semiconductor nanoparticles that exhibit size-dependent optical and electronic properties. Their high quantum yield, photostability, and tunable emission spectra make them attractive for various sensing applications, including calorimetric biosensors (Brown and Taylor 2022). QDs can act as heat-generating agents when exposed to specific wavelengths of light, and their interaction with target molecules can induce measurable thermal changes. In agriculture, QDs have been utilized to detect pesticide residues, soil contaminants, and plant disease markers. For example, QD-based calorimetric biosensors can identify pesticide molecules by generating heat through specific binding interactions. This thermal response can then be correlated to the concentration of the pesticide, enabling accurate and rapid detection. Recent studies, such as those by Chen et al. (2022), highlight the potential of QD-based biosensors in promoting sustainable agricultural practices by minimizing pesticide overuse and ensuring food safety (Yadav and Bhatia 2020).

3.8 Types of Calorimetric Sensors

Calorimetric sensors are typically divided into **differential calorimeters** and **isothermal calorimeters**. Both types measure temperature variations, but they differ in their design and application.

- **Differential Calorimeters**: These sensors measure the temperature difference between the sample and a reference material. They are useful for detecting small temperature changes and can be employed for highly sensitive applications in soil and plant health monitoring (Ahmad and Khan 2021).
- **Isothermal Calorimeters**: These sensors maintain a constant temperature and measure the heat flux in and out of the sample. They are often used in studying biochemical reactions, such as nutrient absorption or pathogen response, where precise temperature control is critical.

3.9 Monitoring Soil Health and Nutrient Levels

Soil health is the cornerstone of agricultural productivity. The availability of nutrients in the soil directly impacts crop yields and the efficiency of fertilizer use. Calorimetric sensors are capable of monitoring nutrient uptake in plants by detecting the heat generated during nutrient absorption. These sensors help in determining optimal fertilization strategies, preventing overfertilization, and minimizing environmental pollution due to excess chemical runoff (Saha et al. 2020).

By integrating calorimetric sensors into precision farming systems, farmers can receive real-time data on nutrient levels in the soil. This data is essential for making timely adjustments to irrigation and fertilization practices, which, in turn, promote sustainable agriculture by minimizing waste and reducing the environmental impact of excessive fertilizer use.

3.10 Assessing Plant Health

Calorimetric nanobiosensors can play a vital role in monitoring plant health. Stress factors such as water scarcity, nutrient deficiencies, disease, and extreme temperatures result in biochemical changes within the plant. These changes can be detected through heat variations that are measured by the calorimetric sensor. For example, when plants undergo oxidative stress due to drought, the metabolic changes lead to heat emissions that can be recorded by the sensors (Chakraborty et al. 2023).

The ability to detect these changes early allows for targeted interventions, such as adjusting irrigation schedules or applying fertilizers, thereby optimizing crop yield. Furthermore, detecting stress responses in plants enables farmers to adjust practices in real time, potentially preventing crop loss.

3.11 Early Detection of Pathogens

Pathogen detection is crucial for preventing crop diseases. Traditionally, detecting plant pathogens involves labor-intensive procedures, including visual inspections and laboratory tests. However, calorimetric nanobiosensors offer a rapid and noninvasive alternative. Pathogen-induced biochemical reactions generate heat, which can be detected by the sensors. This early detection allows farmers to implement disease control measures before pathogens spread extensively, thereby reducing the need for pesticides (Ghosal et al. 2021).

3.12 Environmental Stress and Climate Monitoring

Climate change has exacerbated environmental stress on crops, such as drought, high salinity, and temperature fluctuations. Calorimetric sensors can detect plant responses to these environmental stressors by measuring heat changes resulting from physiological adjustments. For instance, plants may increase respiration or modify their metabolic processes under stress conditions, which can alter their heat signature. By continuously monitoring environmental stress through calorimetric sensors, farmers can better understand how climate change is impacting crop health and adjust their farming practices accordingly (Yuan et al. 2022).

3.13 Recent Advances in Calorimetric Nanobiosensors

3.13.1 Multi-sensor Platforms for Comprehensive Monitoring

Recent research has focused on combining calorimetric biosensors with other types of sensors, such as **optical** and **electrochemical sensors**, to create multi-sensor platforms. These integrated systems allow for more accurate and comprehensive monitoring of multiple agricultural parameters, such as nutrient levels, moisture content, and pathogen presence (Zhou et al. 2023). By combining different sensor types, farmers can obtain holistic data that encompasses both biochemical and environmental factors affecting crop health.

3.13.2 Wireless Technology and IoT Integration

The integration of wireless communication and the Internet of Things (IoT) has allowed for remote monitoring of agricultural fields (Table 3.1). Through the use of IoT-enabled devices, data from calorimetric sensors can be transmitted to central systems for analysis and decision-making. This capability allows farmers to monitor their crops from anywhere and receive real-time data on their mobile devices or computers, which enhances the timeliness of agricultural interventions (Li et al. 2024a, b).

Table 3.1 Wireless technology and IoT integration in agriculture

Aspect	Description	References
Wireless communication	Wireless protocols like Wi-fi, Zigbee, LoRa WAN, and cellular networks enable remote data transmission	Kumar and Kumar (2022)
IoT-enabled devices	Use of IoT devices like sensors (e.g., calorimetric sensors) to monitor environmental parameters in agriculture. These devices collect data and transmit it wirelessly to a central system	Li et al. (2024a, b)
Remote monitoring	Farmers can monitor their fields in real time via mobile devices or computers, enabling immediate interventions	Zhang et al. (2023)
Data transmission and analysis	Data collected by sensors is transmitted to a central system for processing, analysis, and decision-making	Sharma et al. (2023)
Benefits	Timely interventions, resource optimization (water and fertilizer), and improved crop yields	Li et al. (2024a, b)
Decision-making	Data analytics in central systems help in decision-making for irrigation, pest control, and crop management	Singh et al. (2023)
Challenges	Connectivity issues, power requirements, and high setup costs for IoT networks in remote areas	Singh et al. (2023)

3.13.3 Improved Nanomaterials for Enhanced Sensitivity

Advancements in nanomaterial development have led to improved sensor designs with better sensitivity and specificity. For instance, **graphene-based** materials have been explored for use in calorimetric sensors due to their excellent thermal conductivity and biocompatibility (Jiang et al. 2022). Graphene's properties can improve the precision of temperature measurements, allowing for more accurate detection of biochemical reactions (Tian et al. 2021).

3.14 Challenges in the Deployment of Calorimetric Nanobiosensors

3.14.1 High Production Costs

One of the main barriers to the widespread adoption of calorimetric nanobiosensors in agriculture is the high production cost. The synthesis of nanomaterials and the fabrication of sophisticated sensors are resource-intensive processes. As a result, the initial investment in such technologies can be prohibitively expensive for small-scale farmers. While the costs may decrease over time as manufacturing processes improve, the current financial barriers hinder large-scale adoption (Zhao et al. 2021a, b).

3.14.2 Technical Expertise Requirements

Calorimetric sensors require a certain level of technical expertise for both their operation and data interpretation. Many farmers may not have the necessary training to utilize these sensors effectively. To bridge this gap, there is a need for educational programs and training workshops to equip farmers with the knowledge to deploy and interpret sensor data (Saha et al. 2020).

3.14.3 Data Management and Integration Challenges

The vast amounts of data generated by these sensors require efficient data management systems. In addition to hardware challenges, farmers need user-friendly software tools that can integrate and analyze data from various sensor networks. Ensuring compatibility with existing agricultural systems and providing actionable insights from the data remain significant challenges (Yuan et al. 2022).

3.14.4 Environmental Variability and Sensor Calibration

Environmental factors such as temperature, humidity, and soil conditions can affect the performance of calorimetric sensors. Sensors need to be calibrated for specific environmental conditions, which may vary from one agricultural region to another. Furthermore, sensor performance may degrade over time, necessitating regular calibration and maintenance (Chakraborty et al. 2023).

3.15 Future Prospects

Despite the challenges, the future of calorimetric nanobiosensors in agriculture is promising. Advances in nanomaterial development, sensor miniaturization, and integration with IoT are likely to improve the affordability, accuracy, and functionality of these sensors. As these technologies become more accessible and cost-effective, they have the potential to significantly enhance agricultural productivity and sustainability on a global scale. Research in the field is also expected to explore the use of AI and machine learning algorithms for sensor data analysis. These advancements could further optimize farming practices by providing actionable insights from large datasets, allowing for precision farming techniques to be more widely adopted (Li et al. 2024a, b).

3.16 Conclusion

Nanomaterials have revolutionized the field of calorimetric biosensors by enhancing their sensitivity, selectivity, and functionality. The use of materials such as AuNPs, CNTs, AuNRs, and QDs has opened new avenues for real-time monitoring of plant health, soil quality, and pathogen presence in agriculture. Despite the challenges, ongoing research and technological advancements are expected to overcome these limitations, paving the way for the broader application of nanomaterial-based calorimetric biosensors. Their potential to improve agricultural productivity and sustainability makes them a promising tool for the future. Calorimetric biosensors represent a significant step forward in the integration of nanotechnology and agricultural sciences. These sensors offer unparalleled precision in detecting and quantifying key biological and chemical processes in plants and soil. By leveraging the unique properties of nanomaterials, researchers have developed tools capable of addressing complex agricultural challenges, such as nutrient deficiencies, environmental stress, and disease outbreaks. For instance, early detection of pathogens or stress markers enables timely intervention, thereby reducing crop losses and ensuring food security. Moreover, the portability and affordability of nanomaterial-based calorimetric biosensors make them suitable for widespread use, including in resource-limited settings. Farmers can benefit directly from these advancements, using data from biosensors to make informed decisions about irrigation, fertilization, and pest management. The ability to optimize inputs not only improves yield but also contributes to sustainable farming practices by minimizing environmental impacts, such as the overuse of fertilizers and pesticides. Looking ahead, continued innovation in nanomaterial synthesis and functionalization will play a critical role in enhancing the performance and accessibility of calorimetric biosensors. Researchers are exploring the use of hybrid nanomaterials and advanced fabrication techniques to create sensors with even greater sensitivity and specificity. Additionally, integrating these biosensors with digital technologies, such as Internet of Things (IoT) platforms and machine learning algorithms, can further enhance their functionality by enabling real-time monitoring and predictive analytics.

Despite the promising potential, addressing challenges related to cost, scalability, and regulatory approval will be essential for the successful commercialization and adoption of these technologies. Collaborative efforts between researchers, industry stakeholders, and policymakers will be necessary to establish standards, ensure safety, and promote the responsible use of nanotechnology in agriculture. In conclusion, nanomaterial-based calorimetric biosensors are poised to transform modern agriculture by providing accurate, rapid, and cost-effective solutions for monitoring and managing plant and soil health. As research and development efforts continue to push the boundaries of what is possible, these innovative tools will undoubtedly play a pivotal role in achieving sustainable agricultural practices and ensuring global food security.

References

Ahmad R, Khan M (2021) Advances in nanomaterial-enhanced calorimetric sensors for soil health monitoring. J Nanotechnol 2021:8841234

Ahmed S, Yadav N (2020) Challenges in deploying nanobiosensors for field applications. Biosens Bioelectron 176:112945

Ali M, Rahman S (2021) Gold nanoparticle biosensors for pathogen detection in crops. Front Plant Sci 12:689572

Brown RJ, Green D (2019) Biosensors for precision agriculture: current applications and future trends. Sensors Actuators B Chem 305:127845

Brown M, Taylor K (2022) Nanomaterial-based biosensors for real-time soil nutrient monitoring. Soil Biol Biochem 170:108684

Chakraborty S et al (2023) Calorimetric sensing of plant stress: applications in precision agriculture. Sensors Actuators B Chem 323:128467

Chen L, Huang X, Zhao J (2022) Quantum dot-based biosensors for pesticide detection in agriculture. Environ Sci Technol 56(4):2141–2150

Ghosal S et al (2021) Nanobiosensors for detection of plant pathogens: a review. Biosens Bioelectron 171:112696

Gupta V, Singh D (2023) Quantum dots as novel probes in calorimetric detection of pesticides. Food Chem 406:135049

Hossain M, Hasan M (2021) Innovations in calorimetric biosensing using nanomaterials. Curr Nanosci 17(4):486–498

Jiang H et al (2022) Nanomaterial-based calorimetric biosensors: enhancements and applications. J Nanotechnol 2022:671875

Jones MP, White AB (2021) Integration of biosensors with AI for sustainable farming. AI Agric 7(1):67–81. https://doi.org/10.1016/j.aia.2021.067

Kumar R, Kumar M (2022) Wireless IoT solutions in precision farming. J Wirel Netw 25(1):87–98

Kumar S, Roy A (2021) Recent trends in biosensor-based detection of plant stress markers. Plant Biotechnol J 19(12):2431–2445

Lee J, Park S (2021) Synthesis of gold nanorods for agricultural biosensing. ACS Appl Mater Interfaces 13(20):23982–23990

Lee CW, Patel DS (2023) Role of nanomaterials in enhancing biosensor performance for agriculture. Nano Agric J 15(2):189–204

Li X et al (2024a) Integration of calorimetric nano sensors with IoT for smart agriculture: a review. Agric Syst 204:103436

Li Y, Zhang H, Wang Y (2024b) Integration of IoT and wireless technology for smart agriculture monitoring systems. J Agric Technol 36(2):245–256

Liu J, Zhang W, Wang Y (2021) Application of gold nanorods in calorimetric biosensors for soil nutrient analysis. Sensors 21(9):3052

Parker J, Evans D (2023) Nanotechnology-driven biosensors for precision agriculture. Nat Nanotechnol 18(1):21–30

Saha S et al (2020) Advanced biosensors for monitoring soil nutrient dynamics in agricultural systems. Sensors 20(15):4192

Sharma A et al (2020a) Nanomaterials in calorimetric biosensors for agricultural monitoring. Trends Agric Eng 5(3):18–27

Sharma R, Gupta N, Singh P (2020b) Gold nanoparticle-based biosensors for plant pathogen detection. Biosens Bioelectron 145:111722

Sharma A, Gupta R, Shukla S (2023) Data analysis in smart farming: IoT-enabled solutions for agricultural decision-making. Smart Agric J 28(3):155–167

Singh P, Kapoor S, Mehta R (2023) Real-time agricultural monitoring using IoT-enabled devices. Agric Eng Rev 30(1):125–139

Smith L, Johnson R (2021) Advances in nanomaterial-based biosensors for plant disease detection. Plant Pathol J 37(4):312–326

Smith J, Thompson K (2020) Nanotechnology in agriculture: trends and future prospects. J Agric Sci Technol 22(3):45–59

Tian Y et al (2021) Graphene-based calorimetric sensors for agricultural applications. J Agric Food Chem 69(13):4050–4059

Wang H, Zhang Q (2020) Role of plasmonic nanomaterials in thermal biosensors. Biosensors 10(3):45

Yadav R, Bhatia S (2020) Carbon nanotubes for calorimetric biosensing applications in agriculture. J Biosens Bioelectron 8(2):10–20

Yuan L et al (2022) Calorimetric nanobiosensors in assessing environmental stress in crops. Environ Sci Technol 56(12):8037–8045

Zhang T, Wei Z (2022) Functionalized carbon nanotubes in biosensors for plant hormone analysis. Anal Chem 94(7):3467–3475

Zhang S, Chen J, Liu Q (2023) Wireless communication protocols for IoT-enabled agricultural systems. Int J IoT Appl 15(4):200–212

Zhao X, Liang Y (2022) Integration of nanomaterials in agricultural biosensors. Nanotechnology 33(6):065102

Zhao J et al (2021a) Challenges and opportunities in the development of nanobiosensors for agriculture. Nano Today 37:101022

Zhao X, Wang L, Li S (2021b) Advances in precision agriculture technologies: a review. Agric Syst 187:103026

Zhou L et al (2023) Multi-sensor systems for precision agriculture: integration of calorimetric and optical biosensors. Sensors Actuators A Phys 332:113236

Chapter 4
Immuno-nanobiosensors in Precision Agriculture: Methods and Applications

Meskul Zannat, Prodipto Bishnu Angon, Israt Jahan, Sujit Mondal, Shamima Nasrin Asha, and Md. Arif Sakil

Abstract Nanobiosensors have transformed farming through the development of diagnostic techniques, making them essential to agriculture. Nanobiosensors address many environmental and agricultural concerns and are precise and effective. These sensors can be employed to identify fertilizers, moisture content, and various diseases. As a result, plant protectant application may be decreased, crop yields can be improved via effective crop management, and nutrient loss can be minimized. Among the several types, immuno-nanobiosensor is a kind of nanobiosensor that uses a nanomaterial to help detect the creation of an immunocomplex utilizing an antibody or antigen as a bioreceptor. They can identify infections, contaminants, heavy metal ions, and microbial burdens. We review the main approaches used in their creation, such as signal transduction processes, antibody immobilization techniques, and signal amplification tactics based on nanomaterials. The application demonstrates numerous uses of immuno-nanobiosensors for precision agriculture by covering disease detection methods, pathogen surveillance techniques,

---

M. Zannat · I. Jahan
Department of Genetics and Plant Breeding, Bangladesh Agricultural University, Mymensingh, Bangladesh

P. B. Angon
Faculty of Agriculture, Bangladesh Agricultural University, Mymensingh, Bangladesh

Institute of Agronomy, Hungarian University of Agriculture and Life Sciences (MATE), Gödöllő, Hungary

S. Mondal
Faculty of Agriculture, Patuakhali Science and Technology University, Patuakhali, Bangladesh

S. N. Asha
Breeding Division, Bangladesh Sugarcrop Research Institute, Ishwardi Pabna, Bangladesh

M. A. Sakil (✉)
Department of Biochemistry and Molecular Biology, Bangladesh Agricultural University, Mymensingh, Bangladesh
e-mail: arifsakilbmb@bau.edu.bd

J.-T. Chen (ed.), *Nanobiosensors for Crop Monitoring and Precision Agriculture*, Sustainability Sciences in Asia and Africa,
https://doi.org/10.1007/978-981-96-8335-2_4

and pesticide residue testing, alongside soil quality evaluation. We overview the challenges involving the technology's expenses, repeatability, and practical implementation difficulties. These applications lead to future potential for immuno-nanobiosensors to revolutionize agriculture by improving disease detection and boosting crop productivity while supporting sustainable resource practices.

Keywords Immuno-nanobiosensors · Agriculture · Nanotechnology · Biosensors

4.1 Introduction

The agriculture and food industries are major contributors to the global economy. The growth of the agricultural sector is hampered by industrialization and the depletion of natural resources in terms of quantity and quality. Between 2000 and 2019, the condition of approximately 127 million hectares of agricultural land declined globally (Mondal et al. 2022). Every year, the prevalence of hunger and undernourishment rises sharply worldwide. The world's population is expected to reach 8.5 billion by 2030 and grow exponentially to 9.8 billion by 2050 (Thakur et al. 2022). Agriculture is the main food source and nourishment for humans and animals, but traditional agricultural technologies are insufficient to monitor or manage healthy crop production (Dhanaraju et al. 2022). Conventional agriculture faces unprecedented challenges due to environmental degradation and rapid population growth (Kumar et al. 2024). The majority of crop losses every year is caused by biotic and abiotic stresses (Angon et al. 2023). In addition, due to rapid industrialization, the widespread usage of hazardous chemicals like pesticides and inorganic fertilizers has expanded. Fertilizers and insecticides have been crucial in the need for food production, but excessive use of these products threatens the expansion of sustainable agriculture. Also, these substances contribute significantly to the imbalance of the agroecosystem and cause major health issues for the general population (Mondal et al. 2022). The usage of chemical pesticides rose by almost 36% between 2000 and 2019, and about 190 million tons of inorganic fertilizer were spread across agricultural lands in 2019 (FAO 2021).

Nanotechnology is one potential strategy to boost agricultural productivity, enhance crop tolerance, and lessen environmental pollution. Recent years have seen the use of nanotechnology in a variety of technological procedures, such as the synthesis and characterization of nanoparticles, as well as the creation and control of novel materials with broad uses in industry, medicine, pharmaceuticals, and agriculture (Ansari 2023). The term nanosensors, or nanoparticle-integrated biosensors, refers to recent developments in the realm of nanotechnology. The agri-food industry relies heavily on them because of their compact size, portability, efficiency, sensitivity, and specificity. Nanoparticle-based nanobiosensors can be used by researchers in clinical diagnostics and research institutes for smaller delivery systems, plant protection, improved oil health promotion, longer fruit shelf life, and sustainable agriculture (Dar et al. 2020).

Novel nanotechnological techniques are used to create nanocomposites of organic, inorganic, or mixed materials, making nanobiosensors (Romanovski and Periakaruppan 2023). Numerous nanomaterials, including metals, carbon nanotubes (CNTs), magnetic nanoparticles, nanowires, quantum dots (QDs), and nanoprobes, are used in nanobiosensors (Beeguma and Dasb 2022). Piezoelectric nanosensors are used in agriculture to measure temperature, pH, nutrients, and soil moisture content, allowing for real-time crop status monitoring (Dimri et al. 2020). In agricultural crops, the widespread adoption of nanocomposites can prevent plant pests without the need for synthetic chemicals, fungicides, or pesticides. By overcoming biological barriers, small-scale nanocomposites enable effective nutrient transfer via foliar or root treatments (Ijaz et al. 2023).

An immuno-nanobiosensor is a kind of biosensor that uses nanomaterials to help an antibody or antigen function as a bioreceptor in order to detect the creation of an immunocomplex (Zhang and Zhao 2019). Because of the unique antigen–antibody binding, it is very selective. Some of the well-known technologies are electrochemical detection, fluorescence, and ELISA (Enzyme-Linked Immunosorbent Assay)-like techniques. Early disease detection in crops, soil health monitoring, pesticide residue detection, nutrient management in plants, and the detection of mycotoxins and contaminants are some of the crucial roles that immuno-nanobiosensors perform in precision agriculture. The techniques and uses of immuno-nanobiosensors in precision agriculture are highlighted in this chapter.

4.2 Classification and Formation of Biosensor

Analytical devices known as biosensors operate by using biochemical reactions between biomolecules or supramolecular structures. The biological activity functions through an electrical signal recording system, along with a physicochemical transducer and a bioreceptor component. Numerous biosensor types exist, including enzymatic sensors, together with deoxyribonucleic acid (DNA) sensors, alongside cellular and immunosensor devices, and tissular sensors and aptasensors that are classified based on their bioreceptor attributes. Different transducers function in biosensors, including electrochemical and optical, as well as calorimetric, acoustic, and piezoelectric methods (Inshyna et al. 2020).

4.2.1 Enzymatic Biosensors

Enzymatic biosensors developed through nanomaterial integration and microfluidic engineering provide fast inhibitor detection and monitoring capabilities (Amine et al. 2016). Recent research has transformed enzymatic electrochemical biosensors into highly sensitive and stable systems that can detect pesticides in food and environmental specimens (Karadurmus et al. 2021). Both substrate and inhibitor enzymatic biosensors use the Michaelis–Menten equation to detect the concentration of the substrate, whereas inhibitor sensors monitor substances that prevent enzyme activity, such as organophosphate insecticides (Inshyna et al. 2020). The use of nanomaterials and sentinel sensors, alongside permselective membranes, presents potential solutions to expand enzyme biosensor applications and improve selective measurements (Bucur et al. 2021).

4.2.2 Cellular Biosensors

Cellular biosensors are used to investigate metabolic processes like photosynthesis, mutagenesis, and biochemical oxygen consumption. These sensors, which are based on microbial cells, are known as respiratory sensors because of the range of contaminants they can identify. In order for a cell-based biosensor to function, an analyte must be detected and amplified into an optical and electrical signal. These biosensors' efficacy is dependent on the reporter genes found in the cell genome, which are modified by genetic engineering to show the gene's expression or location.

4.2.3 Tissue-Based Biosensors

Biosensors use tissue materials as biocatalytic components because they may provide an environment that is natural for enzymes. Tissue-based biosensors have a greater specific activity and a longer service life than enzymes. The selectivity of the sensor is crucial due to cellular enzymatic activity. The addition of antimicrobial agents is necessary to prevent contamination. The process of optimizing tissular biosensors involves examining many aspects, such as tissue layer immobilization, pH influence, activator and inhibitor concentrations, and biosensor selectivity (Prieto-Simon et al. 2008; Grennan et al. 2006; Shibru et al. 2006). Compared to enzymatic biosensors, tissue-based biosensors provide the advantages of stability, high biocatalytic activity, lower cost, and less inactivation exposure. Plant tissue–based biosensors may be used to assess the levels of flavonoids, catechol, and dopamine (Siedler et al. 2014; Lin and Yeh 2017).

4.2.4 Cellular Organelle–Based Immunosensors

Cellular organelle–based immunosensors and biosensors are used to improve selectivity by either increasing the amount of immobilized enzymes or eliminating interfering enzymes. These organelle-based biosensors, which comprise mitochondria, chloroplasts, liposomes, and microsomes, are used as phytotoxicity sensors to investigate the mechanisms of action of pollutants (Martinkova et al. 2017). A high level of complementarity between the paratope and epitope is necessary for monoclonal antibodies, which are bioreceptors found in immunosensors, to create a long-lasting antigen–antibody pair (Lara and Perez-Potti 2018). Monoclonal antibodies are more specific than polyclonal ones with regard to the associated biomarker. A transducer can detect the signal produced by the biological reaction that creates an antigen–antibody pair (Felix and Angnes 2018). The conditions of the antibody–nanomaterial interaction should get particular consideration since different antibodies have different electrostatic properties.

4.2.5 *DNA Biosensor*

DNA biosensors, often referred to as genosensors, evaluate nucleic acids in clinical diagnostics using hybridization probes. Synthetic bioreceptors known as aptamers are ultra-selective components utilized in aptasensors that attach to a variety of substances. The strong affinity and durability of synthetic bioreceptors, such as polymers with complementary characteristics, make them useful for developing biosensors.

4.2.6 *Optical Biosensor*

A physicochemical transducer is used in optical biosensors to translate biological reactions from bioreceptors into measurable signals. Fluorescence and bioluminescence–based optical biosensors have emerged as crucial instruments in a number of domains, such as food safety, environmental investigation, clinical diagnostics, and cancer medication development (Benito-Peña et al. 2016). Fluorescence, luminescence, bioluminescence, polarization, absorption, scattering, and other optical characteristics are all recorded by these sensors (Kelly and Yang 2024). T-fluorescent protein-based biosensors may identify ions, metabolites, and disease biomarkers; research the effects of anticancer drugs; and visualize biological processes (Damborský et al. 2016).

4.2.7 *Piezoelectric Biosensor*

Piezoelectric biosensors assess mass changes from biomolecule contact, influencing the resonant frequency. They include an antibody as a biorecognition element, with the specificity of the antibody determining the sensor's specificity. These biosensors are ideal for identifying infectious illnesses and high molecular weight analytes. However, they cannot identify low molecular weight analytes. Piezoelectric immunosensors are sensitive, detecting antibodies in the picogram range, making them excellent for microorganism investigation.

4.2.8 Electrochemical Biosensor

Approximately 80% of all electrochemical biosensors are potentiometric, amperometric, or conductometric variations (Pohanka and Skládal 2008; Wang 2022). The conductivity of electrolyte solutions is measured by conductometric biosensors and is influenced by ion mobility and concentration. Potentiometric biosensors determine chemicals quantitatively by comparing the electrical potentials of the reference electrode and the detecting electrode (Wang 2022). They come in three varieties: mediator-free biosensors, mediator biosensors, and direct electron transfer biosensors. Amperometric biosensors detect variations in current density or strength. The sensitivity and selectivity of these biosensors are quite good (Wang 2022). Biosensors that employ optical physicochemical transducers and electrochemical transducers are often used to detect infectious agents and measure metabolite concentrations (Inshyna et al. 2020).

4.3 An Overview of Immuno-nanobiosensors

The term "biosensor" refers to a sensor that uses a transducer to transform changes in a biological analyte into an electrical signal (Wang et al. 2023). The specificity of biosensors depends on recognition molecules, such as antibodies or ligand proteins, referred to as immunosensors, which generate optical or electrical signals for analysis (George et al. 2020). So an immuno-nanobiosensor is a device that uses nanomaterials to detect the formation of an immunocomplex with an antibody or antigen as the bioreceptor (Zhang and Zhao 2019). It consists of three primary parts: a biological recognition element (e.g., antibodies), a transducer, and a detector (Mondal et al. 2022; Kulkarni et al. 2022) (Fig. 4.1).

Antibodies are used to create biosensors because of their highly precise interaction between antigens and their associated antibodies (Sharma et al. 2016). Besides, the use of analyte-specific antibodies in immunosensors allows for straightforward and highly specific interrogation, which has many benefits over other approaches, such as multi-pathogen detection employing sensor platforms with various antibodies (Mendonça and Bhunia 2015). The interaction between the antibody and antigen can be observed as a time-dependent variation in fluorescence signal, which correlates with the ratio of antibody to analyte in the reaction (Bisen 2014). The biological recognition components are often antibodies or antibody derivatives (antigens or haptens), which are either tightly linked to or incorporated into a physicochemical transducer (Gajos et al. 2020). Electrochemical, photonic, and microgravimetric transducer devices use the electronic component to digitalize and amplify the

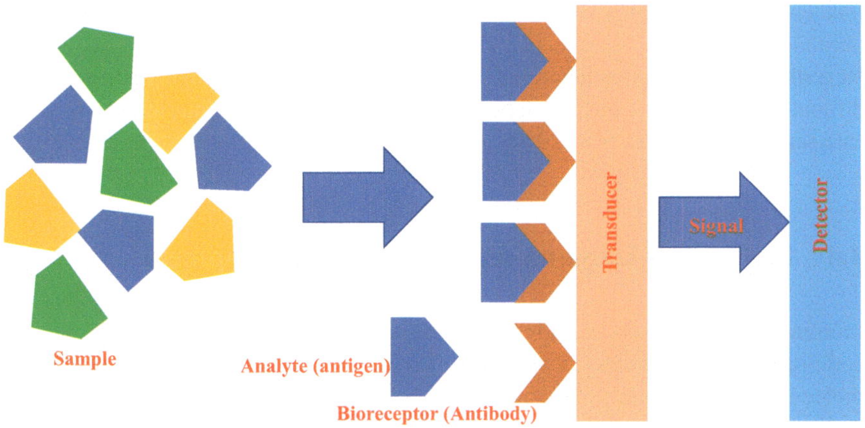

Fig. 4.1 Structure of immuno-nanobiosensor

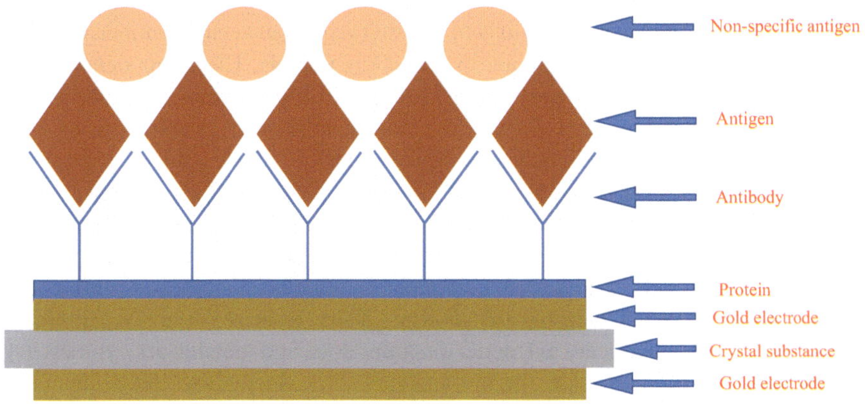

Fig. 4.2 Working of immuno-nanobiosensor

physicochemical output signal (Wang et al. 2008). In a biosensor, the detector coupled to the transducer translates the signal from a biological reaction into an electrical or optical signal that can be quantified and evaluated (Ullah et al. 2016) (Fig. 4.2). An ideal immunosensor design should detect and quantify antigens, transform binding events without the use of external chemicals, allow for repeated measurements on the same device, and detect particular antigen binding in real samples (Chen et al. 2019).

Furthermore, immunosensors may not require complicated sample pretreatment, allowing for faster results. Thus, depending on the sensor design, high-throughput, low-cost sample testing is feasible (Mendonça and Bhunia 2015). High sensitivity, selectivity, and quick analysis times are some of the benefits immunosensors provide as analytical instruments. Due to their high target specificity, which enables the detection of certain analytes, they can be compressed for point-of-care applications

(Okechukwu et al. 2024). Besides, the analytical procedure is made simpler by immunosensors' low sample preparation requirements (Sobhanie et al. 2022). Nevertheless, they can be laborious, necessitate specific knowledge, and encounter difficulties with sensor surface stability and matrix component interference in intricate samples (Nazhand et al. 2020).

4.4 Methods of Immuno-nanobiosensor Development

Immuno-nanobiosensors are advanced biosensors that encompass nanotechnology combined with biosensors (Debnath and Das 2020). The main basis of biosensors is molecular identification and recognition (Christopher et al. 2020). The nanobiosensor is composed of three modules: genetic probe incorporates different substances based on affinities like nucleic acid, cell-based interactions, antibody–antigen, and enzyme–substrate (Kulkarni et al. 2022). Only when the bacteria and the biological receptor share a particular molecular recognition, can the biological receptors identify the bacteria. The best illustrations of the interaction between an antibody and an antigen in molecular recognition are lock and key models. The components of the sensor that communicate with the target are called bioreceptors (Ghafouri et al. 2023). Moreover, another module named a transducer converts the biological signal into an electrical signal and analyzes it as data footage, which loads and transmits the information. Plastics, fibers, metals, ceramics, silicon, and glasses may act as transducers, which convert medium signals into determinate signals such as electrochemical, piezoelectric, optoelectronic, and thermal (Kaur et al. 2022; Chadha et al. 2022). Different kinds of advanced nanomaterials have been employed for the production of novel nanobiosensors (Chen et al. 2020; Kulkarni et al. 2022). Many nanomaterials like graphene, magnetic, quantum dots, and metallic are synthesized for biological probes to sense the bio-analytes. Transducers use electrochemical signals, primarily amperometric, potentiometric, and voltammetric signals, such as optical fibers, fluorescence, and colorimetric signals. The transducer's surface is permanently fixed with bioreceptors that may bind the target entity (enzymes, antibodies, deoxyribonucleic acid (DNA), cells, and aptamers) stably under a variety of storage conditions (Sadani et al. 2022).

The biological recognition element uses various techniques for immobilization, such as trapping, covalent bonding, cross-linking, trapping, and microencapsulation. Immobilization of the nanocomponents is the most difficult step in the creation of nanobiosensors. Some compounds such as synthetic catalysts, ligands, recombinant antibodies, designed artificial proteins, and imprinted polymers (Ghafouri et al. 2023) can be used as alternatives to biological receptors. The function of these receptors mainly determines the selectivity and sensitivity of biosensors (Bhattarai and Hameed 2020). Transducers can detect molecular recognition effects like changes in light, mass, heat, pH, or electroactivity. As an interface, the receptor transforms measurable signals into energy.

In nanobiosensors, the main focus is on transducers, which are modified with nanoparticles, enabling quick detection. In comparison with simple biosensors, nanobiosensors can detect the quantity and presence of analytes (Christopher et al. 2020). The electrical signal of the transducer is amplified by a detector, and a microprocessor measures it. Various types of amplifiers and filters convert analog signals to digital signals. The information can be saved as a tabular, graphic, numeric, or picture file, or it can be shown on the device as concentration units. Smartphone-based detectors have been developed for on-chip or point-of-care analyte detection in nanobiosensors (Seo et al. 2019).

Their performance can be indirectly improved due to the properties of nanobiosensors such as linearity, stability, sensitivity, reproducibility, and selectivity. "Selectivity" is a term that describes the sensor's capacity to differentiate one analyte from several others (Zhang et al., 2020). Fabrication of nanobiosensors is done mainly with glass and silicon. The sensitivity of nanobiosensors, which is correlated with their robustness (Zhang et al., 2020), determines their detection limits. Reliability and repeatability are connected to the accuracy and precision of a nanobiosensor's results. Indicators of linearity or accuracy are working ranges or linear dynamic ranges where concentration is directly proportional to signals. Sensor stability allows for the quantification and detection of analytes in a variety of measurement disruption scenarios without sacrificing accuracy and precision.

4.5 Applications of Immuno-nanobiosensors

The innovative application of immuno-nanobiosensors in precision agriculture is the detection of plant diseases, soil properties, and plant development; control of hormones; and calculation of pesticide residue levels (Dhole and Pitambara 2019). These biosensors use nanotechnology for the improvement of the detection of a wide range of parameters relating to the improvement of agricultural sustainability and productivity.

4.5.1 Monitoring Soil Conditions and Plant Growth Hormones

Immuno-nanobiosensors use nanomaterials, including carbon nanotubes and magnetic nanoparticles, along with quantum dots, for detecting biological markers of plant hormones and soil features (Holzinger et al. 2014). Nanobiosensors help agricultural researchers study soil moisture patterns and examine plant growth requirements for nutrients and pH, together with temperature parameters (Virk et al. 2024). Besides, these sensors identify actionable changes in plant stress behaviors by tracking the amounts of plant hormones and metabolites (Mathivanan 2021). This capability is essential for today's agricultural systems because sensing technologies provide instantly deployable defense against bacterial and environmental challenges.

According to Miguel-Rojas, these sensors detect plant health indicators at their early stages, which allows farmers to initiate preventative actions before damage becomes extensive (Miguel-Rojas and Pérez-de-Luque 2023). Through these nanobiosensors, farmers gain better capabilities to make well-informed choices regarding fertilization practices and irrigation schedules (Ghaffar et al. 2020). The sensors measure plant hormonal substances, including auxins and gibberellins, at exact levels, which leads to better agricultural methods that can both increase output and better control growth environments (Ang et al. 2021). Biological identification elements immobilized on nanoscale transducing elements expand both sensitivity and specificity in nanobiosensing operations. Through enzyme-based nanobiosensors, researchers can directly evaluate hormones in soil and plant tissues to measure plant health and nutritional status (Franca et al. 2011).

4.5.2 Plant Pathogen Detection

The identification of plant diseases stands as the primary essential application of immuno-nanobiosensors in precision agriculture (Ali et al. 2021). The sensors identify pathogen types efficiently through their attachment of nanoparticles to antibodies (Farka et al. 2017). In situ immuno-gold nanoparticle network biosensors enable pathogen detection for early plant disease management, which reduces agricultural losses (Virk et al. 2024). The signal amplification capabilities of these nanomaterials' features boost biosensor sensitivity, thus enabling both new detection techniques and the identification of pathogens at minimal concentrations (Ghafouri, 2023). Electrochemical immunosensors demonstrate ideal suitability for agricultural field applications because they provide both simple operation and quick response times (Li et al. 2020).

Research shows that electrochemical immunosensors offer reliable detection capabilities for distinct plant pathogens. The electrochemical immunosensor successfully detected the coat protein of the papaya ringspot virus, demonstrating that the economic and rapid potential of these devices for pathogen identification (Valekunja et al. 2016). Hu et al. (2023) presented a label-free electrochemical biosensing architecture to detect pathogenic *Escherichia coli* with high sensitivity, according to their research on nanobiosensor development for bacterial pathogen detection. These developments suggest that immuno-nanobiosensors can be customized for particular pathogens, making them more useful in agricultural diagnostic applications.

4.5.3 Harmful Pesticide Residue Detection

Agricultural applications of immuno-nanobiosensors rely crucially on monitoring pesticide residues within farming environments. Instrumental sensors possess the ability to detect problematic pesticide contaminants in agricultural products, alongside agricultural soils, thus upholding both nutritional safety and regulatory protocols. Nanomaterial-based electrochemical biosensors now provide the capability to detect particular pesticides, including organophosphates (Mirres et al. 2022). The combination of nanobiosensor technology and acetylcholinesterase reveals multiple useful attributes regarding response speed and analytic sensitivity, along with minimal sample preparation requirements suitable for agricultural site applications (Sinha and Bachan Upadhyay 2024).

Nanobiosensors based on immunological principles use quantum dots, alongside gold nanoparticles and carbon nanotubes, to boost signal amplification for improved detection sensitivity. Using a magnetoelastic nanobiosensor, atrazine detection accuracy was achieved at 1 ng/mL, which exceeded the US Environmental Protection Agency's (EPA) 3-μg/L permissible threshold (Sang et al. 2018). Nanobiosensors demonstrate their ability to offer rapid, precise contaminant assessments of pesticides across multiple environmental settings from agricultural fields to water sources (Sharma et al. 2021). Biosensors that incorporate nanomaterials accommodate the identification of diverse pesticide types spanning organophosphates and carbamates, among others (Thakur et al. 2022). Besides, a sensor was created containing gold nanoparticles that detect pesticide classes in agricultural products through visual recognition (Morowati and tohidi Moghaddam 2022). Research efforts embrace cost-efficient development and user-friendly, field-deployable biosensing technology that strengthens pesticide residue monitoring systems. Food safety represents an example of the versatile utilization of nanobiosensors that offers multiple applications. The growing use of these sensors in food product safety assessment for toxic chemical detection, including pesticide residues, is due to their reliable identification capabilities (Guruprasath et al. 2024). Furthermore, the combination of nanobiosensors with immunoassays enhances their ability to detect pesticides effectively (Vijayakumar et al. 2022). Nanobiosensors enabled the on-site detection revolution through their combination of nanotechnology and biotechnology approaches for pesticide detection (Srinivasan et al. 2024). Quick and condensed sample-processing methods are essential for combating the analytical roadblocks created by standard analysis systems that need lengthy laboratory testing to deliver results.

4.6 Challenges and Future Prospects

Even if there are already countless inventions and commercialized goods accessible and the findings published in the context of nanomaterials related to farming are becoming more and more optimistic, it is challenging to turn these products into productive uses in the agricultural sector. Studies in this field can occasionally be quite challenging and demand the application of advanced technology and cross-disciplinary teams with experience in physics, chemistry, life sciences, and agricultural practices, among other areas of study. Findings from a particular plant (specifically with regard to NFs (nanoformulations)) could not be instantly pertinent to another since the responses of nanoparticles might vary considerably depending on factors like the variety and morphology of the crop. The cost of employing nanoparticles for extensive use in fields must be somewhat cheap since agriculture sometimes has narrow profitability ratios; otherwise, farmers would not be capable of embracing nanotechnology. An increasing trend toward the application of organic nanoparticles (biological polymers), which are harmless and readily biodegradable, is the consequence of the necessity to also take into account the prospective adverse impacts on the environment and food security associated with these commodities (Miguel-Rojas and Pérez-de-Luque 2023). The expensive price and product responsiveness of detection mechanisms based on nanoparticles are two of their main drawbacks. Nanocomposites are complicated to fabricate on an extensive scale. Minimizing the disruption triggered by atmospheric variables like electromagnetic detectors is a prevalent challenge for nanosensors. This sensitivity may result in inaccurate readings and compromise the preciseness of the data obtained (Virk et al. 2024). Developing nanobiosensors from prototypes to commercialized items is a significant obstacle that has not yet been overcome (Kaushal and Wani 2017). The application of nanomaterials is widespread in both the food and agricultural sectors. However, there are several concerns regarding security with nanoparticles because of their minuscule dimension and capability for getting into the human circulatory system. The development of plants will be hindered by the formation of nanomaterials in the ground, which may also deposit inside plant organs that are consumable (Rajput 2022). Nanocomposites can deteriorate due to atmospheric variables as well as the dispersal of incorporated nanoparticles from polyamide structures into the surrounding atmosphere (Mustafa and Andreescu 2020). The application of pesticides and small-molecule fertilizers for agricultural purposes poses serious health concerns to producers since they can wind up in the ground, water, and atmosphere (Roohinejad and Greiner 2017). Unproductive nanomaterials might be hazardous, and they may trigger damage through oxidation and inflammatory responses (Han et al. 2018).

The goal of smart farming is to support sustainable agriculture by tracking and identifying environmental factors and the health of plant metrics. Nanobiosensors can detect and monitor a variety of characteristics pertaining to both plant health and atmospheric behavior, which can be extremely useful in agriculture. These tiny instruments may identify diseases, pests, and pathogens early on, enabling prompt treatment and reducing damage to crops (Virk et al. 2024). In addition to managing diseases of plants, nanobiosensors improve the security of food and the sector of agriculture by regulating physical, chemical, and biological activities, which enhances prospects of environmentally friendly agri-food development. When nanocomposites are widely used, they can take the place of chemical substances, fungicides, and pesticides in crop cultivation to suppress pests that affect plants. By overcoming biophysical obstacles, modest nanocomposites assure effective nutritional transmission using root or foliar treatments (Ijaz et al. 2023). Scientists and researchers can benefit from nanomaterials due to their capacity as transporters, as well as their tiny magnitude, portability, and functional area on the surface (Hazarika et al. 2021; Dutta et al. 2022). Nanobiosensors can be employed by scientists in therapeutic investigations and research facilities for precision agriculture, prolonged fruit preservation periods, protection of plants, improved lubricant health, and shorter ways to distribute them (Dar et al. 2020). In order to overcome these obstacles and ensure the effective incorporation of nanobiosensors in the agricultural food industry, ongoing advancements in nanomaterials' responsiveness, mobility, lifespan, and resilience are required. Agriculture with precision, resource wastage diminution, and the promotion of ecologically sound methods are some of the ways that these developments in nanobiosensors will help producers (Virk et al. 2024). The fabrication of nanobiosensors for agriculture has a promising future with greater potential, and quantitative biology, including artificial intelligence and advanced understanding, will be essential to realizing that promise (Magarelli et al. 2023). Scientists are highly enthusiastic about utilizing improved variations of nanodiagnostic devices in the future to achieve durability in both farming and the food manufacturing business, despite the fact that simultaneous identification occasionally results in significant interfering concerns (Mondal and Jain 2021).

4.7 Conclusion

The new area of immuno-nanobiosensors, which has the ability to change precise farming methods, was introduced in this chapter. Innovative tools that blend nanomaterials with immune principles may provide unmatched sensitivity, exceptional specificity, and small features for a variety of farming uses. Immuno-nanobiosensors enable quick disease diagnosis while monitoring pathogens, along with pesticide detection and soil evaluation, to deliver optimal crop production with reduced environmental effects and enhance food security standards. Ongoing research focuses on addressing limitations regarding cost-effectiveness, as well as reproducibility and field deployment challenges. Future precision agriculture development relies on further innovation of immuno-nanobiosensors, which grants farmers cleaner choices, alongside better resource management, through informed decisions that produce sustainable agricultural practices.

References

Ali Q, Ahmar S, Sohail MA, Kamran M, Ali M, Saleem MH et al (2021) Research advances and applications of biosensing technology for the diagnosis of pathogens in sustainable agriculture. Environ Sci Pollut Res 28:9002–9019

Amine A, Arduini F, Moscone D, Palleschi G (2016) Recent advances in biosensors based on enzyme inhibition. Biosens Bioelectron 76:180–194

Ang MCY, Dhar N, Khong DT, Lew TTS, Park M, Sarangapani S et al (2021) Nanosensor detection of synthetic auxins in planta using corona phase molecular recognition. ACS Sens 6(8):3032–3046

Angon PB, Mondal S, Akter S, Sakil MA, Jalil MA (2023) Roles of CRISPR to mitigate drought and salinity stresses on plants. Plant Stress 8:100169

Ansari MA (2023) Nanotechnology in food and plant science: challenges and future prospects. Plan Theory 12(13):2565

Beeguma S, Dasb S (2022) Chapter 17: Nanosensors in agriculture. In: Agricultural nanobiotechnology: biogenic nanoparticles, nanofertilizers and nanoscale biocontrol agents. Woodhead Publishing, p 465

Benito-Peña E, Valdés MG, Glahn-Martínez B, Moreno-Bondi MC (2016) Fluorescence based fiber optic and planar waveguide biosensors. A review. Anal Chim Acta 943:17–40

Bhattarai P, Hameed S (2020) Chapter 1: Basics of biosensors and nanobiosensors. In: Wu A, Khan WS (eds) Nanobiosensors: from design to applications. Wiley, pp 1–22

Bisen PS (2014) Laboratory protocols in applied life sciences. CRC Press

Bucur B, Purcarea C, Andreescu S, Vasilescu A (2021) Addressing the selectivity of enzyme biosensors: solutions and perspectives. Sensors (Basel, Switzerland) 21:3038

Chadha U, Bhardwaj P, Agarwal R, Rawat P, Agarwal R, Gupta I, Panjwani M, Singh S, Ahuja C, Selvaraj SK et al (2022) Recent progress and growth in biosensors technology: a critical review. J Ind Eng Chem 109:21–51

Chen Y, Yuan PX, Wang AJ, Luo X, Xue Y, Zhang L, Feng JJ (2019) A novel electrochemical immunosensor for highly sensitive detection of prostate-specific antigen using 3D open-structured PtCu nanoframes for signal amplification. Biosens Bioelectron 126:187–192

Chen YT, Lee YC, Lai YH, Lim JC, Huang NT, Lin CT, Huang JJ (2020) Review of integrated optical biosensors for point-of-care applications. Biosensors 10(12):209

Christopher FC, Kumar PS, Christopher FJ, Joshiba GJ, Madhesh P (2020) Recent advancements in rapid analysis of pesticides using nano biosensors: a present and future perspective. J Clean Prod 269:122356

Damborský P, Švitel J, Katrlík J (2016) Optical biosensors. Essays Biochem 60(1):91–100

Dar FA, Qazi G, Pirzadah TB (2020) Nano-biosensors: NextGen diagnostic tools in agriculture. In: Nanobiotechnology in agriculture: an approach towards sustainability. Springer, Cham, pp 129–144

Debnath N, Das S (2020) Nanobiosensor: current trends and applications. NanoBioMedicine: 389–409

Dhanaraju M, Chenniappan P, Ramalingam K, Pazhanivelan S, Kaliaperumal R (2022) Smart farming: Internet of Things (IoT)-based sustainable agriculture. Agriculture 12(10):1745

Dhole A, Pitambara M (2019) Nanobiosensors: a novel approach in precision agriculture. In: Nanotechnology for agriculture: advances for sustainable agriculture. Springer, pp 241–262

Dimri A, Pathak N, Sharma S (2020) Nanosensors for root zone parameters influencing plant growth. In: Nanomaterials for agriculture and forestry applications. Elsevier, pp 387–406

Dutta P, Sharma V, Bhardwaj H, Agrawal VV, Rajesh, Sumana G (2022) Fabrication of electrochemical biosensor using zinc oxide nanoflowers for the detection of uric acid. Mapan 37(3):585–595

Farka Z, Jurik T, Kovar D, Trnkova L, Skládal P (2017) Nanoparticle-based immunochemical biosensors and assays: recent advances and challenges. Chem Rev 117(15):9973–10042

Felix FS, Angnes L (2018) Electrochemical immunosensors–a powerful tool for analytical applications. Biosens Bioelectron 102:470–478

Food and Agriculture Organization of the United States (FAO) statistical yearbook 2021 2021 Food & agricultural organization https://www.fao.org/3/cb4477en/online/cb4477en.html#

Franca EF, Leite FL, Cunha RA, Oliveira ON Jr, Freitas LC (2011) Designing an enzyme-based nanobiosensor using molecular modeling techniques. Phys Chem Chem Phys 13(19):8894–8899

Gajos K, Szafraniec K, Petrou P, Budkowski A (2020) Surface density dependent orientation and immunological recognition of antibody on silicon: TOF-SIMS and surface analysis of two covalent immobilization methods. Appl Surf Sci 518:146269

George SM, Tandon S, Kandasubramanian B (2020) Advancements in hydrogel-functionalized immunosensing platforms. ACS Omega 5(5):2060–2068

Ghaffar N, Farrukh MA, Naz S (2020) Applications of nanobiosensors in agriculture. Nano:179–196

Ghafouri P, Kasaei B, Aghili S, Monirvaghefi A, Hosseini AM, Amoozegar H et al (2023) Application of Nanobiosensors in detection of pathogenic bacteria: an update. Res Biotechnol Environ Sci 2(4):65–74

Grennan K, Killard AJ, Hanson CJ, Cafolla AA, Smyth MR (2006) Optimisation and characterisation of biosensors based on polyaniline. Talanta 68(5):1591–1600

Guruprasath N, Sankarganesh P, Adeyeye SAO, Babu AS, Parthasarathy V (2024) Review on emerging applications of nanobiosensor in food safety. J Food Sci 89(7)

Han L, Zhao Y, Cui S, Liang B (2018) Redesigning of microbial cell surface and its application to whole-cell biocatalysis and biosensors. Appl Biochem Biotechnol 185:396–418

Hazarika S, Ahmed S, Chamkha AJ (2021) Investigation of nanoparticles Cu, Ag and Fe3O4 on thermophoresis and viscous dissipation of MHD nanofluid over a stretching sheet in a porous regime: a numerical modeling. Math Comput Simul 182:819–837

Holzinger M, Le Goff A, Cosnier S (2014) Nanomaterials for biosensing applications: a review. Front Chem 2:63

Hu C, Wei G, Zhu F, Wu A, Luo L, Shen S et al (2023) Label-free electrochemical biosensing architecture based on a protein-decorated Pt@ BSA nanocomposite for sensitive immunoassay of pathogenic Escherichia coli O157: H7. ACS Sustain Chem Eng 11(20):7894–7907

Ijaz U, Ahmed T, Rizwan M, Noman M, Shah AA, Azeem F et al (2023) Rice straw based silicon nanoparticles improve morphological and nutrient profile of rice plants under salinity stress by triggering physiological and genetic repair mechanisms. Plant Physiol Biochem 201:107788

Inshyna NM, Chorna IV, Primova LO, Hrebenyk LI, Khyzhnia YV (2020) Biosensors: design, classification and application. J Nano Electron Phys 12(3):03033

Karadurmus L, Kaya S, Ozkan S (2021) Recent advances of enzyme biosensors for pesticide detection in foods. J Food Meas Charact 15:4582–4595

Kaur B, Kumar S, Kaushik BK (2022) Recent advancements in optical biosensors for cancer detection. Biosens Bioelectron 197:113805

Kaushal M, Wani SP (2017) Nanosensors: frontiers in precision agriculture. In: Nanotechnology: an agricultural paradigm. Springer, pp 279–291

Kelly T, Yang X (2024) Application of fluorescence-and bioluminescence-based biosensors in cancer drug discovery. Biosensors 14(12):570

Kulkarni MB, Ayachit NH, Aminabhavi TM (2022) Recent advancements in nanobiosensors: current trends, challenges, applications, and future scope. Biosensors 12(10):892

Kumar P, Chugh P, Ali SS, Chawla W, Sushmita S, Kumar R, Raval AV, Shamim S, Bhatia A, Kumar R (2024) Trends of nanobiosensors in modern agriculture systems. Appl Biochem Biotechnol:1–24

Lara S, Perez-Potti A (2018) Applications of nanomaterials for immunosensing. Biosensors 8(4):104

Li Z, Yu T, Paul R, Fan J, Yang Y, Wei Q (2020) Agricultural nanodiagnostics for plant diseases: recent advances and challenges. Nanosc Adv 2(8):3083–3094

Lin Y, Yeh Y (2017) Dual-signal microbial biosensor for the detection of dopamine without inference from other catecholamine neurotransmitters. Anal Chem 89(21):11178–11182

Magarelli G, Freire AM, Silva LP (2023) Electrochemical sensors coupled with machine learning for food safety and quality inspection. In: Food quality analysis. Academic Press, pp 171–200

Martinkova P, Kostelnik A, Valek T, Pohanka M (2017) www. electrochemsci. org. Int J Electrochem Sci 12:7386–7403
Mathivanan S (2021) Perspectives of nano-materials and nanobiosensors in food safety and agriculture. In: Novel nanomaterials. IntechOpen, p 197
Mendonça M, Bhunia AK (2015) Fiber-optic sensors for high throughput screening of pathogens. In: High throughput screening for food safety assessment. Woodhead Publishing, pp 249–262
Miguel-Rojas C, Pérez-de-Luque A (2023) Nanobiosensors and nanoformulations in agriculture: new advances and challenges for sustainable agriculture. Emerg Top Life Sci 7(2):229–238
Mirres ACDM, Silva BEPDMD, Tessaro L, Galvan D, Andrade JCD, Aquino A et al (2022) Recent advances in nanomaterial-based biosensors for pesticide detection in foods. Biosensors 12(8):572
Mondal K, Jain AG (2021) Precision agriculture with nano-based biosensors. Precis Agric 2:15
Mondal R, Dam P, Chakraborty J, Paret ML, Katı A, Altuntas S, Sarkar R, Ghorai S, Gangopadhyay D, Mandal AK, Husen A (2022) Potential of nanobiosensor in sustainable agriculture: the state-of-art. Heliyon 8(12)
Morowati M, tohidi Moghaddam T (2022) Specific and rapid visual detection of organophosphorus and carbamate pesticides in agricultural products by gold nanoparticles conjugated with aptamer
Mustafa F, Andreescu S (2020) Nanotechnology-based approaches for food sensing and packaging applications. RSC Adv 10(33):19309–19336
Nazhand A, Durazzo A, Lucarini M, Souto EB, Santini A (2020) Characteristics, occurrence, detection and detoxification of aflatoxins in foods and feeds. Food Secur 9(5):644
Okechukwu VO, Adelusi OA, Kappo AP, Njobeh PB, Mamo MA (2024) Aflatoxins: occurrence, biosynthesis, mechanism of action and effects, conventional/emerging detection techniques. Food Chem 436:137775
Pohanka M, Skládal P (2008) Electrochemical biosensors—principles and applications. J Appl Biomed 6(2)
Prieto-Simon B, Campas M, Marty JL (2008) Biomolecule immobilization in biosensor development: tailored strategies based on affinity interactions. Protein Pept Lett 15(8):757–763
Rajput JK (2022) Nanomaterial-based sensors as potential remedy for detection of biotoxins. Food Control 135:108686
Romanovski V, Periakaruppan R (2023) Why metal oxide nanoparticles are superior to other nanomaterials for agricultural application? In: Nanometal oxides in horticulture and agronomy. Academic, pp 7–18
Roohinejad S, Greiner R (2017) Nanoscience: relevance for agriculture and the food sector. In: Nanotechnology in agriculture and food science. Wiley-Blackwell, pp 347–362
Sadani K, Nag P, Thian XY, Mukherji S (2022) Enzymatic optical biosensors for healthcare applications. Biosens Bioelectron 12:100278
Sang S, Guo X, Liu R, Wang J, Guo J, Zhang Y et al (2018) A novel magnetoelastic nanobiosensor for highly sensitive detection of atrazine. Nanoscale Res Lett 13:1–8
Seo SE, Tabei F, Park SJ, Askarian B, Kim KH, Moallem G et al (2019) Smartphone with optical, physical, and electrochemical nanobiosensors. J Ind Eng Chem 77:1–11
Sharma S, Byrne H, O'Kennedy RJ (2016) Antibodies and antibody-derived analytical biosensors. Essays Biochem 60(1):9–18
Sharma P, Pandey V, Sharma MMM, Patra A, Singh B, Mehta S, Husen A (2021) A review on biosensors and nanosensors application in agroecosystems. Nanoscale Res Lett 16:1–24
Shibru H, Zhang Y, Cooper KL, Pickrell GR, Wang A (2006) Optimization of layer-by-layer electrostatic self-assembly processing parameters for optical biosensing. Opt Eng 45(2):024401–024401
Siedler S, Stahlhut S, Malla S, Maury J, Neves A (2014) Novel biosensors based on flavonoid-responsive transcriptional regulators introduced into Escherichia coli. Metab Eng 21:2–8
Sinha S, Bachan Upadhyay LS (2024) Biosensing technology for detection and assessment of pathogenic microorganisms. Future Microbiol:1–16

Sobhanie E, Salehnia F, Xu G, Hamidipanah Y, Arshian S, Firoozbakhtian A et al (2022) Recent trends and advancements in electrochemiluminescence biosensors for human virus detection. TrAC Trends Anal Chem 157:116727

Srinivasan S, Raajasubramanian D, Ashokkumar N, Vinothkumar V, Paramaguru N, Selvaraj P et al (2024) Nanobiosensors based on on-site detection approaches for rapid pesticide sensing in the agricultural arena: a systematic review of the current status and perspectives. Biotechnol Bioeng

Thakur A, Kumar A (2022) Recent advances on rapid detection and remediation of environmental pollutants utilizing nanomaterials-based (bio) sensors. Sci Total Environ 834:155219

Thakur M, Wang B, Verma ML (2022) Development and applications of nanobiosensors for sustainable agricultural and food industries: recent developments, challenges and perspectives. Environ Technol Innov 26:102371

Ullah H, Wahid F, Santos HA, Khan T (2016) Advances in biomedical and pharmaceutical applications of functional bacterial cellulose-based nanocomposites. Carbohydr Polym 150:330–352

Valekunja RB, Kamakoti V, Peter A, Phadnis S, Prasad S, Nagaraj VJ (2016) The detection of papaya ringspot virus coat protein using an electrochemical immunosensor. Anal Methods 8(48):8534–8541

Vijayakumar MD, Surendhar GJ, Natrayan L, Patil PP, Ram PB, Paramasivam P (2022) Evolution and recent scenario of nanotechnology in agriculture and food industries. J Nanomater 2022(1):1280411

Virk V, Deepak H, Taneja K, Srivastava R, Giri S (2024) Amelioration in nanobiosensors for the control of plant diseases: current status and future challenges. Front Nanotechnol 6:1310165

Wang Y (2022) Application of electrochemical biosensors for chemical hazards detection. Highlights Sci Eng Technol 3:1–7

Wang H, Shen G, Yu R (2008) Aspects of recent development of immunosensors. In: Electrochemical sensors, biosensors and their biomedical applications, vol 237. Academic

Wang Y, Liu K, Zhang M, Xu T, Du H, Pang B, Si C (2023) Sustainable polysaccharide-based materials for intelligent packaging. Carbohydr Polym 313:120851

Zhang J, Zhao J (2019) Immuno-biosensor. In: Nano-inspired biosensors for protein assay with clinical applications. Elsevier, pp 115–137

Chapter 5
Enzymatic Nanobiosensors in Precision Agriculture: Methods and Applications

Julio C. Anchondo Páez, Esteban Sánchez, Erick H. Ochoa Chaparro, Carlos A. Ramírez Estrada, Cristina L. Franco Lagos, Juan J. Patiño Cruz, and Alan Álvarez Monge

Abstract Modern agriculture is currently facing significant challenges, including climate change, human population growth, increased food demand, reduced agricultural land, and environmental pollution. Precision agriculture has emerged as a vital strategy to optimize resource use, enhance productivity, and decrease environmental impact. In this context, enzymatic nanobiosensors represent an innovative technology with the potential to transform agricultural practices. These devices combine principles of nanotechnology (NT) and molecular biology to accurately and in real time detect various chemical and biological variables, such as nutrients, contaminants, and pathogens. This capability enables more informed and efficient decisions in crop management. The use of advanced sensors in agriculture is essential for addressing global challenges like climate change, water scarcity, and rising food demand. However, traditional monitoring techniques tend to be slow and costly and often require specialized personnel. In contrast, enzymatic nanobiosensors offer high sensitivity and selectivity and can be integrated into smart systems, making them particularly suitable for precision agriculture. These sensors can detect specific levels of key compounds, such as nitrates and phosphates, optimizing fertilization while minimizing environmental impact. Enzymatic nanobiosensors can significantly contribute to precision agriculture by developing next-generation diagnostic tools and techniques. This chapter aims to explore the design, fabrication, and application methods of enzymatic nanobiosensors in precision agriculture, focusing on their ability to effectively detect nutrients and contaminants. It also seeks to identify the primary technological and environmental limitations hindering their widespread adoption and propose strategies for their implementation in smart agricultural systems.

J. C. Anchondo Páez · E. Sánchez (✉) · E. H. Ochoa Chaparro · C. A. Ramírez Estrada
C. L. Franco Lagos · J. J. Patiño Cruz · A. Á. Monge
Centro de Invesstigación en Alimentación y Desarrollo A.C, Chihuahua, Mexico
e-mail: esteban@ciad.mx

J.-T. Chen (ed.), *Nanobiosensors for Crop Monitoring and Precision Agriculture*, Sustainability Sciences in Asia and Africa,
https://doi.org/10.1007/978-981-96-8335-2_5

Keywords Enzymatic nanobiosensors · Precision agriculture · Climate change · Nanotechnology

5.1 Introduction

Modern agriculture is currently encountering major challenges, including climate change, human population growth, and increased food demand (food security), alongside shrinking agricultural land and environmental pollution (Devlet 2021). Precision agriculture, combined with nanotechnology (NT)—specifically, enzymatic nanobiosensors—plays a critical role in addressing these challenges. Precision agriculture encompasses a suite of technologies that integrate sensors, information systems, enhanced machinery, and informed management to optimize production while accounting for the variability and uncertainties inherent in agricultural systems. This approach provides a means to monitor the food production chain and manage both the quantity and quality of agricultural products (Gebbers and Adamchuk 2010; Sahu et al. 2019).

On the other hand, enzymatic nanobiosensors are biological sensing devices that utilize nanotechnology and enzymes to perform specific measurements in biological, chemical, or environmental contexts (Prasad et al. 2017; Solaimuthu et al. 2020). In these sensors, enzymes act as biological recognition elements due to their ability to catalyze specific reactions with target analytes (Singh 2021). Nanotechnology enhances the sensitivity and selectivity of these sensors, as nanometric surfaces (e.g., nanoparticles and nanotubes) facilitate greater interaction between the enzyme and the analyte, ultimately improving detection efficiency. These sensors are widely used in precision agriculture to monitor nutrient levels, contaminants, and stress indicators in plants and soil. The signal transduction methods can include electrochemical, optical, or thermal approaches, allowing for precise quantification of the compounds of interest (Sagadevan and Periasamy 2014).

The application of advanced sensors in agriculture is crucial to tackling global challenges like climate change, water scarcity, and increasing food demand (Singh 2021). However, traditional monitoring techniques are often slow, expensive, and reliant on specialized personnel. In contrast, enzymatic nanobiosensors are distinguished by their high sensitivity, selectivity, and potential for integration into smart systems, making them ideal for precision agriculture. These nanobiosensors facilitate the detection of specific levels of important compounds, such as nitrates or phosphates, thereby optimizing fertilization and reducing environmental impact (Choudhary et al. 2015).

While the potential for nanobiosensors has been explored across various fields, their development and application in precision agriculture remain in the early stages. Current research primarily focuses on detecting nutrients and pesticides, yet studies investigating the stability and performance of these sensors under real field conditions are lacking. Furthermore, there is an ongoing need to integrate these devices into digital platforms that enable remote monitoring and real-time decision-making.

The main objective of this chapter is to examine the design, fabrication, and application methods of enzymatic nanobiosensors in precision agriculture, particularly emphasizing their capacity to effectively detect nutrients and contaminants.

5.2 Fundamentals of Enzymatic Nanobiosensors

5.2.1 Definition and Operating Principles

5.2.1.1 Nanobiosensor

A nanobiosensor is a device that incorporates nanoparticles in its construction, providing high sensitivity and signal transduction performance (Singh 2021). Nanobiosensors use a sensitive bioreceptor to detect certain specific molecules called analytes. The purpose of nanobiosensors is to generate quantitative analytical data with high precision on the molecule of interest. Their operation is achieved by combining physical or chemical transducers with biological recognition components to identify biological products (Bhalla et al. 2016; Mondal et al. 2022). The analytes interact with the biological layers of the nanobiosensors and generate a quantifiable signal that is generally converted into an electrical signal by the effect of the transducers (Huang et al. 2021a, b).

5.2.1.2 Enzymatic Nanobiosensors

Enzymes are organic biomolecules of macro size and complex structure, mainly composed of proteins; their function is to act as biological catalysts and accelerate the transformation of substrates into products (Kumar and Neelam 2016). Enzymatic biosensors are analytical devices in which an enzyme is integrated as a bioreceptor or closely coupled to a physicochemical transducer to provide an electrical signal, which is proportional to the amount of analyte in the sample (Kilic et al. 2023).

5.2.1.3 Operation

Nanobiosensors consist of three essential parts: biological probe elements, a transducer, and a detector.

Within enzymatic nanobiosensors, enzymes serve as biological probe elements and have the role of bioreceptors. The detector has the function of sensing any physical or chemical change, which can be optical, electrochemical, piezoelectric, or thermal (Coronado-Apodaca et al. 2023). The transducer is an element whose function is to convert the signal sent by the detector into an electrical signal. The electrical signal reaches a microcontroller, where the information contained is processed, amplified, and sent to an electronic device, where the information is

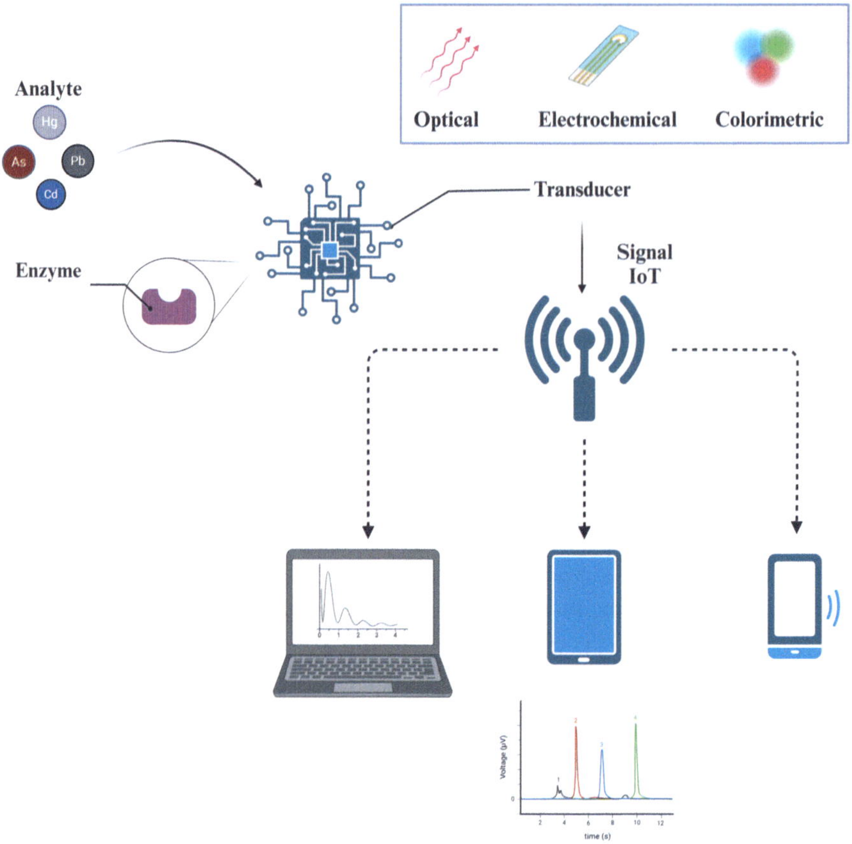

Fig. 5.1 Scheme of operation and parts of an enzymatic nanobiosensor

subsequently extracted for quantification, analysis, and interpretation. Finally, the data are displayed on a screen where the information from the nanobiosensor can be clearly visualized through graphs or figures (Fig. 5.1) (Singh 2021).

5.2.1.4 Role of Enzymes as Bioreceptors

The three-dimensional structure and active site of enzymes are critical to their ability to recognize and bind in a highly specific manner to the analyte. This recognition is achieved by non-covalent interactions, such as hydrogen bonds, electrostatic forces, or other molecular interactions, which facilitate their catalytic reaction. During the reaction, the analyte is transformed into a product that can be quantified and subsequently analyzed. The use of enzymes as bioreceptors offers several advantages, including fast response times, continuous real-time detection, high detection resolution, and reduced operating costs, while maintaining high specificity, reliability, and sensitivity.

5.2.1.5 Transduction Mechanisms

Thermal Transducers

Thermal or calorimetric transducers are used to detect analytes that generate heat release or absorption reactions, i.e., exothermic or endothermic reactions, when interacting with the bioreceptor, making a correlation with analyte concentration and heat (Kaur et al. 2019).

Optical Transducers

Optical transducers detect analytes through the interaction of protons with radiation (UV, visible, and infrared), generating energy transfer through an optical signal that can be detected by different methods such as colorimetry, fluorescence, or luminescence (Damborský et al. 2016; Kaur et al. 2019). They are classified into labeled and label-free detection, the latter being more economical and direct. Although they are highly sensitive and specific, their use is limited by the high cost of the necessary equipment (Chen and Wang 2020).

Piezoelectric Transducers

Piezoelectric transducers measure analytes by voltage changes and vibrations caused by mass variations in the bioreceptor during enzyme–analyte interactions. They use quartz crystal or acoustic devices to record frequency changes, according to the Sauerbrey equation (Pohanka 2018).

Electrochemical Transducers

Electrochemical transducers convert enzyme–analyte reactions into voltage or current changes to quantify contaminants. They are classified into (i) amperometric, which detects electrons by redox reactions; (ii) potentiometric, which measures changes in potential; (iii) conductimetric, which evaluates differences in conductivity; and (iv) impedimetric, which analyzes changes in resistance and reactance. They are the most widely used because of their high sensitivity, low cost, fast response time, and low sample volume required (Singh et al. 2021).

5.2.2 Materials and Applied Nanotechnology

5.2.2.1 Use of Nanomaterials

Nanomaterials have revolutionized the development of enzymatic nanobiosensors, thanks to their unique characteristics such as high surface-to-volume ratio, excellent electrical conductivity, and improved catalytic properties. Usually, carbon is the major element used for biosensors in electroanalysis, as it shows good catalytic properties such as improved chemical stability and biocompatibility. In the case of other types of transducers, metals that show more efficient electron transfer capabilities, such as gold, platinum, or palladium, are the preferred sensors to generate the signals of nanobiosensors.

The properties of nanomaterials allow a more efficient immobilization of enzymes, which increases the selectivity and sensitivity of nanobiosensors. In addition, their versatility has allowed the diversification of applications, from the monitoring of environmental conditions to the detection of pathogens, nutrients, and water quality, specifically in precision agriculture (Cavalcante et al. 2021).

5.2.2.2 Advantages of Miniaturization and High Sensitivity

The miniaturization of biosensors constructed with nanomaterials offers advantages such as higher sensitivity, faster response times, and smaller sample quantities. This is mainly due to the ability of nanomaterials to facilitate more efficient electron transfer and improve biocompatibility, which maximizes the catalytic activity of immobilized enzymes. In addition, their nanoscale size allows for integration into portable devices and real-time monitoring systems.

These technological advances have made nanotechnology-based enzymatic biosensors very efficient tools in fields of study such as environmental control, food safety, and precision agriculture (Cavalcante et al. 2021).

5.3 Design and Fabrication Methods for Enzymatic Nanobiosensors

For a sensor to be called an enzymatic nanobiosensor, the fabrication process must combine enzymes, nanomaterials, and transduction components that together form a system capable of detecting the desired compounds or analytes (Ghaffar et al. 2020). The key point for the development of enzyme-based nanosensors depends on a fabrication methodology capable of converting the biocatalytic activity of enzymes into signals that are measurable and interpretable (Soy et al. 2022). Likewise, in order to ensure that the sensor retains its high sensitivity, selectivity, and stability under operating conditions, the fabrication procedure is essential (Putzbach and Ronkainen 2013). This manufacturing process starts with the selection of the enzyme to be conjugated.

5.3.1 *Enzyme Selection*

One of the critical steps in designing a highly specific and efficient enzyme nanobiosensor is the selection of the enzyme to be used. Likewise, due to the large number of enzymes available and the factors that affect their performance, such as substrate specificity, catalytic activity, operational stability, and compatibility, choosing one that is adapted to be conjugated to nanomaterials is a challenge (Kilic et al. 2023).

Generally, the most common signal transduction platforms use electron exchange between redox proteins and an electrode. Thus, the candidate enzymes best suited to perform this process are globular proteins composed of linear chains of amino acids, which, when unfolded, form three-dimensional structures that provide the specificity required by the system (Verma 2017). In addition, authors have mentioned that the enzymes that are best suited for sensor assembly are those with a wide range of heat tolerance, which is why enzymes isolated from thermophilic bacteria or genetically engineered enzymes are preferred (Cavalcante et al. 2021).

Some of the most commonly used enzymes for the fabrication of nanobiosensors are glucose oxidase (GOx), peroxidase, laccase, lipase, amylase, and urease, among others (Arora 2018). Glucose oxidase (GOx) (E.C. 1.1.3.4.) has been widely studied for glucose sensing since, when this enzyme is immobilized, glucose can be easily attached to the electrode surface for detection. Likewise, peroxidase (E.C. 1.11.1.7) has been used as a nanobiosensor for plant pathogens because of its high specificity in detecting single-nucleotide changes in some groups of viruses and bacteria (Kaushal and Wani 2017). For its part, the enzyme laccase (E.C. 1.10.3.2) is currently the enzyme with the best properties for the detection of phenolic compounds due to its ability to catalyze the reaction without the help of additional cofactors. Therefore, nanobiosensors for the detection of harmful compounds, especially phenols, based on the enzymatic activity of laccase, have increased in the past decade (Kadam et al. 2022).

5.3.2 *Integration with Nanomaterials*

One of the most widely accepted methodologies for the application of enzymes is immobilization (Razzaghi et al. 2022). Immobilization allows support, often metallic, to function as a matrix, allowing for subsequent recovery and reuse, as well as increased traceability when used as a sensor in living systems such as plants (Kumar and Arora 2020).

Similarly, when developing a methodology for the application of enzymes as sensors, the selection process must be considered. Given the variety of enzymes, the materials that can serve as support, and the conditions for immobilization, the selection must be carried out appropriately. Therefore, both the selection and the immobilization method, in combination, represent a key factor for the correct application of nano-bioenzymes as sensors (Table 5.1) (Meena et al. 2021).

Table 5.1 Main enzymes used for conjugation with nanomaterials in the fabrication of enzymatic nanobiosensors

Immobilized enzyme	Nanomaterial used	Objective	Reference
Esterase	Multi-walled carbon nanotubes	Fabrication of screen-printed electrodes as biosensors for the detection of pesticides	Abdullah et al. (2020)
Cysteine protease	Single-walled carbon nanotubes	Detection of Huanglongbing bacteria, responsible for citrus greening	Tran et al. (2020)
Glucose	Spherical gold nanoparticles (AuNPs) stabilized with chitosan and plant-type zinc oxide (ZnO)	Glucose detection	Tian et al. (2015)
Plant-derived esterase	Composite of chitosan/gold nanoparticles and graphene nanosheets	Organophosphate (OP) pesticide detection	Bao et al. (2015)
Tyrosinase	Quantum dot conjugate and chitosan	Detection of phenolic compounds	Han et al. (2015)
Tyrosinase	Nanocrystalline cellulose	Detection of phenolic compounds	Abd-Manan et al. (2019)

Some conventional nanomaterials to be used as supports are nanoparticles, nanowires, nanotubes, and graphene derivatives (Cavalcante et al. 2021). The properties of these materials that make them stand out from other nanomaterials are higher conductivity and tunable surfaces, suitable for optical enzymatic nanobiosensors (Singh 2019). Similarly, their ability to increase electron transfer, enhance signal amplification, and provide stable support systems ensures that enzyme activity is maintained or even increased during sensor operation, making these materials the perfect candidates to be selected as enzyme-coupling nanomaterials (Ashkan et al. 2021).

5.3.3 *Transduction Platforms*

Another key element for the fabrication of enzyme nanobiosensors is the signal transduction mechanism, also called transduction platforms. Achieving robust immobilization of enzymes, maintaining their biological activity, and improving the interface between the enzyme and the signal transducer are important factors in this process (Bharti et al. 2024). In addition, both the spatial arrangement and orientation of enzymes can be precisely controlled by advanced fabrication procedures, which improve signal generation and ensure efficient interaction with analytes (Kucherenko et al. 2019).

Some of the most popular methods are advanced lithographic processes, chemical bonding, and physical deposition (Sharma et al. 2021). Likewise, techniques such as electrochemical deposition, inkjet printing, and layer-by-layer assembly have become more popular for nanoscale biosensors because of their precision and suitability for nanomaterials. Functionalized nanomaterials, such as graphene, nanoparticles, or nanotubes, can be integrated, thanks to these methods (Tonelli et al. 2019).

Once different transduction methods were elaborated, categorization according to their functionality was resorted to, the most common being optical, electrochemical, piezoelectric, pyroelectric, electronic, and gravimetric nanobiosensors. However, despite the large number of methods used for signal transduction, the common denominator is the detection of events that can be transformed into measurable signals (Sharma et al. 2021).

5.4 Applications in Precision Agriculture

Enzymatic nanobiosensors offer better advantages than conventional methods and are becoming an essential tool for sustainable and precision agriculture. While conventional methods can process a small number of samples, need specialized personnel, and require extended periods of time for their analysis, nanobiosensors provide a faster, more specific, and enhanced sensitivity for the detection of target analytes (Thakur et al. 2022). Moreover, the information recollected for obtained by these nanobiosensors can be uploaded to devices, machines, or software in order to help farmers improve yields and crop quality (Quy et al. 2022).

5.4.1 Soil Nutrient Monitoring

Soil is an important natural resource essential for growing the food necessary for humans' and terrestrial animals' needs. Thus, it is important to identify potential risks to soil quality by continuously monitoring soil parameters such as pH, salinity, macronutrient and micronutrient concentrations, or pollutant presence (Fan et al. 2022; Kim and Lee 2022). Real-time monitoring could provide important information and updates for decision-making and reduce fertilizer use and wastage while also providing economic benefits.

5.4.1.1 Detection of Nitrates and Phosphates

Nitrate (NO_3^-) is an important compound necessary for plant growth and development, owing to its role in many biochemical and physiological activities within the plant (Matimati et al. 2014; Leghari et al. 2016), and is commonly applied as a fertilizer to improve crop yields. However, disproportionate use of N-fertilizers increases the concentration of NO_3^- in soils, and it can accumulate in leafy vegetables and fruits, affecting human and animal health (Singh et al. 2022). Kundu et al. (2022) developed three nanobiosensors for NO_3^- monitoring in soil extracts. Among these, the nanobiosensor consisting of hematite nanostructures, nitrate reductase, and bovine serum albumin demonstrated the best performance, high sensitivity, high accuracy, and better shelf life of all the nanobiosensors in the study.

Phosphate (PO_4^{3-}) and copper (Cu^{2+}) ions are essential for living organisms in adequate quantities but can represent a significant source of pollution if used excessively. Thus, Amalraj et al. (2024) developed a colorimetric low-cost paper strip sensor with a cobalt and bi-organic–based metal framework (Co-MOF) that acts as a nanozyme. The nanozyme functions like peroxidase and is PO_4^{3-}-specific and also shows high affinity to tetramethylbenzidine (TMB) and H_2O_2. The cobalt nodes catalyze TMB into oxidized TMB, and the Co-MOF seizes PO_4^{3-}, limiting electron transfer between Co-MOF and H_2O_2. This causes a color change, from blue to colorless, due to the reduction of oxidized TMB. However, the flow of electrons can be reactivated when Cu^{2+} complexes with PO_4^{3-} form $Cu_3(PO_4)_2$, changing H_2O_2 to OH. This restores the original blue color because TMB oxidation occurs, thus enabling monitoring by colorimetric assay of PO_4^{3-} and Cu^{2+} ions.

5.4.2 *Water Quality Control*

Water is a vital compound necessary for the survival of humans, plants, and animals. However, the quality of water used to support these communities has deteriorated because of industrialization, urbanization, mining, and agriculture (Gupta and Chinnasamy 2022). Water contaminated with pesticides, heavy metals (HMs), and fertilizer residues poses serious dangers to human health, especially if used for crop irrigation, due to the persistence of these compounds throughout the food chain. Therefore, precise and timely monitoring of pollutants is imperative for water quality monitoring.

5.4.2.1 Pesticide Detection

Pesticides are chemical compounds extensively used to manage pests and boost agricultural productivity. They commonly include organophosphates, carbamates, organochlorines, and pyrethroids as active ingredients (WHO 2020; Kumar et al. 2021). Organophosphates (OPs) inhibit acetylcholinesterase (AChE), leading to high concentrations of acetylcholine, impairing the nervous system of pests. These compounds can induce respiratory, reproductive, nervous, and other health concerns in humans and animals when toxic levels are reached (Sidhu et al. 2019). In a similar way, carbamates also inhibit acetylcholinesterase but with low mammalian toxicity and low persistence. However, they can still be a threat to the nervous system, and some carbamates are presumed to be carcinogenic and mutagenic.

Organochlorines are very stable, persistent, and potent compounds, widely used in emerging countries for farming activities and other purposes. Long-term exposure to organochlorines can cause their bioaccumulation in living organisms, increasing the risk of developing cancer and immune system disorders (da Silva Júnior et al. 2024). Pyrethroid pesticides are designed to modify the sodium channels of the nervous system, abnormally exciting the nerve cells and provoking paralysis and death (Wakeling et al. 2012). While most pyrethroids are much deadlier to insects than humans, they can induce dizziness, headaches, anorexia, and even convulsive episodes in high quantities (Soderlund 2010).

The excessive use of pesticides in agriculture poses a serious threat to the population, environment, and food security. Thus, a timely, precise, and cost-effective detection of these compounds in water is necessary to mitigate adverse health risks. By using the catalytic ability of enzymes, enzymatic nanobiosensors can monitor pesticide residues in the environment. In these systems, several enzymes are used: acetylcholinesterase (AChE), organophosphorus hydrolase, choline esterase, acid phosphatase, tyrosinase, ascorbate oxidase, acetolactate synthase, and aldehyde dehydrogenase (Samal et al. 2023). For example, Maanaki et al. (2023) designed a nanobiosensor that works by the inhibition of AChE by organophosphate pesticides present in food and water. The enzymatic reaction of AChE with acetylcholine results in choline and acetic acid, with the latter dissociating into acid residues and a proton. The transit of this proton throughout polyaniline nanofibers causes a change in the conductance in a gold-interdigitated electrode, which can be registered by a digital meter. When OP is added, the generation of protons is slowed, reducing the change of conductance or even registering no change if AChE is fully inhibited by OP.

5.4.2.2 Heavy Metal Detection

Large amounts of heavy metals (HMs) are released by industrial activities, traffic emissions, sewage sludge, and widespread agrochemical applications. Their presence in agricultural soil and water endangers food security and human health, owing to their toxicity, persistence, non-biodegradability, and bioaccumulation in the food chain (Oladoye et al. 2022). Examples of these HMs include chromium (Cr), arsenic (As), nickel (Ni), cadmium (Cd), lead (Pb), mercury (Hg), zinc (Zn), and copper (Cu) (Angon et al. 2024).

Some of these elements, including Zn, Ni, and Cu, are required in trace amounts for plant growth and development (Siddiqi and Husen 2020; Khan et al. 2023; Jalal et al. 2024). However, high quantities of HMs in the soil can impair several physiological processes in plants, such as seed germination, photosynthetic activity, nutrient uptake, and damage to proteins, lipids, and DNA (Dobrikova et al. 2021; Abdullah et al. 2024). An abundance of HMs in fruits, vegetables, and grains exposes consumers to cancer and cardiovascular, neurological, bone, and kidney issues (Kumar et al. 2024a, b; Sweta and Singh 2024). Thus, timely and accurate detection is necessary to control the accumulation of HMs in the environment and prevent human health issues.

The use of enzymatic nanobiosensors could help in the identification of a wide array of HMs, such as Pb, Ni, and Cd, in agricultural soils. By using the highly selective nature of enzymes for molecules, several enzyme-based nanobiosensors can be designed. For instance, Attaallah and Amine (2022) developed an enzyme-based (horseradish peroxidase) immunokinetic membrane to detect Cd^{2+} in drinking water. This membrane can monitor Cd^{2+} in a precise manner in water without significant obstruction by other cations, such as Cu^{2+}, due to dielectric exchange technology. Cd^{2+} can act as an irreversible inhibitor, immobilizing horseradish peroxidase. The authors confirmed the immobilization by cyclic voltammetry, showing that small quantities of Cd can be successfully monitored. In another study, Rigo et al. (2020) designed a cantilever nanobiosensor with urease for the detection of HMs in water. If HMs are present in water, they bind to the active sites of urease by reacting with the sulfhydryl groups, triggering stress tension of the cantilever surface, causing a deflection that can be measured by voltage changes in the cantilever nanobiosensor.

5.4.3 Disease Monitoring

Plant diseases are a major source of losses in agricultural production and are often caused by fungi, bacteria, and viruses. Approximately 16% of the global harvest is lost due to plant pathogens. Early diagnosis of infections often requires several days and is done by visual inspection of crops (Dyussembayev et al. 2021). However, by the time visual symptoms are detected, the pathogen has already established itself and spread throughout the host population. Therefore, the development of pathogen monitoring has been driven by the need for greater sensitivity, accuracy, and identification speed. One approach is the detection of methyl salicylate, a volatile organic compound (VOC) emitted by plants when infected by pathogens. In their study, Fang et al. (2016) created a bi-enzyme electrochemical sensor capable of detecting methyl salicylate. Recognition of the VOC begins by hydrolysis into methanol, which then transforms into formaldehyde by alcohol oxidase, at the same time reducing oxygen to hydrogen peroxide. Finally, hydrogen peroxide is reduced by horseradish peroxidase, producing an amperometric signal by direct electron transport. However, as VOC liberation is a general response to pathogen infection and since many causal species trigger similar symptoms, discrimination of the pathogens is necessary for accurate diagnosis. For example, Tang et al. (2024) devised a portable sensing device consisting of Au-Co/Zn ZIF nanozymes, together with Au-DNA clusters, for the detection of sugarcane smut pathogen. These clusters are then caught by an H_2 probe, adsorbing an electron with methylene blue. The nanozyme component then catalyzes glucose oxidation, which produces a large flow of electrons and causes methylene blue to change color. As a result, this system emits a dual-signal output (current/colorimetric), allowing early and precise diagnosis, which can be monitored in real time with a smartphone. Another serious threat to sugarcane is pokkah boeng disease, for which Song et al. (2024) developed a dual-signal biosensor for early detection. This platform conjugates Au/Mn_3O_4 nanozymes with nitrogen-doped graphdiyne (N-GDY), together with targeted-induced DNA space arrays, for the precise and specific detection of pokkah boeng bacteria. When Au/Mn_3O_4 nanozymes on a bioanode are oxidized, the generation of electrons flows through a biocathode, reducing $[Ru(NH_3)_6]^{3+}$ to $[Ru(NH_3)_6]^{2+}$ and leading to an increase in an open-circuit voltage proportional to the concentration of pokkah boeng bacteria.

5.4.4 Integrated Systems and Precision Agriculture

Nowadays, agriculture faces significant challenges such as scarce resources, increasing population, and climate change (Yazdinejad et al. 2021). To address such issues, farmers have employed intelligent technologies such as remote sensing, Global Positioning Systems (GPS), Geographic Information Systems (GIS), big data analysis, and the Internet of Things (IoT) with the objective of maximizing crop productivity, reducing input costs, and minimizing environmental pollution (Zain et al. 2024). This approach, known as precision agriculture (PA), prioritizes data collection, decision-support tools, and data-driven equipment to facilitate diagnosis, decision-making, and execution (McFadden et al. 2023). For instance, farmers can manage their entire field, only some square meters, or even single plants based on the information received (Yazdinejad et al. 2021).

5.4.4.1 Use of Nanobiosensors in IoT Networks for Remote Monitoring Systems

The concept of IoT refers to the collective network of devices connected to technologies that simplify communication and information transfer between devices and the cloud (Karunathilake et al. 2023). The IoT can greatly aid precision agriculture by gathering data via interconnected sensors (like nanobiosensors) in real time. Different sensors can collect information on temperature, nutrient concentrations, and pathogen infestations and upload it to cloud support systems (Hundal et al. 2023). Moreover, improved communication and wireless technologies have increased coverage, bandwidth, connection density, and end-to-end latency, enhancing the reliability of on-site and remote monitoring systems in agriculture (Table 5.2).

Table 5.2 Enzymes used in successful cases of nanobiosensors in precision agriculture

Nanobiosensor	Enzyme used	Working procedure	Reference
Current/ colorimetric	Au-Co/Zn ZIF nanozyme and Au-DNA nanocluster	Dual-signal output allows for the precise detection of sugarcane smut pathogen; this signal can be read in real time by a smartphone	Tang et al. (2024)
Current	AuNPs/Mn_3O_4 nanozymes with nitrogen-doped graphdiyne	Dual-electrode signal bio-detection for ultrasensitive and specific detection of pokkah boeng disease. Real-time detection enabled by Bluetooth and smartphone	Song et al. (2024)

5.4.4.2 Advantages

High Specificity and Sensitivity

Biosensors can achieve remarkable sensitivity, selectivity, and response times (Pratama and Rahayu 2024). Nanoscale dimensions of different nanomaterials extend sensitivity through signal amplification and integration of various transduction principles into nanosensors (Srivastava et al. 2018). Proper use of nanobiosensors can contribute to sustainable agriculture that improves crop productivity (Ghaffar et al. 2020). Nanobiosensors and nanoparticle-based systems possess unique characteristics such as portability, specificity, and sensitivity that make them necessary for food and agriculture (Dar et al. 2020). Low-cost portable nanobiosensors are the key to rapid in situ assessment of plant health. To achieve reliable detection analysis, these nanobiosensors require nanoparticles to amplify the signal and the sensitivity of the sensor to reduce the detection limit. Also, the high sensitivity and selectivity of these devices ensure early detection and management of specific anomalies (Mondal et al. 2022).

Real-Time Response and Continuous Monitoring Capabilities

The integration of nanotechnology and the Internet of Things (IoT) has revolutionized precision agriculture, and this offers unprecedented opportunities for real-time monitoring and decision-making (Pratama and Rahayu 2024). The various nanobiosensors: electrochemical, optical, wireless, and nanobarcode technology, have revolutionized sensing in the agricultural field with multiplex and real-time sensing capabilities (Srivastava et al. 2018). The importance of nanobiosensors highlights their ability to influence the future of agriculture by enabling real-time monitoring of decision-making and crop management (Gangopadhyay et al. 2024). Nanobiosensors can play a key role in real-time quality improvement by ensuring the safety of resources used in agriculture. These biosensors allows us to face this challenge with increased accuracy, mobility, and sensitivity (Rana et al. 2024). In real time, nanobiosensors contribute significantly to agriculture and promote sustainable food production by detecting infections, pathogens, fertilizers, herbicides, pesticides, nutrient levels (micronutrient and macronutrient uptake), metabolites, heavy metal ions (e.g., Hg^{2+}, As^{3+}, and Cu^{2+}), toxins, pH, chlorophyll content, photosynthetic content, protein content, temperature, and soil moisture. This ensures the precision of agricultural techniques and maximizes the use of resources (Bharti et al. 2024) (Kumar et al. 2024a, b).

Reduction of Chemical Use and Minimization of Environmental Impact

After the green revolution, farmers started to indiscriminately apply chemical fertilizers, pesticides, insecticides, and herbicides, which led to the loss of soil biodiversity and caused resistance to pathogens and pests (Dhole and Pitambara 2019). Due to climate change, the world's agricultural systems face remarkable challenges to increase production while decreasing negative environmental impact with the use of agricultural nanotechnology. Nanobiosensors help detect abiotic and biotic stress in plants before production is affected, while nanoformulations (NFs) make agrochemicals more effective and less polluting. NFs facilitate the controlled and localized administration of agrochemicals by increasing their efficacy and reducing the dose (Miguel-Rojas and Pérez-de-Luque 2023). Knowing the threshold level of biological interactions in the rhizosphere at a very small scale helps the farmer obtain maximum yield by delivering material to plants through nanoparticles and advanced biosensors (Dhole and Pitambara 2019). Biosensors have the ability to decrease nutrient loss as well as the likelihood of removing phytosanitary substances, improving yield by effective nutrient management (Rana et al. 2024).

5.4.5 Limitations

5.4.5.1 Costs and Stability in Real Environments

In agriculture and the food industry, the application of nanotechnology (NT), such as nanobiosensors, is a topic that is currently receiving a lot of attention. Investment in nanotechnology applied to agriculture and the food industry is on the rise due to the motive of significantly improving the quality and safety of agricultural inputs, as well as the processing and nutritional value of food. While investments in NT for agriculture and food are highest in developed countries, current research reveals a high potential for applications in marginalized regions of the world. This could have a significant impact on the food security and welfare of rural populations in developing countries (Bodkhe et al. 2024).

The increasing accumulation of pollutants and toxic agents in the environment requires their rapid and accurate detection and quantification. Nanobiosensors, especially those of the enzymatic type, offer an attractive alternative in the monitoring of different pollutants in the environment in a fast, efficient, automated, and economical way (Ramírez and Albores 2014).

To manufacture a nanobiosensor, it is necessary to have separate knowledge of each component that makes it up; this allows it to be built as an integrated device, i.e., a device in which the components exist as if they were one. In the case of enzymatic nanobiosensors, the element that does the recognition is an enzyme (Ramírez and Albores 2014).

The enzyme is the central component of an enzyme nanobiosensor; enzyme-based nanobiosensors prioritize electrochemical transducers during instrumentation. Therefore, enzyme electrodes integrate the excellent specificity of enzymes with the benefits of electrochemical sensing, i.e., accurate and rapid estimation of nutrient concentrations in soil, increased productivity, reduced costs for farmers, and easy monitoring (Mondal et al. 2022).

5.4.5.2 Enzyme Stability Under Adverse Conditions (Heat, Humidity, and Extreme pH)

The stability of enzymes in nanobiosensors is crucial to ensure the effectiveness and accuracy of measurements under adverse conditions, such as high temperatures, humidity, and extreme pH, in precision agriculture. These nanobiosensors are widely used for their ability to distinguish a specific compound in a complex mixture. During the enzymatic reaction, changes such as variations in the number of protons or electrons, manifestation of light or heat, and the formation of electroactive compounds are generated, resulting in amplified signals detectable by the monitor at the end of the reaction (Mondal et al. 2022). Despite their benefits, in agricultural environments, enzyme stability is affected by several factors.

Heat: Heat can denature enzymes, affecting their catalytic activity, so it is essential that nanobiosensors are designed to maintain stability under adverse thermal conditions. In addition, electrochemical nanobiosensors, which are the most common, have several advantages, such as minimization of sample preparation and high sensitivity. These devices also stand out for their ability to directly convert electrochemical signals into electrical signals, which facilitates their rapid use, portability, and miniaturization, contributing to stability in hot conditions (Kumar et al. 2024a, b).

Moisture: Moisture can affect enzyme activity and stability in the device. Enzyme biosensors require a fast and selective response to low concentrations of the analytes, and the presence of moisture could interfere with the accuracy of these measurements. To maintain the effectiveness of the nanobiosensor in humid environments, the use of stabilizing materials that protect the enzymes is essential. In addition, electrochemical biosensors are advantageous because they provide direct conversion of the electrochemical signal, which can improve cost feasibility and ensure accuracy and reproducibility of measurements, even under variable humidity conditions (Yadav et al. 2023).

Extreme pH: Enzymes are also sensitive to changes in pH, which can alter their structure and activity. Enzyme biosensors must be able to maintain their specificity and activity, even in a variable pH range, which is common in agricultural soils. The selectivity of these biosensors allows them to detect a specific analyte despite the presence of other interfering compounds. To achieve this stability under extreme pH conditions, immobilization of enzymes in protective materials or on stable supports can be an effective solution. In this way, the enzymes remain active and stable, ensuring accurate measurements, even when the pH of the sample changes (Yadav et al. 2023).

Nanobiosensor Requirements: The nanobiosensor must be small, portable, and easy enough to use so that operators with even a basic level of skill can make accurate and rapid measurements. In addition, the response time must be extremely short, allowing farmers to make real-time decisions. Electrochemical nanobiosensors excel in their ability to directly convert electrochemical signals into electrical signals, which simplifies device design and operation and ensures that measurements are precise, accurate, and reproducible, unaltered by factors such as temperature, pH, and agitation (Bodkhe et al. 2024).

Enzymatic nanobiosensors are easy to use, sensitive at very low concentrations, and highly accurate and, even when coupled with nanoparticles, show great potential for miniaturization and real-time diagnostic capability, as well as requiring minimal sample preparation and promoting high throughputs.

5.4.5.3 Need for Integration with Existing Agricultural Systems

Nanobiosensors are designed to monitor the physicochemical characteristics of plants when exposed to comparably sized biological and chemical analytes (Mohammad et al. 2022). Nanoparticles are frequently used as transducer materials that enhancing signal transduction systems while recognizing and combining analytes. Nanobiosensors offer promising capabilities for precision agriculture, enabling farmers to monitor soil quality, detect plant diseases, and optimize resource management to improve crop yields and environmental sustainability (Huang et al. 2021a, b).

In the field of nanotechnology, nanobiosensors can be used in various agricultural applications, including precision agriculture, plant physiology monitoring and management, pesticide residue detection in plants, pathogen detection, disease management, and environmental monitoring (Kumar et al. 2024a, b).

The main challenges to ensuring the sustainability of the agricultural and food sectors include a growing world population, finite natural resources, rapid depletion of arable land, and rising production costs. Environmental conditions, such as moisture levels, insects, plant viruses, weeds, and temperature, restrict the agricultural sector from producing needed products. Technology must be used in agriculture to reduce yield loss and meet the nutritional needs of a growing world population (Kumar et al. 2024a, b).

5.5 Future Perspectives

The accelerated growth of the global population demands agricultural production that meets quality and quantity standards at a low cost and with minimal environmental impact (Daria et al. 2023). In this context, the emergence of precision agriculture and the use of nanobiosensors present themselves as key solutions to the various current challenges facing agricultural production, promising to resolve critical issues. These innovations aim to optimize resources and increase productivity while enabling precise, real-time monitoring of various agricultural parameters efficiently and sustainably. This section explores the growth of these technologies and their potential development in the agricultural sector.

5.5.1 Development of More Robust and Cost-Effective Biosensors

Advancements in technologies like nanobiosensors aim to create more affordable devices (Marín-Bustamante et al. 2021). Current trends lean toward device miniaturization and the implementation of mass production technologies, strategies designed to reduce costs, improve manufacturing reproducibility, and expand potential applications (Gálan-Vidal et al. 2010). Among various analytical techniques, electroanalytical methods have garnered special attention due to the use of metallic nanoparticles. For example, gold nanoparticles provide a natural environment for biomolecule immobilization, enabling greater stability and longevity (Fernández et al. 2020). Therefore, implementing nanotechnology is vital for developing scalable detection and monitoring processes with high sensitivity and multifunctionality (Salek-Maghsoudi et al. 2021).

5.5.2 Integration with Artificial Intelligence and Machine Learning for Predictive Analysis

The application of technology to the agricultural sector, combined with precision agriculture, has given rise to digital agriculture (Liakos et al. 2018). This new scientific field leverages data-intensive approaches to enhance agricultural productivity by collecting and analyzing information generated during agricultural operations. This improves the understanding of plant–soil–climate relationships, enabling more precise and faster decision-making. The integration of digital agriculture, artificial intelligence, and nanotechnology offers promising prospects for food production (Ashique et al. 2024). Artificial intelligence's ability to process complex data, synergized with the precision and efficiency of nanotechnology, facilitates real-time data analysis of key parameters. These include disease forecasting, detection of nutrient deficiencies, and identification of water requirements for different crops. These innovations transform agricultural productivity by reducing operating costs and mitigating environmental impact, contributing to more sustainable and resilient practices (Barreto-Mendoza et al. 2024).

5.5.3 *Advances in Multiparametric Detection*

Nanobiosensors have shown significant evolution, enabling the simultaneous and detailed detection of parameters due to their high sensitivity. These new technologies have been employed as diagnostic tools for identifying microbial activity, assessing diseases, monitoring soil conditions, and measuring nutrient levels (Virk et al. 2024). These parameters are crucial for comprehensive crop management, positioning nanobiosensors as essential agents for promoting more sustainable agricultural practices. These advances facilitate the implementation of more adaptive and personalized management strategies, improving agricultural efficiency and productivity.

5.5.4 *Potential Impact on Global Sustainability and Food Security*

The agricultural sector has undergone a massive transformation, tackling one of the greatest challenges of the twenty-first century: Producing quality food in sufficient amounts through the use of innovation and technology (Gyamfi et al. 2024). Devices such as nanobiosensors applied in precision agriculture have the potential to transform sustainability and global food security. This integration offers significant potential for improving agriculture by enabling the development of nanoscale agrochemicals that balance safety and functionality, optimizing agricultural inputs such as water and fertilizers to ensure optimal yields in agricultural systems (Ashique, S. et al., 2024). Therefore, it is evident that the implementation of new technologies like nanobiosensors in precision agriculture is a promising field of interdisciplinary and multidisciplinary research aimed at improving the productivity of agroecosystems (Lira Saldivar et al. 2018).

5.6 Conclusions

Enzymatic nanobiosensors represent a groundbreaking innovation in precision agriculture, offering high sensitivity and specificity in detecting essential nutrients, contaminants, and stress factors within agricultural environments. By utilizing nanotechnology and enzymatic biorecognition, these sensors enable real-time monitoring and improved resource optimization, which significantly reduces the environmental impact of farming practices. Their integration into smart systems has the potential to enhance crop yields, improve soil health, and promote overall sustainability.

However, challenges related to stability in field conditions, cost-effectiveness, and large-scale implementation remain critical barriers that must be addressed.

The future of enzymatic nanobiosensors lies in enhancing their robustness, affordability, and adaptability to various agricultural environments. Research should focus on developing more stable enzymes and materials that maintain their functionality under extreme conditions. Additionally, integrating artificial intelligence and machine learning will improve predictive analytical capabilities, allowing for more precise crop and resource management.

Furthermore, designing multiparametric sensors that can simultaneously monitor various environmental factors will be essential for comprehensive agricultural diagnostics. Global efforts should also aim to democratize access to this technology, ensuring that its benefits reach smallholder farmers in developing regions, helping effectively address food security challenges.

These advancements have the potential to revolutionize agricultural practices, paving the way for a more sustainable and resilient food production system.

References

Abd-Manan FA, Hong WW, Abdullah J, Yusof NA, Ahmad I (2019) Nanocrystalline cellulose decorated quantum dots based tyrosinase biosensor for phenol determination. Mater Sci Eng C 99:37–46. https://doi.org/10.1016/j.msec.2019.01.082

Abdullah NA, Yusof F, Ahmad FB (2020) Preliminary study on immobilization of plant esterase on functionalized multi-walled carbon nanotubes (MWCNTs) for biosensor application. IOP conference series: materials science and Engineering 932(1):012040. https://doi.org/10.1088/1757-899X/932/1/012040

Abdullah, Wani KI, Naeem M, Jha PK, Jha UC, Aftab T et al (2024) Systems biology of chromium-plant interaction: insights from omics approaches. Front Plant Sci 14:1305179. https://doi.org/10.3389/fpls.2023.1305179

Amalraj A, Ayyanu R, Pavadai R, Govindaraj TS, Aham EC, Li X et al (2024) Smartphone assisted paper strip-based colorimetric sensing of phosphate and copper ions utilizing bi-ligand intercalated cobalt-MOF as a dual functional nanozyme. J Environ Chem Eng 12(5):113522. https://doi.org/10.1016/j.jece.2024.113522

Angon PB, Islam MS, Kc S, Das A, Anjum N, Poudel A et al (2024) Sources, effects and present perspectives of heavy metals contamination: soil, plants and human food chain. Heliyon. https://doi.org/10.1016/j.heliyon.2024.e28357

Arora K (2018) Advances in Nano based biosensors for food and agriculture. In: Nanotechnology, food security and water treatment. Environmental chemistry for a sustainable world. Springer, Cham, pp 1–52. https://doi.org/10.1007/978-3-319-70166-0_1

Ashique S, Raikar A, Jamil S, Lakshminarayana L, Gajbhiye SA, Kumar S (2024) Artificial intelligence integration with nanotechnology: a new frontier for sustainable and precision agricultura. Curr Nanosci 21(2):242–273. https://doi.org/10.2174/0115734137275111231206072049

Ashkan Z, Hemmati R, Homaei A, Dinari A, Jamlidoost M, Tashakor A (2021) Immobilization of enzymes on nanoinorganic support materials: an update. Int J Biol Macromol 168:708–721. https://doi.org/10.1016/j.ijbiomac.2020.11.127

Attaallah R, Amine A (2022) An ultrasensitive and selective determination of cadmium ions at ppt level using an enzymic membrane with colorimetric and electrochemical detection. Biosensors 12(5):310. https://doi.org/10.3390/bios12050310

Bao J, Hou C, Chen M, Li J, Huo D, Yang M et al (2015) Plant esterase–chitosan/gold nanoparticles–graphene nanosheet composite-based biosensor for the ultrasensitive detection of organophosphate pesticides. J Agric Food Chem 63(47):10319–10326. https://doi.org/10.1021/acs.jafc.5b03971

Barreto-Mendoza DA, Cedeño-Mendoza RD, Chavarria-Mendoza EJ, Caicedo-Mero SG, Villa Palma CM (2024) Aplicación de la inteligencia artificial en la producción de sembrío de ciclo corto en el sector agro productivo. Biotech Eng 4(2):102–114. https://doi.org/10.52248/eb.Vol4Iss2.172

Bhalla N, Jolly P, Formisano N, Estrela P (2016) Introduction to biosensors. Essays Biochem 60(1):1–8. https://doi.org/10.1042/EBC20150001

Bharti A, Jain U, Chauhan N (2024) From lab to field: nano-biosensors for real-time plant nutrient tracking. Plant Nano Biol 100079. https://doi.org/10.1016/j.plana.2024.100079

Bodkhe M, Chalke T, Kulkarni S, Goswami A (2024) A review on sustainable applications of nanobiosensors in various fields and future potential. BioNanoScience 1–21. https://doi.org/10.1007/s12668-024-01344-w

Cavalcante FT, de A. Falcão IR, da S. Souza JE, Rocha TG, de Sousa IG, Cavalcante AL et al (2021) Designing of nanomaterials-based enzymatic biosensors: synthesis, properties, and applications. Electrochem 2(1):149–184. https://doi.org/10.3390/electrochem2010012

Chen C, Wang J (2020) Optical biosensors: an exhaustive and comprehensive review. Analyst 145(5):1605–1628. https://doi.org/10.1039/C9AN01998G

Choudhary MK, Singh M, Saharan V (2015) Application of nanobiosensors in agriculture. Popular Kheti 3:130–135

Coronado-Apodaca KG, González-Meza GM, Aguayo-Acosta A, Araújo RG, Gonzalez-Gonzalez RB, Oyervides-Muñoz MA, Martínez-Ruiz M, Melchor-Martínez EM, Barceló D, Parra-Saldívar R, Sosa-Hernández JE (2023) Immobilized enzyme-based novel biosensing system for recognition of toxic elements in the aqueous environment. Top Catal. https://doi.org/10.1007/s11244-023-01786-8

da Silva Júnior AH, De Oliveira CRS, Leal TW, Mapossa AB, Fiates J, De Souza AAU et al (2024) Organochlorine pesticides remediation techniques: technological perspective and opportunities. J Hazard Mater Lett 5:100098. https://doi.org/10.1016/j.hazl.2023.100098

Damborský P, Švitel J, Katrlík J, Estrela P (2016) Optical biosensors. Essays Biochem 60(1):91–100. https://doi.org/10.1042/EBC20150010

Dar FA, Qazi G, Pirzadah TB (2020) Nano-biosensors: NextGen diagnostic tools in agriculture. In: Nanobiotechnology in agriculture: an approach towards sustainability. Springer, Cham, pp 129–144. https://doi.org/10.1007/978-3-030-39978-8_7

Daria T, Valeriy S, Aleksander L, Aleksandr B, Roland G, Kirill K et al (2023) Climate change and future food security: predicting the extent of cropland gain or degradation. arXiv preprint arXiv 2310.03261

Devlet A (2021) Modern agriculture and challenges. Front Life Sci Relat Technol 2(1):21–29. https://doi.org/10.51753/flsrt.856349

Dhole A, Pitambara M (2019) Nanobiosensors: a novel approach in precision agriculture. In: Nanotechnology for agriculture: advances for sustainable agriculture. Springer, Cham, pp 241–262. https://doi.org/10.1007/978-981-32-9370-0_13

Dobrikova AG, Apostolova EL, Hanć A, Yotsova E, Borisova P, Sperdouli I et al (2021) Cadmium toxicity in Salvia sclarea L.: an integrative response of element uptake, oxidative stress markers, leaf structure and photosynthesis. Ecotoxicol Environ Saf 209:111851. https://doi.org/10.1016/j.ecoenv.2020.111851

Dyussembayev K, Sambasivam P, Bar I, Brownlie JC, Shiddiky MJ et al (2021) Biosensor technologies for early detection and quantification of plant pathogens. Front Chem 9:636245. https://doi.org/10.3389/fchem.2021.636245

Fan Y, Wang X, Funk T, Rashid I, Herman B, Bompoti N et al (2022) A critical review for real-time continuous soil monitoring: advantages, challenges, and perspectives. Environ Sci Technol 56(19):13546–13564. https://doi.org/10.1021/acs.est.2c03562

Fang Y, Umasankar Y et al (2016) A novel bi-enzyme electrochemical biosensor for selective and sensitive determination of methyl salicylate. Biosens Bioelectron 81:39–45. https://doi.org/10.1016/j.bios.2016.01.095

Fernández L, Espinoza-Montero P, Rodríguez A (2020) Electrodos modificados con nanopartículas de oro y su aplicación en biosensores electroquímicos que utilizan peroxidasa de rábano como sistema de bioreconocimiento biológico: detección de peróxido de hidrógeno. InfoANALÍTICA 8(1):101–132. https://doi.org/10.26807/ia.vi.179

Gálan-Vidal CA, Pacheco-Hernández ML, Álvarez-Romero GA, Páez-Hernández ME, Mendoza-Huizar LH, Sierra-Zenteno A (2010) Desarrollo de biosensores miniaturizados de bajo costo en configuración plana. Cienc Univ:73–81

Gangopadhyay M, Chakravarti S, Sanyamat S, Dewanjee S (2024) Biosensors in agriculture: pioneering the future of precision farming and sustainability. In: Biosensors: developments, challenges and perspectives. Springer, pp 259–301. https://doi.org/10.1007/978-981-97-3048-3_13

Gebbers R, Adamchuk VI (2010) Precision agriculture and food security. Science 327:828–831. https://doi.org/10.1126/science.1182768

Ghaffar N, Farrukh MA, Naz S (2020) Applications of nanobiosensors in agriculture. In: Nanoagronomy. Springer, Cham, pp 179–196. https://doi.org/10.1007/978-3-030-41275-3_10

Gupta M, Chinnasamy P (2022) Trends in groundwater research development in the South and Southeast Asian countries: a 50-year bibliometric analysis (1970–2020). Environ Sci Pollut Res 29(50):75271–75292. https://doi.org/10.1007/s11356-022-21163-4

Gyamfi EK, ElSayed Z, Kropczynski J, Yakubu MA, Elsayed N (2024) Agricultural 4.0 leveraging on technological solutions: study for smart farming sector. arXiv preprint arXiv 2401.00814

Han E, Yang Y, He Z, Cai J, Zhang X, Dong X (2015) Development of tyrosinase biosensor based on quantum dots/chitosan nanocomposite for detection of phenolic compounds. Anal Biochem 486:102–106. https://doi.org/10.1016/j.ab.2015.07.001

Huang X, Aguilar ZP, Xu H, Lai W, Xiong Y, Sun J (2021a) Biosensors for analysis of pollutants in agriculture. Anal Chem 93(7):3309–3318. https://doi.org/10.1021/acs.analchem.0c05254

Huang X, Zhu Y et al (2021b) Nano biosensors: properties, applications and electrochemical techniques. J Mater Res Technol 12:1649–1672. https://doi.org/10.1016/j.jmrt.2021.03.048

Hundal GS, Laux CM, Buckmaster D, Sutton MJ et al (2023) Exploring barriers to the adoption of internet of things-based precision agriculture practices. Agriculture 13(1):163. https://doi.org/10.3390/agriculture13010163

Jalal A, Júnior EF et al (2024) Interaction of zinc mineral nutrition and plant growth-promoting bacteria in tropical agricultural systems: a review. Plan Theory 13(5):571. https://doi.org/10.3390/plants13050571

Kadam AA, Saratale GD, Ghodake GS, Saratale RG, Shahzad A, Magotra VK et al (2022) Recent advances in the development of laccase-based biosensors via nano-immobilization techniques. Chemosensors 10(2):58. https://doi.org/10.3390/chemosensors10020058

Karunathilake EMBM, Le AT, Heo S, Chung YS et al (2023) The path to smart farming: innovations and opportunities in precision agriculture. Agriculture 13(8):1593. https://doi.org/10.3390/agriculture13081593

Kaur J, Choudhary S, Chaudhari R, Jayant RD, Joshi A (2019) Enzyme-based biosensors. In: Bioelectronics and medical devices. Elsevier, pp 211–240

Kaushal M, Wani SP (2017) Nanosensors: frontiers in precision agriculture. In: Nanotechnology. Springer, Singapore, pp 279–291. https://doi.org/10.1007/978-981-10-4573-8_13

Khan A, Khan AA et al (2023) Effects of different concentrations of nickel (ni) on the vegetative and reproductive growth parameters of Nigella sativa L. Gesunde Pflanzen 75(3):677–686. https://doi.org/10.1007/s10343-022-00733-z

Kilic NM, Singh S, Keles G, Cinti S, Kurbanoglu S, Odaci D (2023) Novel approaches to enzyme-based electrochemical nanobiosensors. Biosensors 13(6):622. https://doi.org/10.3390/bios13060622

Kim MY, Lee KH (2022) Electrochemical sensors for sustainable precision agriculture—a review. Front Chem 10:848320. https://doi.org/10.3389/fchem.2022.848320

Kucherenko IS, Soldatkin OO, Kucherenko DY, Soldatkina OV, Dzyadevych SV (2019) Advances in nanomaterial application in enzyme-based electrochemical biosensors: a review. Nanosc Adv 1(12):4560–4577. https://doi.org/10.1039/C9NA00491B

Kumar V, Arora K (2020) Trends in nano-inspired biosensors for plants. Mater Sci Energy Technol 3:255–273. https://doi.org/10.1016/j.mset.2019.10.004

Kumar H, Neelam R (2016) Enzyme-based electrochemical biosensors for food safety: a review. Nanobiosens Dis Diagn 5:29–39. https://doi.org/10.2147/NDD.S64847

Kumar M, Yadav AN, Saxena R, Paul D et al (2021) Biodiversity of pesticides degrading microbial communities and their environmental impact. Biocatal Agric Biotechnol 31:101883. https://doi.org/10.1016/j.bcab.2020.101883

Kumar A, Kumar V, Rodríguez-Seijo A, Setia R, Singh S, Kumar A et al (2024a) Appraisal of heavy metal (loid) s contamination in rice grain and associated health risks. J Food Compos Anal 106215. https://doi.org/10.1016/j.jfca.2024.106215

Kumar P, Chugh P, Ali SS, Chawla W, Sushmita S, Kumar R et al (2024b) Trends of nanobiosensors in modern agriculture systems. Appl Biochem Biotechnol 1-24. https://doi.org/10.1007/s12010-024-05039-6

Kundu M, Krishnan P, Chobhe KA, Manjaiah KM, Pant RP et al (2022) Fabrication of electrochemical nanosensor for detection of nitrate content in soil extract. J Soil Sci Plant Nutr 22(3):2777–2792. https://doi.org/10.1007/s42729-022-00845-5

Leghari SJ, Wahocho NA, Laghari GM, Hafeez Laghari A, MustafaBhabhan G, HussainTalpur K et al (2016) Role of nitrogen for plant growth and development: a review. Adv Environ Biol 10(9):209–219

Liakos KG, Busato P, Moshou D, Pearson S, Bochtis D (2018) Machine learning in agriculture: a review. Sensors 18(8):1–29. https://doi.org/10.3390/s18082674

Lira Saldivar RH, Méndez Argüello B, Santos Villarreal GDL, Vera Reyes I (2018) Nanotechnology potential in sustainable agriculture. Acta Univ 28(2):9–24. https://doi.org/10.15174/au.2018.1575

Maanaki H, Xu T, Chen G, Du X, Wang J (2023) Development of integrated smartphone/resistive biosensor for on-site rapid environmental monitoring of organophosphate pesticides in food and water. Biosens Bioelectron: X 15:100402. https://doi.org/10.1016/j.biosx.2023.100402

Marín-Bustamante MQ, Hernández-Flores A, Cásarez-Santiago RG (2021) Nanotecnología y agricultura: detección, monitoreo y remediación de contaminantes. Salud y Administración 8(23):29–35

Matimati I, Verboom GA et al (2014) Nitrogen regulation of transpiration controls mass-flow acquisition of nutrients. J Exp Bot 65(1):159–168. https://doi.org/10.1093/jxb/ert367

McFadden J, Njuki E, Griffin T (2023) Precision agriculture in the digital era: recent adoption on U.S. Farms, vol 248. US Department of Agriculture, Economic Research Service. Available online: https://www.ers.usda.gov. Accessed 13 Jan 2025

Meena J, Gupta A, Ahuja R, Singh M, Panda AK (2021) Recent advances in nano-engineered approaches used for enzyme immobilization with enhanced activity. J Mol Liq 338:116602. https://doi.org/10.1016/j.molliq.2021.116602

Miguel-Rojas C, Pérez-de-Luque A (2023) Nanobiosensors and nanoformulations in agriculture: new advances and challenges for sustainable agriculture. Emerg Top Life Sci 7(2):229–238. https://doi.org/10.1042/ETLS20230070

Mohammad ZH, Ahmad F, Ibrahim SA, Zaidi S (2022) Application of nanotechnology in different aspects of the food industry. Discov Food 2(1):12. https://doi.org/10.1007/s44187-022-00013-9
Mondal R, Dam P, Chakraborty J, Paret ML, Katı A, Altuntas S et al (2022) Potential of nanobiosensor in sustainable agriculture: the state-of-art. Heliyon 8(12). https://doi.org/10.1016/j.heliyon.2022.e12207
Oladoye PO, Olowe OM et al (2022) Phytoremediation technology and food security impacts of heavy metal contaminated soils: a review of literature. Chemosphere 288:132555. https://doi.org/10.1016/j.chemosphere.2021.132555
Pohanka M (2018) Overview of piezoelectric biosensors, immunosensors and DNA sensors and their applications. Materials 11(3):448. https://doi.org/10.3390/ma11030448
Prasad R, Bhattacharyya A, Nguyen QD (2017) Nanotechnology in sustainable agriculture: recent developments, challenges, and perspectives. Front Microbiol 8:1014. https://doi.org/10.3389/fmicb.2017.01014
Pratama A, Rahayu S (2024) Biosensors enabled by nanotechnology: advancing precision agriculture through IoT monitoring. J Intell Connect Emerg Technol 9(1):25–39. https://questsquare.org/index.php/JOUNALICET/article/view/47
Putzbach W, Ronkainen NJ (2013) Immobilization techniques in the fabrication of nanomaterial-based electrochemical biosensors: a review. Sensors 13(4):4811–4840. https://doi.org/10.3390/s130404811
Quy VK, Hau NV, Anh DV, Quy NM, Ban NT, Lanza S et al (2022) IoT-enabled smart agriculture: architecture, applications, and challenges. Appl Sci 12(7):3396. https://doi.org/10.3390/app12073396
Ramírez ET, Albores AM (2014) Biosensores enzimáticos. http://www.revista.unam.mx/vol.15/num12/art97/index.html
Rana A, Thakur K, Thakur M (2024) Smart biosensors for precision agriculture. In: Innovative methods in horticultural crop improvement: biosensors and nanosensors. Springer, pp 237–259. https://doi.org/10.1007/978-3-031-61095-0_9
Razzaghi M, Homaei A, Vianello F (2022) Industrial applications of immobilized nano-biocatalysts. Bioprocess Biosyst Eng 45:237–256. https://doi.org/10.1007/s00449-021-02647-y
Rigo AA, Cezaro AMD, Muenchen DK, Martinazzo J, Brezolin AN, Hoehne L et al (2020) Cantilever nanobiosensor based on the enzyme urease for detection of heavy metals. Braz J Chem Eng 36(4):1429–1437. https://doi.org/10.1590/0104-6632.20190364s20190035
Sagadevan S, Periasamy M (2014) Recent trends in nanobiosensors and their applications-a review. Rev Adv Mater Sci 36(2014):62–69
Sahu B, Chatterjee S, Mukherjee S, Sharma C (2019) Tools of precision agriculture: a review. Int J Chem Stud 7(6):2692–2697
Salek-Maghsoudi A, Hassani S, Mirnia K, Abdollahi M (2021) Recent advances in nanotechnology-based biosensors development for detection of Arsenic, Lead, Mercury, and Cadmium. Nanomedicine 2021(16):803–832. https://doi.org/10.2147/IJN.S294417
Samal S, Mohanty RP, Mohanty PS, Giri MK, Pati S et al (2023) Implications of biosensors and nanobiosensors for the eco-friendly detection of public health and agro-based insecticides: A comprehensive review. Heliyon 9(5). https://doi.org/10.1016/j.heliyon.2023.e15848
Sharma P, Pandey V, Sharma MMM (2021) A review on biosensors and nanosensors application in agroecosystems. Nanoscale Res Lett 16:136. https://doi.org/10.1186/s11671-021-03593-0
Siddiqi KS, Husen A (2020) Current status of plant metabolite-based fabrication of copper/copper oxide nanoparticles and their applications: a review. Biomater Res 24(1):11. https://doi.org/10.1186/s40824-020-00188-1
Sidhu GK, Singh S, Kumar V, Dhanjal DS, Datta S, Singh J (2019) Toxicity, monitoring and biodegradation of organophosphate pesticides: a review. Crit Rev Environ Sci Technol 49(13):1135–1187. https://doi.org/10.1080/10643389.2019.1565554
Singh S (2019) Nanomaterials exhibiting enzyme-like properties (nanozymes): current advances and future perspectives. Front Chem 7:46. https://doi.org/10.3389/fchem.2019.00046

Singh RP (2021) Recent trends, prospects, and challenges of nanobiosensors in agriculture. In: Pudake RN, Jain U, Kole C (eds) Biosensors in agriculture: recent trends and future perspectives, Concepts and strategies in plant sciences. Springer, Cham, pp 3–13. https://doi.org/10.1007/978-3-030-66165-6_1

Singh A, Sharma A, Ahmed A, Sundramoorthy AK, Furukawa H, Arya S, Khosla A (2021) Recent advances in electrochemical biosensors: applications, challenges, and future scope. Biosensors 11(9):336. https://doi.org/10.3390/bios11090336

Singh S, Anil AG, Kumar V, Kapoor D, Subramanian S, Singh J et al (2022) Nitrates in the environment: a critical review of their distribution, sensing techniques, ecological effects and remediation. Chemosphere 287:131996. https://doi.org/10.1016/j.chemosphere.2021.131996

Soderlund DM (2010) Toxicology and mode of action of pyrethroid insecticides. In: Hayes' handbook of pesticide toxicology. Academic, pp 1665–1686. https://doi.org/10.1016/B978-0-12-374367-1.00077-X

Solaimuthu A, Vijayan AN, Murali P, Korrapati PS (2020) Nanobiosensores y su relevancia en la ingeniería de tejidos. Opinión actual en ingeniería biomédica 13:84–93. https://doi.org/10.1016/j.cobme.2019.12.005

Song Y, Wang Z, Liao J, Zhang X, Yan J, Luo H et al (2024) Dual-electrode signal amplification self-powered biosensing platform based on nanozyme boosting target-induced DNA nanospace array for ultrasensitive detection of sugarcane Pokkah Boeng disease pathogenic bacteria. Int J Biol Macromol 281:136423. https://doi.org/10.1016/j.ijbiomac.2024.136423

Soy S, Sharma SR, Nigam VK (2022) Bio-fabrication of thermozyme-based nano-biosensors: their components and present scenario. J Mater Sci Mater Electron 33(8):5523–5533. https://doi.org/10.1007/s10854-022-07741-9

Srivastava AK, Dev A, Karmakar S (2018) Nanosensors and nanobiosensors in food and agriculture. Environ Chem Lett 16:161–182. https://doi.org/10.1007/s10311-017-0674-7

Sweta, Singh B (2024) A review on heavy metal and metalloid contamination of vegetables: addressing the global safe food security concern. Int J Environ Anal Chem 104(16):4762–4783. https://doi.org/10.1080/03067319.2022.2115890

Tang D, Wang Z, Che R, Liao J, Yan M, Zhang X et al (2024) Design of a portable bioassay system based on highly efficent nanozyme and Au-DNA nanocluster: toward current/colorimetric dual-signal detection of sugarcane smut. Chem Eng J 500:156912. https://doi.org/10.1016/j.cej.2024.156912

Thakur M, Wang B, Verma ML (2022) Development and applications of nanobiosensors for sustainable agricultural and food industries: recent developments, challenges and perspectives. Environ Technol Innov 26:102371. https://doi.org/10.1016/j.eti.2022.102371

Tian K, Alex S, Siegel G, Tiwari A (2015) Enzymatic glucose sensor based on au nanoparticle and plant-like ZnO film modified electrode. Mater Sci Eng C 46:548–552. https://doi.org/10.1016/j.msec.2014.10.064

Tonelli D, Scavetta E, Gualandi I (2019) Electrochemical deposition of nanomaterials for electrochemical sensing. Sensors 19(5):1186. https://doi.org/10.3390/s19051186

Tran TT, Clark K, Ma W, Mulchandani A (2020) Detection of a secreted protein biomarker for citrus Huanglongbing using a single-walled carbon nanotubes-based chemiresistive biosensor. Biosens Bioelectron 147:111766. https://doi.org/10.1016/j.bios.2019.111766

Verma ML (2017) Enzymatic nanobiosensors in the agricultural and food industry. In: Nanoscience in food and agriculture, vol 24. Springer, Cham, pp 229–245. https://doi.org/10.1007/978-3-319-53112-0_7

Virk V, Deepak H, Taneja K, Srivastava R, Giri S (2024) Amelioration in nanobiosensors for the control of plant diseases: current status and future challenges. Front Nanotechnol 6:1–24. https://doi.org/10.3389/fnano.2024.1310165

Wakeling EN, Neal AP, Atchison WD et al (2012) Pesticides—advances in chemical and botanical pesticides. Arcler Education Incorporated

World Health Organization (2020) The WHO recommended classification of pesticides by hazard and guidelines to classification 2019. World Health Organization

Yadav A, Yadav K, Ahmad R, Abd-Elsalam KA (2023) Emerging frontiers in nanotechnology for precision agriculture: advancements, hurdles and prospects. Agrochemicals 2(2):220–256
Yazdinejad A, Zolfaghari B, Azmoodeh A, Dehghantanha A, Karimipour H, Fraser E et al (2021) A review on security of smart farming and precision agriculture: security aspects, attacks, threats and countermeasures. Appl Sci 11(16):7518. https://doi.org/10.3390/app11167518
Zain M, Ma H, Rahman SU, Nuruzzaman M, Chaudhary S, Azeem I et al (2024) Nanotechnology in precision agriculture: advancing towards sustainable crop production. Plant Physiol Biochem 206:108244. https://doi.org/10.1016/j.plaphy.2023.108244

Chapter 6
Apta-Nanobiosensors in Precision Agriculture: Methods and Applications

Rahul Gogoi, Fung Swrangshee Daimari, Hridesh Harsha Sarma, Abhisek Rath, Madhurjya Ranjan Sharma, Anshu, Madhumita Barooah, and Sudipta Sankar Bora

Abstract Aptamer-based nanobiosensors are changing the landscape of precision agriculture (PA) by providing cutting-edge tools for monitoring and managing agricultural systems in real time. In this chapter, we review the methodologies and applications of aptamer-nanobiosensors, which utilize the high affinity and specificity of aptamers to bind to target molecules, combined with the unique properties of nanomaterials. We first provide a basic background on aptamer technology and then discuss their coupling with nanomaterials, which include materials like gold nanoparticles (AuNPs), carbon nanotubes (CNTs), quantum dots (QDs), and others. Nanomaterials increase sensor sensitivity and specificity, ensuring their broad utilization for environmental and agricultural detection of a variety of parameters. Sensor development is described, including aptamer selection, strategies for coupling aptamers to nanomaterials, and methods for developing sensors. Applications to precision agriculture, including monitoring for soil-based nutrients, detecting plant diseases, and identifying pests and pollutants, are highlighted. Each application is illustrated with case studies that focus on how these sensors can help increase crop health and improve the use of resources while developing sustainability. In

R. Gogoi · M. R. Sharma · M. Barooah
Department of Agricultural Biotechnology, Assam Agricultural University, Jorhat, Assam, India

F. S. Daimari
Department of Botany, Boroland University, Deborgaon, Assam, India

H. H. Sarma
Department of Agronomy, Assam Agricultural University, Jorhat, Assam, India

A. Rath
Department of Plant Pathology, Assam Agricultural University, Jorhat, Assam, India

Anshu
Department of Medical Oncology, All India Institute of Medical Sciences, Delhi, India

S. S. Bora (✉)
Silchar Regional Forensic Science Laboratory, Ghungur, Silchar, Assam under the Directorate of Forensic Science, Guwahati, Assam, India

J.-T. Chen (ed.), *Nanobiosensors for Crop Monitoring and Precision Agriculture*, Sustainability Sciences in Asia and Africa,
https://doi.org/10.1007/978-981-96-8335-2_6

addition, the real-world challenges are discussed, including sensor robustness, scaling, and sensor integration with digital agriculture platforms. Finally, we highlight prospects for future research and development that include sensor development and forcing the integration of novelly developed aptamer-nanobiosensors into data science methodology to assist in fine-tuning agricultural practice and yield. The chapter concludes with the overall importance of increased tracking of targeted pest detection and sustainability through the use of aptamer-nanobiosensors.

Keywords Precision agriculture · Nanomaterials · Sensor technology · Soil nutrients · Agricultural innovation

6.1 Introduction

The global population is projected to reach 8.5 billion by 2030 and increase to 9.8 billion by 2050, leading to challenges in meeting the demand for natural resources and food supply (Alvarado et al. 2019). Reduced available land, food and crop shortages, resource competitiveness, and the need to boost crop yields despite unfavorable weather are other major obstacles that further aggravate this situation. Food contamination by bacteria during preparation and storage is a serious worldwide issue. To maximize the utilization of natural resources, plant hormones, fertilizers, metal ions, beneficial microorganisms, and other agricultural and food products, it is imperative that new and sustainable solutions be developed. Sustainable agricultural intensification is the sole option for augmenting food output on existing land without inflicting detrimental environmental effects. One strategy for sustainable agricultural intensification involves integrating nanotechnology and its distinctive features into already-established conventional agricultural crop handouts (Kaphle et al. 2018). The potential of nanotechnology in agriculture can be harnessed to minimize nutrient losses, increase yields through better nutrient management, accelerate pest detection, and reduce the use of plant protection chemicals. Moreover, the incorporation of nanomaterials into biosensors, referred to as nanobiosensors, has greatly improved the biosensors' capacity to detect environmental stimuli. For the quick and precise identification of pests and diseases of plants, deficiencies in nutrients, mineral levels, pH levels, moisture levels, soil quality, groundwater accessibility, and other changes at the microclimatic level, powerful, real-time sensing technologies that combine the special qualities of nanomaterials with highly specific biological factors provide an excellent substitute (Kaphle et al. 2018).

Nanobiosensors are extremely sensitive, noninvasive, and purpose-built contemporary monitoring tools for the continuous, real-time detection and measurement of biological chemicals across diverse environments (Srivastava et al. 2018). These cutting-edge analytical tools work by translating biological signals in real time into quantifiable optical, thermal, or electrical outputs. In order to interpret and present the results, this transformation process depends on three essential elements: a transducer that transforms the biological response into a measurable signal, a bioreceptor that interacts specifically with the target analyte, and an electronic system that includes a processor, display, and signal amplifier. A key part of the biosensor, the bioreceptor is in charge of guaranteeing a very precise one-to-one connection between the sensor and the target analyte. This selectivity is a key component of biosensor operation since it permits precise detection and monitoring. Any biological material with a strong affinity for the target analyte can serve as a bioreceptor, including enzymes, antibodies, nucleic acids, and cells. The biosensor can detect analytes even in complicated mixtures and at extremely low concentrations because of the very selective nature of this interaction. In these processes, nanomaterials of all kinds—including nanotubes, nanowires (NWs), nanoparticles (NPs), nanocrystals, and nanocomposites—are essential. Their distinct physicochemical characteristics improve signal transduction, sensitivity, and selectivity, making them crucial for the effective operation of nanobiosensors (Kashyap et al. 2019). For example, Rai et al. (2015) and Singh Sekhon (2014) have measured growth hormone, moisture, and fertility levels using nanobiosensors, giving accurate and timely information for evaluating soil productivity. Khaledian et al. (2017) employed gold nanoparticles (AuNPs) biofabricated with single-stranded (ss) oligonucleotides to identify the genomic DNA of the potato bacterial wilt pathogen, *Ralstonia solanacearum*. Furthermore, Talan et al. (2018) created a bioreceptor-mediated electrochemical immunoassay technology based on chl-Ab that can detect the organophosphate pesticide chlorpyrifos in fruits and vegetables with an impressive limit of detection (LOD) of up to 10 fM. In this way, a network of small, compact, and transportable nanodevices could enable farmers to keep on-site tabs on microclimate, topsoil, and subsoil conditions, as well as plant health. Conversely, nanobiosensors are also finding increasing use in the food industry to detect allergens and/or pathogens in food during production and packaging, thereby ensuring food safety (Kim and Karadeniz 2012). In particular, aptamer-based nanobiosensors have lately garnered a lot of attention due to their exceptional functionalities, such as high stability across a broad spectrum of pH levels and temperatures, as well as their low development costs. These aptamer constructs are single-stranded oligonucleotides or peptide molecules, ranging from 15 to 40 bases in length, that adopt three-dimensional conformations to selectively bind to their complementary regions via van der Waals forces, hydrogen bonding, and electrostatic interactions (Zhao et al. 2017). The study estimates that the aptamer market will reach 342 million USD by 2026, representing a compound annual growth rate (CAGR) of 17.7% over the forecast period. The market was valued at about 151 million USD in 2021. In agriculture, apta-nanobiosensors have been extensively investigated as biorecognition components in lateral flow assays (LFAs), which serve as quick, on-site detection instruments for toxins, crop-specific biomarkers, plant growth regulators, plant pathogens, and pesticides (herbicides, insecticides, and fungicides). The four primary parts of an LFA are typically a conjugate pad, a sample pad, a nitrocellulose membrane, and an absorbent pad. Capillary action guides the specimen through specific analytical zones as it is applied, causing it to distribute evenly across the sample pad. The sample pad is pretreated with active reagents and buffer salts to control the flow and guarantee sample stability. Following the positioning of the conjugate pad, the biorecognition molecule with the label is added. A detectable signal is produced by certain interactions between the target analyte and biorecognition components on the nitrocellulose membrane during the detection process. The absorbent pad is essential to preserving the assay's effectiveness since it stops backflow and makes sure the sample moves continuously across the test strip. The exact detection of biological molecules is made possible by optical (colorimetric) and electrochemical (impedimetric) sensors (Huang et al. 2021). They offer a thorough solution for comprehending, identifying, and examining the presence of substances, thus aiding in environmental sustainability and the protection of crop health. As technology progresses, these nanodevices will assume a progressively crucial role in protecting our environment.

Apta-nanobiosensors are thus examined in this chapter, with particular attention paid to their design, the nanomaterials used, the underlying mechanisms, and their most recent uses in precision agriculture (PA) to support sustainable development. It also covers these sensors' benefits, drawbacks, and potential future developments in terms of environmental factors, stability, reproducibility, and biocompatibility. Both academic and industrial researchers can get important insights from their incorporation into new smart multifunctional systems.

6.2 The Role of Precision Agriculture in Modern Farming

Precision agriculture is the application of technology that enables farmers to manage their fields more precisely and accurately. Profitability, sustainability, and environmental protection are the three main objectives of PA (Huang et al. 2023). A data-driven, production-oriented agricultural method, precision agriculture aims to maximize productivity while reducing negative impacts on the environment and wildlife. This approach optimizes resource utilization, boosts long-term productivity, and increases overall farm profitability by utilizing site-specific management strategies (Fig. 6.1).

It all comes down to acting appropriately at the appropriate time, in the appropriate location, and in the appropriate manner. Precision agriculture technologies (PATs) are revolutionizing the agricultural sector by enhancing productivity, efficiency, and sustainability. These advanced tools, including GPS, the Internet of Things (IoT), drones, and data analytics, enable real-time monitoring and precise management of crop performance, optimizing resource use and improving overall farm yields (Duguma and Bai 2025). Precision agriculture minimizes waste, lessens its impact on the environment, and promotes sustainable farming methods by making the most of inputs like water, fertilizers, and pesticides. In addition to increasing crop yields, this strategy protects natural resources and reduces pollution, improving the general health of the ecosystem (Padhiary et al. 2024).

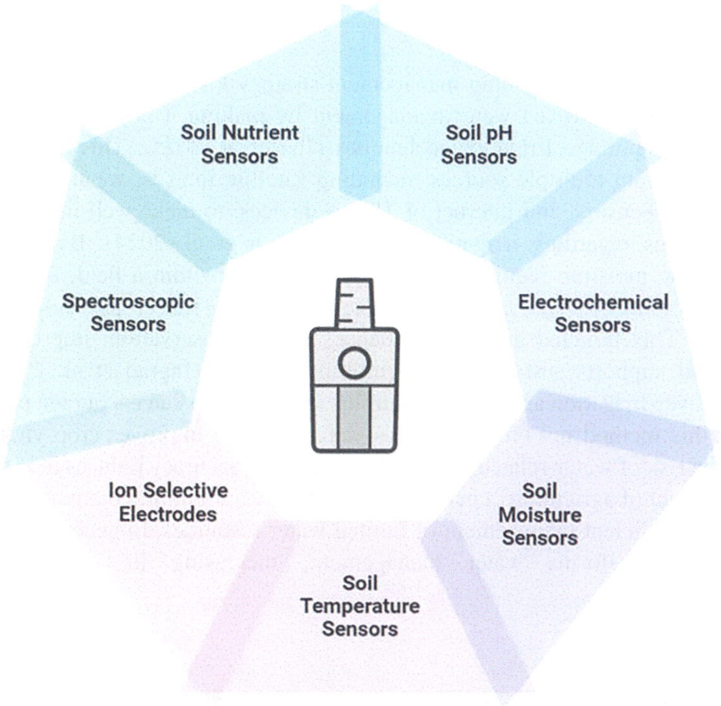

Fig. 6.1 Different agricultural sensors used for soil nutrient management

6.2.1 Soil Management

Precision agriculture, which uses advanced technologies like GPS, remote sensing, and IoT devices, transforms soil management by allowing farmers to adopt site-specific interventions (Soussi et al. 2024). This approach allows for detailed mapping of soil parameters, including pH, nutrient levels, and moisture content, across different field regions. By leveraging this data, farmers can tailor soil management practices to the specific needs of each area, optimizing the use of fertilizers and water for improved efficiency and sustainability (Rajak et al. 2023). This technique improves crop quality and agricultural yields while lowering environmental effects, including groundwater pollution and fertilizer runoff, and increasing resource efficiency. Furthermore, precision agriculture makes it possible for ongoing data gathering and monitoring, which supports proactive management techniques and offers real-time insights into soil health. Farmers can take prompt corrective action and save any losses by using drones and sensors to identify early indicators of nutrient deficits, pest infestations, or soil degradation (Getahun et al. 2024). Thus, precision agriculture promotes long-term soil health and soil productivity, which helps ensure food security and sustainable farming methods.

6.2.2 *Water Management*

The technology-driven farming management strategy known as precision agriculture has greatly improved water management by making it possible to precisely monitor and optimize irrigation techniques (Jha et al. 2025). This method uses information from multiple sources, including satellite images, weather forecasts, soil moisture sensors, and Internet of Things devices, to make well-informed recommendations regarding irrigation techniques (Qi et al. 2024). By accurately assessing the moisture requirements of specific areas within a field, farmers can apply water more efficiently, reducing waste and ensuring that crops receive optimal hydration. This targeted approach enhances water conservation, improves crop health, and supports sustainable agricultural practices (Ingrao et al. 2023). By avoiding over-irrigation and reducing surplus moisture that can encourage pathogen growth, this method not only conserves water but also improves crop yields and lowers the risk of water-related diseases. This level of accuracy reduces the environmental impact of agriculture, encourages sustainable agricultural methods, and aids in the more efficient management of limited water resources. In general, precision agriculture transforms water management, increasing its efficiency and sustainability.

6.2.3 *Crop Monitoring*

Crop monitoring has been transformed by precision agriculture, which incorporates cutting-edge technologies like GPS, remote sensing, and Internet of Things devices (Alahmad et al. 2023). These technologies allow farmers to gather comprehensive data about their fields, such as crop status, soil health, and moisture levels, with remarkably high accuracy. In order to detect pests, illnesses, and nutrient deficits early on, farmers can use satellite photography and drones with multispectral sensors to track crop health in real time. This proactive strategy minimizes input waste, maximizes yields, and enables prompt responses (Rejeb et al. 2022). Targeted interventions, including variable rate fertilizer and pesticide application, input optimization, and environmental impact reduction, are made possible by this granular data (Getahun et al. 2024). As a result, precision agriculture reduces excessive chemical use and conserves resources, increasing productivity and encouraging sustainable agricultural methods.

Moreover, data analytics and machine learning (ML) algorithms are frequently integrated into precision agriculture systems in order to predict crop performance and recognize trends (Elbasi et al. 2023). By analyzing historical data, in addition to current observations, these techniques can provide farmers with valuable information to improve production projections and schedule planting timings (Mgendi 2024). Predictive models can examine weather trends and offer information on the best times to plant in order to lessen adverse situations (Jaseena and Kovoor 2022).

Finally, precision agriculture improves decision-making, increases production, and strengthens the sustainability and profitability of modern farming.

6.2.4 Nutrient Management

Utilizing cutting-edge technologies such as GPS, remote sensing, and data analytics, precision agriculture revolutionizes nutrient management in agriculture, guaranteeing effective resource utilization, increased crop yields, and less environmental impact (Xu et al. 2024). This technique makes it possible to apply fertilizers and soil amendments precisely, guaranteeing optimal nutrient use efficiency, cutting waste, and increasing crop yield by using thorough, site-specific data on soil nutrient levels and crop requirements (Sabir et al. 2024). Through the integration of data from crop yield monitors, soil tests, and aerial imagery, farmers may create variable rate application (VRA) maps. A cheap and nonintrusive way to evaluate crop health and find nutrient deficits in plants or soil is remote sensing. Through the integration of remote-sensing data into geographic information systems (GIS), Indian farmers may apply fertilizer and soil amendments at precise sites, increasing crop productivity, decreasing waste, and increasing efficiency (Farooqui et al. 2024). As a result, nitrogen losses would be reduced and fertilizer use efficiency would increase. To improve efficiency and crop productivity, precision agriculture combines information from soil tests, remote sensing, and past yield records to create nutrient management plans that are specifically suited to certain field zones.

Furthermore, precision agriculture enhances environmental sustainability by minimizing the negative impacts of conventional farming. Targeted fertilizer application reduces nutrient runoff into water bodies, helping prevent algal blooms and other ecological disturbances (Kushwaha et al. 2024). Furthermore, it reduces greenhouse gas emissions associated with fertilizer production and application by optimizing nutrient use, decreasing excess fertilizer inputs, and lowering environmental impact. The continuous improvement of nutrient management tactics is made possible by sophisticated monitoring and feedback systems, which let farmers modify their methods based on historical trends and real-time data (Mgendi 2024). The sustainability of agricultural ecosystems is supported by this adaptive method, which guarantees economical and environmentally friendly nutrient management.

6.2.5 Pest and Disease Management

Precision agriculture improves crop health and sustainability by using data-driven techniques and cutting-edge technologies to manage pests and diseases. This allows for early identification, focused interventions, and less chemical use (Anastasiou et al. 2023). By combining geographic information systems (GIS), GPS, and remote sensing, farmers may accurately monitor and manage crop health, allowing for early

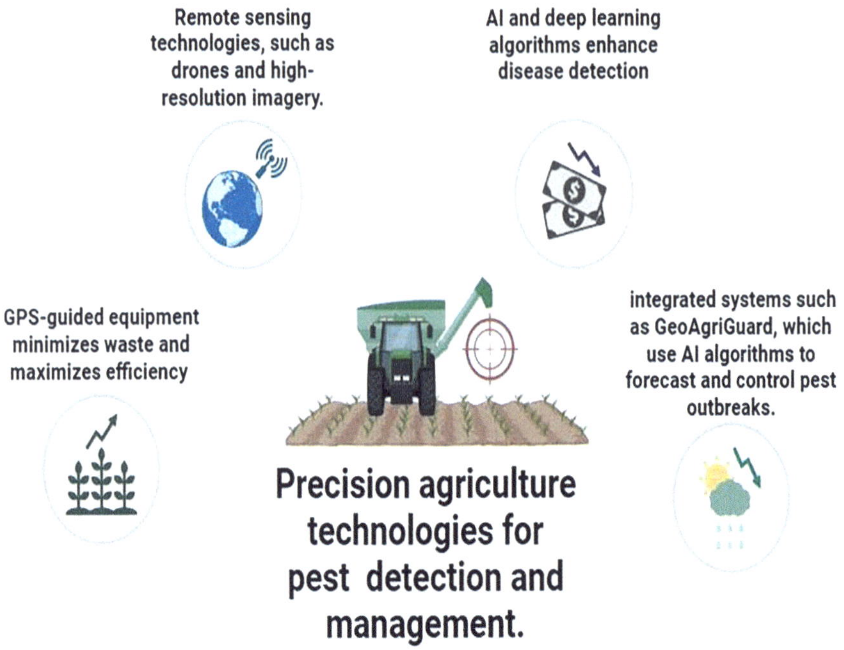

Fig. 6.2 Different precision agriculture technologies for pest and disease detection and management

problem diagnosis and focused treatments for increased yield. These technologies, which provide detailed spatial and temporal data, enable the early detection of pest infestations and disease outbreaks (Yang et al. 2025). For example, remote sensing can detect minute changes in crop vigor and color that may be invisible to the human eye but are suggestive of infection or stress (Kouadio et al. 2023). By using this information, farmers may apply targeted treatments, decreasing the need for widespread pesticide applications and encouraging more ecologically friendly and sustainable farming methods (Fig. 6.2).

Precision agriculture's capacity to provide more effective and efficient treatments, lower input costs, lessen environmental impact, and enhance crop health is a major benefit in the management of pests and diseases. Variable rate technologies (VRT) permit the exact application of insecticides and fertilizers to different parts of a field based on their individual demands (Pramod Pawase et al. 2024). This strategy minimizes environmental effects and maximizes resource efficiency by avoiding soil pollution and chemical runoff. In order to foresee pest and disease threats and enable proactive and focused management methods, precision agriculture also makes use of predictive modeling, which combines historical data with current conditions (Akhter and Sofi 2022).

6.2.6 Planting and Harvesting

By incorporating cutting-edge technologies to increase production, reduce resource waste, and improve efficiency, precision agriculture has transformed planting and harvesting processes, resulting in more economical and sustainable farming methods (Marinello et al. 2023). Precision agriculture makes use of cutting-edge planting methods, such as GPS-guided equipment and variable rate technology (VRT), to ensure consistency throughout the field. This maximizes crop establishment and yield potential by precisely controlling seed depth and spacing (Sharma and Shivandu 2024). Farmers can tailor planting plans to optimize the potential of each crop by evaluating soil data, such as moisture content and nutrient levels. This focused strategy optimizes the use of seeds and fertilizers to increase yields while lowering input costs. Furthermore, harvesting processes are streamlined by precision agriculture, increasing production and efficiency (Javaid et al. 2022). By using remote sensing and yield monitoring devices, farmers may track crop health and maturity in real time, allowing them to choose the best time to harvest in order to maximize production efficiency and quality (Alahmad et al. 2023). Farmers can improve yield quality and overall efficiency by maximizing harvest time and minimizing losses from extreme weather or crop stress by combining weather forecasts with soil data (Marinello et al. 2023). Additionally, using sophisticated equipment with sensors and automation technology improves harvesting efficiency, lowering labor costs and raising total yield. All things considered, by increasing productivity, sustainability, and profitability, precision agriculture transforms planting and harvesting. Farmers can maximize every phase of the agricultural process, from planting seeds to harvesting, by utilizing data-driven insights and cutting-edge technologies. With increased yields, less of an adverse effect on the environment, and larger financial returns, this method guarantees a more effective and sustainable farming system.

6.2.7 Sustainable Land Management

A crucial strategy for sustainable land management, precision agriculture maximizes yields and optimizes inputs to increase resource efficiency while reducing environmental impact (Mgendi 2024). By using GPS, sensors, drones, and machine learning algorithms, precision agriculture allows farmers to monitor and manage their fields with previously unheard-of accuracy, improving yields, reducing environmental impact, and using resources more efficiently (Xu et al. 2024). Because of this accuracy, resources like water, fertilizers, and pesticides can be applied precisely, which minimizes waste and its negative effects on the environment. By allowing farmers to customize management strategies to the unique requirements of each field zone, field variability mapping is a crucial component of precision agriculture that promotes soil health and conservation (Huang et al. 2023). In order to achieve more effective and sustainable land management, this strategy may entail modifying irrigation schedules in accordance with soil moisture levels, using conservation tillage to reduce soil disturbance, or applying fertilizers only where and when required (Al-Shammary et al. 2024). Moreover, precision agriculture makes it easier to implement regenerative agricultural methods, which support sustainability in the long run. Farmers can make well-informed decisions to improve soil fertility and resilience over time by routinely evaluating soil health indicators, including compaction, nitrogen levels, and organic matter content. The advancement of sustainable land management techniques depends heavily on precision agriculture, which maintains a balance between environmental stewardship, agricultural productivity, and the demands of future generations.

6.3 Aptamer Technology: An Overview

6.3.1 Definition and Origin

Aptamers are single-stranded (ss) oligonucleotides that are either DNA or RNA and, due to their unique three-dimensional structures, bind to their target molecules with high affinity and selectivity. Aptamers were initially discovered in 1990 by means of a repeating in vitro selection procedure that was separately established by three laboratories (Adachi and Nakamura 2019). This procedure is now popularly known as Systematic Evolution of Ligands by Exponential Enrichment (SELEX). Under the premise that the best sequences will be more frequently enhanced throughout choosing, SELEX entails putting together a sizable first archive of randomly assigned DNA or RNA sequences (together with a total range that includes up to 10^{15} unique members) through a number of mechanically implemented picking limitations. SELEX procedures were used to develop the first aptamers with binding affinity–based selection pressures. This was accomplished by repeatedly screening a nucleic acid bank versus the ligand that is important and retaining only the members of the library with the highest affinity.

The accepted method for choosing aptamers is still the Systematic Evolution of Ligands by Exponential Enrichment (SELEX) method, which was originally used in 1990. The procedure has now been improved by important developments, though, and chemically modified aptamers with increased binding affinities and more effective, adaptable workflows have been produced. Recent developments in sequence-defined aptamers and flexible amplification of synthetic oligonucleotides enhance specificity, stability, and application potential in a variety of domains, including biomedicine, diagnostics, and therapeutics (Amraee et al. 2017). The development of aptamers incorporating expanded DNA alphabets (Jiang et al. 2021) and the introduction of slow off-rate modified aptamers (SOMAmers) (Yang et al. 2021) have significantly advanced the field, while the automation of the SELEX process has further improved the scalability and efficiency of aptamer selection. Chemically altered aptamers are now being investigated as synthetic ligands for a number of uses, such as drug delivery, biomarker detection, diagnostics, and drug discovery in therapeutic development. Aptamers function as chromatographic media in process chemistry, allowing for the effective and traceless purification of biomolecules while maintaining their biological integrity. As such, they are useful instruments in the production of pharmaceuticals and biotechnology (Aljohani et al. 2022).

Aptamers have evolved the capacity to bind proteins involved with neurologic diseases; inhibit neurodegenerative processes triggered by PTM, like phosphorylation; and/or halt aggregation. Aptamers are single-stranded (ss) oligonucleotides (DNA or RNA) that attach themselves through their unique three-dimensional structures to their unique substrates with excellent selectivity. Systematic Evolution of Ligands by Exponential Enrichment (SELEX) is commonly employed to select aptamers. First introduced in 1990, the Systematic Evolution of Ligands by Exponential Enrichment (SELEX) technique is widely used for the selection of aptamers (Yan et al. 2019) . However, great improvements have led to the production of chemically modified aptamers with enhanced binding affinities in workflows that are more effective and flexible since then. Adaptable proliferation of artificial oligonucleotides (evolution of sequence-defined aptamers) (Amraee et al. 2017), aptamers made from enlarged DNA "alphabets" (Jiang et al. 2021), slow off-rate modified aptamers (SOMAmer) (Yang et al. 2021), and the high-throughput automation of the SELEX process are a few examples of advancements. In recent years, several methods have been developed to produce aptamers more effectively and consistently. These consist of the exponential enrichment method for the systematic evolution of ligands (SELEX) and its variants, including cap-SELEX, cell-SELEX, capillary electrophoresis–SELEX (CE-SELEX), atomic force microscopy–SELEX (AFM-SELEX), immunoprecipitation-coupled SELEX (IP-SELEX), and artificially expanded genetic information system–SELEX (AEGIS-SELEX) (Zhuo et al. 2017).

6.3.2 Properties and Advantages

6.3.2.1 High Specificity, Versatility, and Stability Under Diverse Conditions

Aptamers are widely used in environmental monitoring, food safety, and fundamental research. Their excellent specificity and affinity for disease-related targets make them useful tools in modern medicine, and they also hold considerable potential as therapeutic agents and in clinical diagnostics. Because of their numerous benefits over traditional protein antibodies, such as their small size, increased attachment affinity, accuracy, adaptability, capacity for chemical synthesis and modification, good biocompatibility, extended stability, and low immune response, aptamers are frequently employed in the biomedical industry (Zhang et al. 2019) and have a larger target potential that can range from ions to live animals (Allemailem et al.

2020). Aptamers are perfect for use in drug administration, diagnostics, and targeted therapeutics because they have a number of benefits over antibodies, including increased permeability, less immunogenicity, and smaller size (Zhang et al. 2019). Nucleic acid molecules called aptamers can bind selectively to target molecules, much like antibodies, since they can fold into intricate three-dimensional shapes. While the majority of aptamers are made in vitro and customized for particular targets, others are found in nature as riboswitch ligand-binding components. Aptamers are superior to antibodies in a number of ways, including their low immunogenicity, reversible folding, low batch-to-batch variability, ease of synthesis, and reduced production costs. But their greatest strength is their adaptability—aptamers are easily designed into sensors, actuators, and other useful parts—which makes them indispensable for developing medicinal, diagnostic, and biotechnological applications (Byun 2021).

Due to their capacity to attach to target molecules selectively through the creation of three-dimensional structures, nucleic acid aptamers—single-stranded DNA or RNA molecules that range in length from 20 to 100 nucleotides—have garnered significant scientific attention. The efficacy of aptamers is that they can withstand several denaturation/renaturation cycles (Fu and Xiang 2020). Stable at room temperature: resistant to temperature changes, extended shelf life (many years), is lyophilized, susceptible to nucleases, impervious to proteases. Potential targets range from ions and tiny molecules to entire cells and living creatures. Without losing their affinity, aptamers are easily and quickly modifiable. Ions and small compounds as well as whole cells and living things are examples of potential targets. The aptamers can be readily and rapidly modified without losing their affinity. High and enhanced particularity of affirmation multivalent aptamers Single-point mutations are identifiable.

Systematic Evolution of Ligands by Exponential Enrichment (SELEX) is an in vitro selection and amplification technique that uses repeated cycles of binding, recovery, and amplification processes to produce aptamers, which are target-specific ssDNA, RNA, or peptide sequences. With high specificity and selectivity, aptamers can attach to a wide range of targets, including medications, proteins, heavy metals, and infections (Chai et al. 2011). In terms of binding affinities, aptamers are comparable to monoclonal antibodies; nevertheless, they have several benefits over current antibody-based detection techniques, making them intriguing analytical and treatment instruments for upcoming analytical and biological applications.

6.3.3 Target Molecule Interactions

6.3.3.1 Mechanisms of Binding to Proteins, Ions, and Other Molecules

The interactions between proteins and nucleic acids, which form the foundation of cell life, are crucial to basic cellular functions. Aptamers are special single-stranded oligonucleotides that can interact selectively with proteins. They are produced by in vitro evolution techniques. Aptamer structure modification will significantly alter how they interact with proteins and, furthermore, impact associated cellular behaviors (Cozzolino et al. 2021). Aptamers fold into distinctive secondary and tertiary structures based on single-stranded designs, allowing them to demonstrate high binding affinity and selectivity for a wide range of substrates, including proteins, metal ions, tiny molecules, and even entire cells (Zhang et al. 2021). Aptamer research has advanced significantly during the past few decades. Pegaptanib, also known as Macugen, was the first aptamer-based medication to be licensed for clinical use by the Food and Drug Administration (FDA). By selectively targeting vascular endothelial growth factor (VEGF165), pegaptanib has shown promise in treating age-related macular degeneration (AMD) and preventing aberrant blood vessel formation in the retina, hence preserving vision. This achievement demonstrates how aptamers can be used therapeutically in contemporary medicine (Sun et al. 2015). In addition, a number of other aptamers are in various phases of clinical testing and might have application in medicine (Ni et al. 2011).

Nuclear magnetic resonance (NMR) spectroscopy and X-ray crystallography are currently two of the most crucial methods for researching binding complexes, offering in-depth understanding of their three-dimensional structures, chemical interactions, and binding mechanisms. The properties of the targets and the adaptability of aptamers determine aptamer–target interactions (Ruigrok et al. 2012). Aptamers frequently produce a broad range of structural forms, including G-quadruplexes, hairpins, bulges, and pseudoknots (Yoon and Rossi 2018). These structural configurations allow aptamers to attach to their targets by a variety of non-covalent interactions, including the hydrophobic effect, π–π stacking, van der Waals forces, electrostatic interactions, hydrogen bonds, or a mix of these. These various binding interactions provide aptamers with their great specificity and affinity, which make them useful tools for biotechnology, medicine, and diagnostics (Caporale et al. 2021). Peptide aptamers are a different class of aptamers in addition to nucleic acid aptamers that are composed of oligonucleotides. Combinatorial protein molecules known as peptide aptamers are designed to have a very particular intracellular binding affinity for target proteins. These compounds are more stable and target-specific because they have a changeable peptide loop that is bound by a scaffold protein (Reverdatto et al. 2015).

The small molecules that aptamers can attach to include metal ions, hormones, antibiotics, adenosine monophosphate (AMP) and adenosine triphosphate (ATP), colors, pesticide residues, and even active ingredients in herbal medicine used in traditional Chinese medicine (Mahmoudian et al. 2024). A thorough understanding of the aptamotif, which is the binding site region occupied by specific aptamers, might be helpful in guiding post-SELEX modification. Sekhon et al. (2017) assessed 11 aptamers that selectively target copper (Cu) and had binding affinity values ranging from 10^{-11} to 10^{-8} M. Four high-affinity aptamers (Cu-A1 through Cu-A4) were chosen for additional study in order to uncover possible binding sites, underscoring the potential of aptamers in environmental monitoring and metal ion detection. Additionally, aptamers have the ability to attach to macromolecules, the majority of which are proteins, including antibodies, antigens, viruses, cytochromes, neurotoxins, and interleukins (Caporale et al. 2021). For the first time, Morris and coworkers (Morris et al. 1998) produced RNA aptamers that were able to attach to a variety of possible targets on the membrane of human red blood cells. The Tan Group then developed the cell-SELEX technique in 2006 (Shangguan et al. 2006).

Many aptamers have since been chosen in vitro to target particular cells, such as bacteria, viruses, and different types of cancer cells (Lyu et al. 2016). For instance, following 12 rounds of cell-SELEX-based selection, Wang et al. (2017) chose the aptamer W786-1, which has a high affinity (KD of 9.4 ± 2.0 nM) for binding to renal cell carcinoma, or RCC, cells. The aptamer has a lot of potential to be employed as an investigative molecular expander to detect and target RCC in clinical tests (Wang et al. 2017). In a similar manner, Amraee et al. (2017) used *Escherichia coli* (*E. coli*) O157 as a whole-cell target and successfully screened DNA aptamers that could attach to their target selectively after nine rounds of selection. This paper shows how aptamers can be used to identify pathogens, providing a promising tool for applications in biosensing, diagnostics, and food safety. The aptamers that were developed can be used to identify related foodborne infections (Amraee et al. 2017). Techniques for examining the aptamer–target binding mechanism include circular dichroism spectroscopy (CD), isothermal titration calorimetry (ITC), surface plasmon resonance (SPR), truncation and mutation assays, fingerprinting assay, X-ray crystallography, nuclear magnetic resonance spectroscopy (NMR), and simulation of molecular docking.

6.4 Nanomaterials in Biosensing

6.4.1 Key Nanomaterials

The biosensor is "a device enabling identifying the presence of a substance that generates an identifiable electrical signal by coupling a bio-recognition component to a signal transducer." Three generations of biosensor design innovations are then introduced: the enzyme-dependent biosensor, the mediator-based biological sensor, and the catalyst electrode-dependent biosensor. The third generation attracts the most attention among these. The primary goal of third-generation sensors was to co-immobilize the metallic enzymatics as well as mediators on the electrode surface, thus helping make the biorecognition component a crucial part of the electrode transducer. When added to a catalytic framework, metal nanoparticles may present chances for cost reduction as well as catalytic fine-tuning (Hou et al. 2016). These analytes frequently have biological origins, such as bacterial or viral DNA or proteins produced by the immune response (antibodies and antigens) of contaminated or infected living things. The target analytes might be anything from basic substances like glucose to complicated pollutants when a biological sensor component with great specificity is included. However, guaranteeing effective signal transduction is a significant obstacle in the development of biosensors since it has a direct impact on detection accuracy, sensitivity, and dependability. To overcome this obstacle, improvements in transducer technology, nanomaterials, and signal amplification techniques are needed to improve biosensor performance in a range of applications, such as food safety, environmental monitoring, and medical diagnostics (Holzinger et al. 2014).

The inability of transduction technology to directly detect the recognition event between a bio-specific element and its target is a major obstacle in biosensing devices. This problem is especially pertinent to affinity biosensors, when binding by itself does not produce a visible signal, like DNA hybridization or antigen–antibody interactions. Bio-specific components, including DNA strands or secondary antibodies, are frequently functionalized with labels for optical or electrochemical transmission in order to address this. These changes based on nanomaterials improve signal amplification and detection, increasing the sensitivity and dependability of biosensors for use in biotechnology, environmental monitoring, and medical diagnostics. Certain nanomaterials' unique characteristics either directly aided in the creation of "label-free" transduction methods or, when applied as labels, clearly enhanced signals (Grieshaber et al. 2008).

6.4.2 Gold Nanoparticles

Due to their remarkable optical and electrical characteristics, biocompatibility, and simplicity of synthesis and functionalization, gold nanoparticles (AuNPs), one of the noble metal nanoparticles, are frequently employed in biosensor applications. Their superior conductivity improves electrochemical biosensors, and their distinctive surface plasmon resonance (SPR) effect allows for extremely sensitive visual detection (Biju 2014). The optic properties of gold surfaces, wherein exposure to light of a certain wavelength results in resonant surface plasmons—a vibration of electrons that occurs in the conduction band—is particularly intriguing. Unlike standard surface plasmon resonance (SPR) systems, which propagate oscillating electrons along a surface, gold nanoparticles (AuNPs) behave differently when their size is much smaller than the incident wavelength. In this instance, the electron density becomes polarized on the opposite side of the particle, oscillating in time with the reflected light frequency—a phenomenon known as localized surface plasmon resonance (LSPR). The environment of the gold sheet's propagating surface plasmons and its dielectric constant is typically changed in this method, which allows for the recording of the analyte's detection in a variety of ways, such as variations in the direction, magnitude, or wavelength of the light that reflects (Guo 2012).

6.4.3 Carbon Nanostructures

In biosensor devices, nanostructured carbons, such as graphene or carbon nanotubes (CNTs), are commonly used as either electronic or electrochemical transducers due to their beneficial properties (Ayanda et al. 2024). Consequently, carbon nanotube interfaces show significantly enhanced capabilities, including the capacity to contact the active portions of a redox enzyme coated with aminosilica and wire it to the bulk electrode. Furthermore, the straightforward and proven organic functionalization of nanostructured electrodes enables them to perform additional functions such as specific docking sites for biomolecules or redox mediation of bioelectrochemical interactions. CNT films also naturally form exceptionally porous three-dimensional networks, which give them highly electroactive surface areas suitable for anchoring a lot of bioreceptor units and, hence, excellent responses (Lee et al. 2018). However, the examples of carbon nanomaterials offered focus on their electrical or electrochemical properties for biosensing applications, while increasing specific surface area is not explored here. As was previously indicated, CNTs' nanowire shape makes it possible to reach redox enzyme active sites, resulting in quick and effective electron transfers. In rare instances, the enzyme's adsorption on the nanocarbon causes this enzyme wiring to happen on its own.

6.4.4 Quantum Dots

Luminous semiconducting nanocrystals known as quantum dots (QDs) are another well-known example of a nanomaterial used in bioanalytics. Cadmium chalcogenides (S, Se, and Te) are the basis for the most researched colloidal QDs (Reiss et al. 2009). These compounds offer a limited emission spectrum that varies with size and a very wide absorption spectrum. The reason behind this phenomenon is that different nanocrystal sizes result in varied band gaps for the semiconductor material (the larger the particle, the smaller the band gap), which causes different wavelengths for emission from the electron-hole exciton's recombination (Ghobadi 2013). Core/shell QDs' exceptional photochemical stability has made them an effective alternative to organic fluorophores (Resch-Genger et al. 2008). According to Biju et al. (Biju 2014), QDs are available with biocompatible or inert coatings to provide functional components for bioreceptor immobilization and foresee any toxicity difficulties. The duration of the photophysical recombination process is unaffected; practically any type of biomolecule can be bonded to these nanocrystals.

6.4.5 Magnetic Nanoparticles

Magnetic nanoparticles appear to offer a promising alternative to fluorescent labeling in biosensor devices. Nano-sized magnetic nanoparticles have different magnetic behaviors from their bulk counterparts, a phenomenon known as superparamagnetic behavior, because they have fewer magnetic domains, which are regions of parallel-oriented magnetic moments caused by an atom's interacting unpaired electrons (Vats et al. 2024). This implies that superparamagnetic materials have zero net magnetization on average when there is no external magnetic field present and that their magnetic moments fluctuate arbitrarily over a very brief period of time, referred to as the Néel relaxation time. Nevertheless, this temperature-dependent behavior is eliminated when an external magnetic field is provided because these magnetic moments align. One important difference is that, although their behavior may seem similar, superparamagnets have a far higher magnetic susceptibility than conventional paramagnetic materials. Superparamagnetic materials are especially helpful in applications like magnetic biosensors, targeted drug administration, and biomedical imaging because of their increased susceptibility (Bishop et al. 2009). Iron oxide nanoparticles' distinct magnetic characteristics and biocompatibility make them, along with other ferromagnetic materials, popular in bioanalytical applications. The main benefit of magnetic nanoparticles is that they can be controlled and concentrated in a particular area prior to the detection event, greatly increasing biosensing sensitivity and efficiency. Simply mixing receptor unit–modified magnetic nanoparticles with the analyte solution allows them to interact with the target in a targeted manner. Effective separation procedures are made possible by the aggregation and easy extraction of nanoparticles from the solution upon the application of an external magnetic field. Even in complicated surroundings, core/shell nanoparticles coated with gold or silica have proven to be an excellent method for quickly and selectively isolating DNA strands (Min et al. 2014).

6.4.6 Properties That Enhance Sensing

6.4.6.1 Optical, Electrical, and Chemical Characteristics

Nanomaterials have been shown to be suitable for biosensing applications, as they are in numerous other technical domains. Sensitivities were raised, and detection limits were lowered by multiple orders of magnitude as a result of the clever utilization of such nano-objects. All nanoparticles share the advantage of having a large specific surface area, which already enables the immobilization of more bioreceptor units (Holzinger et al. 2014). The process of immobilization needed to firmly attach the biorecognition element (such as enzymes, antibodies, or aptamers) to the surface of magnetic nanoparticles is one of the main obstacles to employing them for biosensing. The biosensor's sensitivity, stability, and specificity are all directly impacted by how well this immobilization works. The development of a dependable and effective immobilization technique is therefore essential to the production of a high-performance biosensor (Putzbach and Ronkainen 2013). In summary, non-covalent techniques represent entrapment in polymers, van der Waals forces, π–π stacking, or electrostatic interaction between the biological entity and the nanomaterial. These guidelines maintain every unique characteristic of both biomolecules and nanomaterials.

An electrical signal can be produced from a biological reaction via a biosensor, which is an incorporated receptor–transducer device. Over the past decade, researchers and scientists have concentrated on creating and developing biosensors because of their many uses, including medicine delivery, surveillance of the environment, food and water integrity monitoring, and healthcare and illness detection (Naresh and Lee 2021). (i) The main obstacles in the development of biosensors are the effective transmission of biorecognition signals and their transformation into electrochemical, electrical, optical, gravimetric, or auditory signals (transduction process). Even for the detection of individual molecules, (ii) enhancing transducer performance—such as sensitivity, reaction time, repeatability, and low detection limits—and (iii) using micro- and nanofabrication technologies to reduce the size of biosensing devices.

A potent remedy for the difficulties in biosensor creation is the integration of sensing technologies with nanomaterials, which can have zero- to three-dimensional structures. These nanoparticles are very useful for biosensing applications because of their high surface-to-volume ratio, superior conductivity, mechanical stability, and adjustable optical characteristics. Quantum dots (QDs) have tunable optical properties for improved detection, carbon nanotubes (CNTs) offer a large surface area with superior electrical and thermal conductivity, nanowires (NWs) and nanorods (NRs) enable ultrasensitive detection, and nanoparticles (NPs) offer high stability and large carrier capacity. Apart from these advantages, nanomaterials can also operate as transduction elements, directly transforming biological interactions into quantifiable signals. This greatly increases the biosensors' sensitivity, effectiveness, and adaptability (Naresh and Lee 2021).

Field-effect transistors (FETs) are extremely sensitive transduction devices that are frequently used in biosensors that make use of electrical characteristics. A FET is a three-terminal device that operates between a semiconductor's source and drain terminals to regulate current flow using an electric field. The gate terminal modulates the semiconductor's impedance, enabling accurate target analyte detection (Presnova et al. 2017). FET-based biosensors have drawn a lot of interest lately because of their capacity to convert analyte–FET surface interactions directly into quantifiable electrical signals, improving sensitivity and real-time detection capabilities (Arya et al. 2019).

To enable extremely sensitive and selective detection, electrochemical biosensors take advantage of the electrochemical characteristics of the analyte and transducer. These biosensors work by promoting an electrochemical reaction at the transducer surface, where quantifiable electrochemical signals are produced by the interaction of the bioreceptor and analyte. These signals offer a flexible platform for precise and real-time analysis since they can be identified in terms of voltage, current, impedance, or capacitance. Electrochemical biosensors are frequently utilized in industries including food safety, environmental monitoring, and medical diagnostics because of their great sensitivity, selectivity, and quick reaction times (Malhotra and Ali 2018).

6.4.7 Synergistic Effects with Aptamers

6.4.7.1 Improved Sensitivity, Specificity, and Real-Time Detection

A "synergistic effect" occurs with aptamers when an aptamer, in combination with another medicinal product or molecules, elicits a therapeutic effect that is noticeably more potent than what either agent alone could produce. In other words, the two agents work together in order through supplementary mechanisms of action to increase the desired outcome (Nimjee et al. 2009).

It is thus possible to design aptamers that bind at several points of a protein or multiple proteins along the pathway in combination with another therapeutic drug, which could interfere with yet another pathway. The simultaneous blocking of many key steps can thereby be achieved. An aptamer combined with another therapeutic agent can produce synergistic effects if they target the same molecule through different mechanisms, such as blocking protein–protein interactions while simultaneously disrupting enzymatic activity (Nimjee et al. 2009). Aptamers can deliver drugs to specific cells, making them more effective when combined with other delivery methods (Xiao et al. 2024).

The advantages of aptamer interaction include:

A single aptamer has a lower inhibitory effect than aptamer cocktails, which are collections of aptamers that bind to different epitopes on a target molecule (Nimjee et al. 2009). Drugs can be delivered to particular cells with aptamer–drug conjugates, increasing their effectiveness and lowering their side effects (Xiao et al. 2024). In terms of focusing on distinct disease progression pathways, aptamers may be used together with other medications, like immunotherapy or chemotherapy, for their enhanced efficiency (Soule et al. 2022).

6.5 Designing and Developing Aptamer-Nanobiosensors

The identification, linker, localization, signal receptor, signal amplification, and signal transducer components are among the many crucial parts that make up a biosensor. Of these, two elements are essential: the recognition element, which permits accurate identification and binding of the target analyte, and the signal transduction element, which transforms the analyte interaction into a signal that can be measured. Biosensors are useful instruments in environmental monitoring, biotechnological applications, and medical diagnostics because of these essential elements that cooperate to guarantee sensitive, accurate, and efficient detection. The selectivity, sensitivity, stability, and potential applications of biosensors are directly determined by recognition components, which are usually enzymes, antibodies, or nucleic acids. Aptamers are DNA or RNA single-stranded oligonucleotides that are highly selective and adapt to their targets (Zhou and Rossi 2017). Aptamers are known to interact with their targets and create unique three-dimensional structures. Targets can be categorized into two types based on their relative spatial locations in target–aptamer complexes: embedded and outside-binding (Kara et al. 2023). Aptamers' distinct oligonucleotide sequences frequently encapsulate ligands in minuscule binding pockets, guaranteeing great affinity and specificity for target molecules. Using a variety of techniques, including mass-sensitive, optical, and electrochemical biosensors, numerous investigations have been conducted on the creation of aptamer-based biosensors. Even though these biosensors have well-thought-out designs, many of them use flexible or analogous components. Four different modes—target-induced structural switching, sandwich or sandwich-like, target-induced dissociation or displacement, and competitive replacement—have been proposed to better categorize these design efforts. By streamlining the construction of biosensors, these classification models increase the accuracy, versatility, and efficiency of detecting a variety of analytes.

6.5.1 Sensor Development Workflow

Aptamer-nanobiosensors for agricultural use are developed using a rigorous scientific procedure intended to meet the highest performance requirements in terms of effectiveness, sensitivity, and specificity. The four distinct processes that make up the procedural framework are both required and crucial.

6.5.1.1 Aptamer Selection

The SELEX technology provides a systematic methodology intended to isolate aptamers characterized by robust binding capabilities to certain agricultural targets, such as diseases, heavy metals, pesticides, and soil nutrients. The approach comprises several cycles of iterative selection processes, which contribute to strengthening the specificity and binding strength of the aptamers, thereby ensuring their efficacy in identifying target molecules (Didarian et al. 2024). Furthermore, while the intrinsic properties of aptamers are vital, the adoption of chemical modifications greatly aids in their structural stability. This improvement resulted in improved durability under tough agricultural situations, thereby boosting their utility for biosensing and monitoring duties (Chinchilla-Cárdenas et al. 2024).

6.5.1.2 Coupling Aptamers to Nanomaterials

The conjugation of aptamers to nanoparticles is a vital step in the production of biosensors characterized by high sensitivity and selectivity for applications in agriculture. Achieving this coupling can be performed by several means, including covalent bonding, such as thiol-gold contacts, electrostatic adsorption, and affinity interactions involving biotin and streptavidin (Yang et al. 2011). The intended functionality of the biosensor determines which strategy is best because each one has unique benefits in terms of stability, sensitivity, and specificity (Odeh et al. 2019). The establishment of stable and reproducible binding is critical, since it directly affects the accuracy and operational longevity of the sensor. Ultimately, this reliability is important to the efficacy of aptamer-based detection systems utilized in the agricultural economy.

6.5.1.3 Biosensor Fabrication and Testing

The prototype of biosensors is undertaken in line with the planned detection technology, which may be optical, electrochemical, or fluorescence-based. Following the construction of these biosensors, rigorous laboratory testing is done to evaluate critical performance parameters, including reaction time, sensitivity, and specificity (Carpenter et al. 2018). Additionally, detection limits must be defined to reduce the frequency of false positives and false negatives in the identification of agricultural analytes within a sample while guaranteeing the overall accuracy and reliability of the detection process (Vigneshvar et al. 2016).

6.5.1.4 Field Deployment and Validation

Field testing is done to assess the practicality and stability of the biosensor across diverse agricultural situations. Data validation is necessary to confirm that the outputs supplied by the sensors remain trustworthy and consistent under diverse scenarios. The integration of biosensors with online platforms increases monitoring capabilities and encourages real-time decision-making, thereby enhancing the effectiveness of precision agriculture and resource management.

6.5.2 Detection Methods

6.5.2.1 Optical Detection

Optical detection technologies constitute an essential strategy within the field of aptamer-based biosensing in agriculture. For example, gold nanoparticle–based colorimetric assays encourage rapid and straightforward identification by exhibiting a visible color change upon engagement with the target, therefore making them appropriate for on-site deployment (Kou et al. 2020). Additionally, surface plasmon resonance (SPR) techniques allow real-time analytical capabilities defined by great sensitivity, thereby permitting the exact identification of agricultural chemicals without the requirement for labeling or complex sample preparation (Khan and Song 2020).

6.5.2.2 Electrochemical Detection

Electrochemical detection methods give a particularly sensitive and portable solution for the application of aptamer-based biosensing in agriculture. For instance, carbon nanotube sensors may detect changes in impedance that occur upon binding to the target, thus allowing for a reliable assessment of soil nutrients and contaminants (Suhito et al. 2020). Furthermore, conductivity-based sensing offers a robust and practical alternative for field applications, offering real-time monitoring of environmental variables with minimal sample preparation. These properties make electrochemical biosensors particularly well-suited for agricultural applications, thereby facilitating effective resource management and environmental monitoring (Sajeevan et al. 2025).

6.5.2.3 Fluorescence-Based Techniques

Fluorescence detection techniques offer sensitive and real-time monitoring capabilities for agricultural biosensing applications. The interaction of quantum dots with target molecules, mediated by aptamers, results in the generation of fluorescence signals, which permits precise monitoring of plant health and environmental pollutants (Nath et al. 2023). In addition, multiplexed fluorescence assays enable simultaneous detection of various analytes, thereby boosting monitoring efficiency and contributing to a more thorough understanding of agricultural conditions. Such techniques are crucial for early illness diagnosis, nutrient content analysis, and the implementation of precision agricultural programs (Camarca et al. 2021).

6.5.3 Challenges in Sensor Development

6.5.3.1 Environmental Robustness

The successful deployment of aptamer-nanobiosensors in agriculture requires their capacity to retain stability despite variations in environmental parameters, such as temperature, humidity, and soil pH (Hanif et al. 2019). Variations in these parameters may considerably affect the operational efficiency and durability of the sensors; consequently, the development of robust designs is crucial. The application of protective shell coatings on nanomaterials, such as biocompatible coatings or polymers, may boost durability and resistance to environmental pressures (Janghorban et al. 2022). These developments are vital for ensuring that biosensors perform properly over extended periods, hence facilitating their usage in real agricultural contexts.

6.5.3.2 Reproducibility and Standardization

Ensuring reproducibility and standardization of aptamer-nanobiosensors is vital to demonstrating their dependability for agricultural applications. Variability in the synthesis of aptamers and the functionalization techniques of nanomaterials could influence the sensitivity, specificity, and stability of the sensors, resulting in variable outcomes in the final biosensing products (Kangarshahi and Naghib 2024). The implementation of standardized production methods can promote the creation of sensors of uniform quality in repeated manufacturing cycles, thereby enhancing dependability and expediting regulatory approval processes. The trustworthy functioning of these biosensors is vital for their widespread adoption, since they must deliver consistent and accurate results across various agricultural circumstances (Hanif et al. 2019).

6.5.3.3 Cost and Scalability

To render aptamer-nanobiosensors suited for widespread application in agriculture, it is important to achieve mass production at cheap rates while keeping excellent sensitivity and dependability. The lowering of production costs through the optimization of manufacturing processes and the building of scalable nanomaterial synthesis technologies is vital for their economic viability (Hanif et al. 2019). Moreover, connecting these biosensors with IoT-based platforms can enhance their accessibility for farmers, particularly small-scale farmers, enabling them to access and comprehend real-time data. Such technological innovations enhance precision agriculture by giving cost-effective and user-friendly diagnostic tools that promote sustainable farming methods (Sen et al. 2021).

6.6 Applications in Precision Agriculture

Aptamers are crucial for nanobiosensors in precision agriculture due to their ability to rapidly detect pathogens and monitor soil health. Their enhanced sensitivity and specificity improve nutrient management, reduce chemical inputs, and increase crop yields, contributing to sustainable farming practices and food safety.

6.6.1 Soil Nutrient Monitoring

The analysis of chemical species in particular habitats, especially the origin and interactions of cell metabolites in the agricultural rhizosphere, has shown great promise with aptamer-based nanosensors. Real-time biochemical change monitoring is made possible by these nanosensors, which offer important new information about the interactions between microbes and plants. By measuring microbial activity, they also play a vital part in soil diagnostics. Nanosensors can efficiently measure microbial populations by comparing the oxygen consumption levels of advantageous and detrimental bacteria during respiration, providing a potent tool for soil health evaluation, precision agriculture, and sustainable farming methods (Kaushal and Wani 2017). Both macronutrients and micronutrients in soil can be detected with aptamer-based nanosensors, which are incredibly sensitive and selective instruments. They provide precise, real-time nutrient analysis and enhance soil management techniques for maximum agricultural yield by binding selectively to target molecules. With the ability to precisely monitor biochemical processes in living cells, aptamer-based nanobiosensors are becoming increasingly potent instruments for examining the origins and interactions of metabolites in plant rhizospheres. They also have a lot of potential for researching controlled-release agrochemicals, which could increase the effectiveness of pest control and nutrient delivery. Layer-by-layer polymer films' potential for focused applications is further increased by the incorporation of nanodevices, including aptamers, which provide site-specific action, controlled release, and enhanced stability in environmental monitoring and precision agriculture (Kanjana 2017).

6.6.2 Disease and Pest Detection

Aptamers, which are substitutes for traditional antibodies and provide more reliable biosensors for pesticide detection, have been introduced by innovative developments in the field of pesticide analysis. Aptamer-based nanobiosensors are used to detect diseases and pests because aptamers are short DNA or RNA sequences that bind preferentially to target molecules such as proteins or pathogens. A quantifiable reaction is triggered by this binding, signifying the existence of pathogens or pests. Lately, QDs have been employed as signal reporters to detect acetamiprid, which is an insecticide, using DNA aptamers that bind to it specifically. Finding and selecting the appropriate aptamer for the target pesticide can be challenging, though, which may limit their application as recognition molecules (Nsibande and Forbes 2016).

Fei et al. (2015) successfully used a label-free direct impedimetric aptasensor to measure the femtomole level of acetamiprid. Two different research studies tried to use aptasensors to detect acetamiprid in order to detect pesticides. The acetamiprid-specific S-18 aptamer and the nanozyme activity of gold nanoparticles (AuNPs) were combined in 2014 to create an aptasensor. By using the aptamer's target-specific binding capability, this creative design was able to inactivate the nanozyme surface and thus prevent its enzymatic activity when it came into contact with acetamiprid. The AuNP nanozyme is reactivated when such a target is present because aptamer molecules separate from the AuNP surface. Having a ppm or so detection limit, this reversible nanozyme activation/inhibition technique produced extremely quick, sensitive, and specific acetamiprid detection (Weerathunge et al. 2014).

6.6.3 Environmental Monitoring

Heavy metals, including Ag^{+}, Cd^{2+}, As^{3+}, Pb^{2+}, Zn^{2+}, and Hg^{2+}, are acknowledged as chemical pollutants because of their toxicity and durability in the environment. Biosensors have been created to identify these dangerous substances by combining genetically altered bacterial cells with a green fluorescent signal amplifier (Pola-López et al. 2018). The biosensors with aptamer and DNA-based properties can detect these heavy metals on nanoscale levels and on very large scales that are suitable for screening and monitoring of food safety (Gupta et al. 2020).

According to Ye et al. (2022), Pb^{+2} and Hg^{+2} detection was made easier by multiplex sensors with a particular locus-centered aptamer function that allied AuNRs. Different responses of individual AuNRs with distinctive spectra changes in the scattering spectrum were employed to maximize the detection of Pb and Hg, two heavy elements. Using a spectral imaging dark field microscope to examine bound Pb and Hg, the response of several single plasmonic nanosensors was determined with high precision and time resolution. By tracking the concentration-specific response of position-specific nanosensors for mixtures containing Hg^{+2} and Pb^{+2}, the assay's limit of detection (LOD) was established. Sensitivities of 5 nM (Pb^{+2}) and 1 nM (Hg^{+2}) were found during inspection, suggesting that more heavy metal ions (similar to periodic table groups) might be screened. It is quite easy to expand the mercury-specific aptamer to an electrochemical sensing platform in the future. Lead ions (Pb^{2+}) may be detected with high sensitivity and selectivity using an electrochemical DNAzyme sensor that amplifies the signal using DNA-Au bio-bar codes. This invention made use of a DNAzyme's catalytic activity, which was activated when it bound to Pb^{2+}, allowing for extremely specific detection. This biosensor is a potential instrument for environmental monitoring and heavy metal detection because of the large increase in signal sensitivity brought about by the combination of DNAzyme catalysis with gold-based bio-bar codes. Despite being shown to detect Pb^{2+}, this DNAzyme sensor might be used as a universal platform to create sensors for various heavy metal ions and small molecules (Shen et al. 2008). For the extremely sensitive and specific detection of Cu^{2+}, an electrochemical aptasensor based on gold nanoparticles was created. This sensor proved to be a useful instrument for identifying Cu^{2+} in lake water samples due to its quick response, simplicity of use, and excellent selectivity. For environmental monitoring and water quality evaluation, precise and trustworthy measurements are essential, and the use of gold nanoparticles improved electrical conductivity and signal amplification.

6.6.4 Water Quality Assessment

To avoid bacterial diseases and preserve human health, it is crucial to quickly identify harmful bacteria in drinking water. However, unavoidable problems, including low levels of pathogens, pathogenic and nonpathogenic bacterial strains, and difficulty differentiating between closely related subspecies, provide a significant obstacle in methodology development. Finding the disease's genesis depends heavily on the identification, reporting, and characterization of waterborne outbreaks. A targeted organism was bio-captured from surface water using a DNA aptamer specific to *Salmonella typhimurium*, and then, its DNA was extracted and quantified using Quantitative Polymerase Chain Reaction (qPCR) in molecular beacon format. Aptamer and qPCR could be used to treat bacterial illnesses and assess the microbiological purity of water. It was discovered that the test improved the sensitivity of the qPCR assay for the detection of *Salmonella*, which was otherwise inhibited by the presence of PCR inhibitors. Aptamers, due to their peculiar properties, have been used as biosensing tools for the detection of pathogenic bacteria (Hayat and Marty 2014). Jo et al. (2011) created a sol-gel biochip aptasensor to identify water samples that contain bisphenol A (BPA). Additionally, an aptamer-based technology has been used to build electrochemical aptasensors for BPA detection. Then, using a modified glassy carbon electrode and a graphene nanocomposite sheet sprinkled with gold nanoparticles, a label-free electrochemical aptasensor was created. This method demonstrated its promise for environmental monitoring and safety evaluations by offering a quick, easy, and economical way to detect BPA accurately and efficiently (Zhou et al. 2014).

Aptamer and nanotechnology are combined to develop a biosensor with molecular recognition elements that selectively attach to analytes and generate a visual signal with the help of a transducer (signaling element) (Sharma 2014). For instance, by combining aptamer and silver nanoparticles, an electrochemical immunosensor was created to identify *Staphylococcus aureus* (Abbaspour et al. 2015). Two aptamers specific to *S. aureus*, one coupled with streptavidin-coated magnetic beads and the other with silver nanoparticles, create a sandwich complex when targeted bacteria are present. With a sensitivity as low as 1 CFU/mL, this aptamer-based assay allowed for the electrochemical stripping voltammetry detection of *S. aureus* in a variety of environmental samples. Furthermore, drinking water containing algae toxins is becoming a bigger risk to human and environmental health, which emphasizes the necessity of efficient monitoring and detection techniques. Aptamer-functionalized hybrid nanomaterials have been used to detect algae toxins, including lipopolysaccharides, okadaic acid (OA), tetrodotoxin, brevetoxin, and microcystins (Bilibana et al. 2022).

6.7 Integration into Agricultural Systems

Aptamer-nanobiosensors are making precision farming possible by implementing smart farming, which uses technologies that interface with data analytics to monitor crop conditions and soil health in real time. This allows for resource-optimized yield gain and targeted intervention while also minimizing environmental damage and promoting agricultural sustainability.

6.7.1 Smart Farming Technologies

6.7.1.1 Sensors and IoTs

Based on reduced graphene quantum dots (rGQDs) and multi-walled carbon nanotubes (MWCNTs), an optical apta-nanosensor was created and utilized to detect diazinon, one of the most common organophosphorus pesticides, selectively. It enables the selective detection of diazinon over other pesticides in contaminated areas using a sensitive, affordable, quick, and portable technology (Talari et al. 2021). The effectiveness of the aptamer-based nanosensor in identifying diazinon pesticide in actual samples, such as farm runoff, tap water, urine, and river water, was assessed. Its low binding affinity for the diazinon-specific aptamer allowed the proposed apta-nanosensor to detect diazinon quickly and reliably, according to the results. Because of this selectivity, emission recoveries were not impacted by the presence of additional pesticides, demonstrating the sensor's excellent specificity and dependability for real-world uses. The new method that was introduced might be used for other pesticides as well, providing a practical on-site detection method for real samples of several pesticides. A new FRET-based nanobiosensor was created to detect aflatoxin B1 in rice and peanut samples by combining aptamer-conjugated quantum dots (QDs) with gold nanoparticles. With a linear detection range of 10–400 nM and a detection limit of 0.34 nM, this biosensor proved to be very sensitive and selective for applications involving food safety.

By immobilizing a glyphosate-specific aptamer onto a gold electrode that has been treated with catechol, a catechol-based biosensor for glyphosate detection was created. The excellent sensitivity and specificity of this biosensor made it useful for glyphosate detection in food, water, and soil samples. Its linear range was between 0.1 and 100 ng/mL, while its detection limit was 0.1 ng/mL. Food safety could be improved, and customers could be shielded from glyphosate exposure, thanks to this development (Wang et al. 2022).

6.7.2 Data Utilization

Aptamer-based nanobiosensors, which offer real-time data on a range of environmental and plant health factors, can greatly improve predictive analysis and agricultural decision-making. For a wide variety of non-nucleic acid target molecules, such as peptides, proteins, medications, organic and inorganic chemical cues, and even entire cells, aptamers demonstrate great affinity and selectivity (Mairal et al. 2008). By detecting illnesses, pesticide residues, soil moisture, and nutrient levels, these sensors assist farmers in making informed crop management and resource allocation decisions. The data produced can be examined to forecast agricultural results, maximize inputs, and reduce risks related to biotic and abiotic stressors by incorporating machine learning techniques. This results in increased food safety, decreased environmental impact, and higher crop yields. Compared to immunoassays for environmental monitoring, aptamer-based assays are still in the early stages of research. Quick developments in new aptamer sequences and the incorporation of novel nanomaterials into aptamer-based tests are enhancing current protocols (Hayat and Marty 2014).

6.8 Challenges and Future Insights

Significant obstacles still stand in the way of the practical deployment of biosensing devices, especially apta-nanobiosensors for food and agricultural applications, despite encouraging published results, a large number of patents, and currently available commercial goods. For these industries to remain viable and sustainable, these challenges must be overcome (Prasad et al. 2017). First, research in this field is intrinsically complex and calls for teams of experts from many different fields to work together, including chemistry, physics, biology, and agronomy, alongside the use of state-of-the-art technology such as particle accelerators and synchrotrons. Second, rendering data from one crop cannot be readily transferable to another because nanomaterial behavior can be highly dependent on parameters such as plant type and physiology (Pérez-de-Luque 2017). Third, farmers may not be able to adopt nanodevices for crops if such adoption restricts their already constrained profit margins; hence, the cost of these devices should be comparatively small for large-scale field applications. Thus, it is essential to reduce the cost of fabricating nanobiosensors. This can be accomplished by employing low-cost biological elements, such as cells or enzymes, in conjunction with novel matrices for the stabilization or immobilization of nanomaterials. Chitosan and other materials can improve the durability and reusability of nanobiosensors, increasing their potential for broad use (Kah et al. 2019). Moreover, these devices present greater challenges in applications, necessitating the development of cost-effective, sustainable, and targeted methods for their deployment on crops. In various scenarios, such as noninvasive electronic systems on leaves, they may be ineffective or prohibitively expensive, unless applied to perennial crops (e.g., fruit trees).

Given their special properties, including size, structure, content, and a high surface-to-volume ratio, it is also critical to take into account the possible hazards connected to the usage of nanomaterials, especially their toxicity and ecotoxicity. To achieve this objective, it is necessary to precisely evaluate their environmental impact in relation to factors such as size, dosage, accumulation characteristics, immunological response, and retention time, which has led to an increasing inclination toward the use of natural nanomaterials (biopolymers) that are nontoxic and readily degradable. Another problem that needs to be addressed is the fabrication of a miniaturized form to improve portability. Therefore, the agricultural and food industries will greatly benefit from the successful integration of information technology, including IoT, ML, deep learning (DL), and cyber-physical systems (CPS), particularly in remote locations. These advantages include lower expenses, higher output, better prediction of disease outbreaks, and more effective use of natural resources such as soil, water, and climate. Artificial intelligence (AI)–powered apta-nanobiosensors are anticipated to have transdisciplinary uses in the near future, improving sensing capabilities and lowering costs (Alvarado et al. 2019).

Given the aforementioned challenges, transitioning nanobiosensors from prototype to commercial product may seem formidable. As a result, developing apta-nanobiosensors with excellent stability and reproducibility is crucial for achieving the highest levels of accuracy and precision. The use of effective nano-raw materials that can withstand high temperatures and moisture while simultaneously increasing the sensitivity of biosensors through different signal amplification methods can accomplish this (Fathi-Karkan et al. 2023). This also requires manufacturing optimization of natural and artificial biorecognition components, including antimicrobial agents, enzymes, complementary DNA (cDNA), dendrimers, aptamers, and thin films, by rigorous monitoring and high-throughput methods in the development process (Fu et al. 2024). Furthermore, greater emphasis should be given to real-field trials in agronomy, which have been insufficient and primarily rely on extrapolating results obtained in laboratory or controlled environments (such as greenhouses or growth chambers) (Kah et al. 2019). This is because, although these nanotechnology products perform well in laboratory/controlled settings, their performance in real-world conditions is not assured, and there may be obstacles in their manufacturing and/or distribution.

Therefore, with modernization and the mitigation of the aforementioned issues, these new and future nano-enabled technologies can expand the use of nano-based products that are nurtured through specialization and government policies, which may optimize production in the agricultural sector and offer innovative and cost-effective pathways for global human nutrition.

6.9 Conclusion

Recent breakthroughs in apta-nanobiosensors in contemporary agri-food industries have fostered significant trust and improved productivity by optimizing the use of natural resources, thus promoting long-term sustainability in both agricultural and food sectors. These nanobiosensors possess numerous potential applications, beginning with the agri-food industry and continuing through soil analysis, natural resource assessment, moisture evaluation, soil pH measurement, disease management, quick detection of dangerous substances and adulterants, and the identification of infectious organisms and, finally, commercialization, where professionals make informed decisions based on preliminary test results. Companies like Roche, Nippon, and IBM are among the limited entities involved in the commercial production of apta-nanobiosensors due to their broad range of applications. While there are some articles exploring the application of apta-nanobiosensors in medical diagnostics, there are few reports on their commercialization in the agri-food industry. Significant challenges are presented by the high expenses associated with fabrication, automation trials, outcome evaluation, and field trial validation. The miniaturization of prototypes for industrial production is further hampered by regulatory and public concerns about the safety and dependability of these technologies, which prevent their broad acceptance. Moreover, there is no market to alleviate and absorb these expenses. Because of that, there might be reduced availability of commercial apta-nanobiosensors on the market. One potential and affordable way to overcome some of the financial obstacles related to apta-nanobiosensors is to extract novel nanomaterials from waste biomass. The versatility of these sensors in many applications is another important issue that needs more research. The development of a wide variety of nanomaterials designed to detect different compounds through bioassays could lead to increased commercialization of portable apta-nanobiosensors. The performance and adaptability of biosensors could be greatly improved by materials including carbon nanotubes, graphene, quantum dots, and nanocomposites, which would make them appropriate for a variety of industrial and agricultural uses. The commercialization process and the production of affordable, durable apta-nanobiosensors are both expected to be accelerated through a collaboration between academic studies and industrial enterprises. The transition from innovation of versatile, robust, and dynamic apta-nanobiosensors, stemming from foundational research to commercial products, is likely to be widespread to explore their multifaceted potential through tailored and rapid recognition systems for use in agroecosystems, with an emphasis on integration, automation, and miniaturization. Additionally, to bring the product to market, it is necessary to incorporate and employ modern technical components, including IoT, cloud computing, AI, data analysis, deep learning, and cyber-physical systems. Improving a comprehensive database and fostering international cooperation in policy formulation, strategic planning, and legislative processes are also essential for the effective management and global dissemination of such knowledge. Furthermore, the appropriate authorities ought to set precise guidelines and criteria for reducing the hazards associated with using these nanotechnological items.

References

Abbaspour A, Norouz-Sarvestani F, Noori A, Soltani N (2015) Aptamer-conjugated silver nanoparticles for electrochemical dual-aptamer-based sandwich detection of staphylococcus aureus. Biosens Bioelectron 68:149–155. https://doi.org/10.1016/J.BIOS.2014.12.040

Adachi T, Nakamura Y (2019) Aptamers: a review of their chemical properties and modifications for therapeutic application. Molecules 24(23):4229. https://doi.org/10.3390/MOLECULES24234229

Akhter R, Sofi SA (2022) Precision agriculture using IoT data analytics and machine learning. J King Saud Univ Comput Inform Sci 34(8):5602–5618. https://doi.org/10.1016/J.JKSUCI.2021.05.013

Alahmad T, Neményi M, Nyéki A (2023) Applying IoT sensors and big data to improve precision crop production: a review. Agronomy 13(10):2603. https://doi.org/10.3390/AGRONOMY13102603

Aljohani MM, Cialla-May D, Popp J, Chinnappan R, Al-Kattan K, Zourob M (2022) Aptamers: potential diagnostic and therapeutic agents for blood diseases. Molecules 27(2):383. https://doi.org/10.3390/MOLECULES27020383

Allemailem KS, Almatroudi A, Alsahli MA, Basfar GT, Alrumaihi F, Rahmani AH, Khan AA (2020) Recent advances in understanding oligonucleotide aptamers and their applications as therapeutic agents. 3 Biotech 10(12):1–20. https://doi.org/10.1007/S13205-020-02546-1/METRICS

Al-Shammary AAG, Al-Shihmani LSS, Fernández-Gálvez J, Caballero-Calvo A (2024) Optimizing sustainable agriculture: a comprehensive review of agronomic practices and their impacts on soil attributes. J Environ Manag 364:121487. https://doi.org/10.1016/J.JENVMAN.2024.121487

Alvarado K, Bolaños M, Camacho C, Quesada E, Vega-Baudrit J (2019) Nanobiotechnology in agricultural sector: overview and novel applications. J Biomater Nanobiotechnol 10:120–141. https://doi.org/10.4236/jbnb.2019.102007

Amraee M, Oloomi M, Yavari A, Bouzari S (2017) DNA aptamer identification and characterization for E. coli O157 detection using cell based SELEX method. Anal Biochem 536:36–44. https://doi.org/10.1016/J.AB.2017.08.005

Anastasiou E, Fountas S, Voulgaraki M, Psiroukis V, Koutsiaras M, Kriezi O, Lazarou E, Vatsanidou A, Fu L, Di Bartolo F, Barreiro-Hurle J, Gómez-Barbero M (2023) Precision farming technologies for crop protection: a meta-analysis. Smart Agric Technol 5:100323. https://doi.org/10.1016/J.ATECH.2023.100323

Arya A, Gangwar A, Kumar A (2019) Biosensors in animal biotechnology. In: Nanotechnology in modern animal biotechnology: concepts and applications. Springer, pp 75–95. https://doi.org/10.1016/B978-0-12-818823-1.00006-5

Ayanda OS, Mmuoegbulam AO, Okezie O, Durumin Iya NI, Mohammed SE, James PH, Muhammad AB, Unimke AA, Alim SA, Yahaya SM, Ojo A, Adaramoye TO, Ekundayo SK, Abdullahi A, Badamasi H (2024) Recent progress in carbon-based nanomaterials: critical review. J Nanoparticle Res 26(5):1–42. https://doi.org/10.1007/S11051-024-06006-2

Biju V (2014) Chemical modifications and bioconjugate reactions of nanomaterials for sensing, imaging, drug delivery and therapy. Chem Soc Rev 43(3):744–764. https://doi.org/10.1039/C3CS60273G

Bilibana MP, Citartan M, Fuku X, Jijana AN, Mathumba P, Iwuoha E (2022) Aptamers functionalized hybrid nanomaterials for algal toxins detection and decontamination in aquatic system: current progress, opportunities, and challenges. Ecotoxicol Environ Saf 232. https://doi.org/10.1016/J.ECOENV.2022.113249

Bishop KJM, Wilmer CE, Soh S, Grzybowski BA (2009) Nanoscale forces and their uses in self-assembly. Small (Weinheim an Der Bergstrasse, Germany) 5(14):1600–1630. https://doi.org/10.1002/SMLL.200900358

Byun J (2021) Recent progress and opportunities for nucleic acid Aptamers. Life 11(3):193. https://doi.org/10.3390/LIFE11030193

Camarca A, Varriale A, Capo A, Pennacchio A, Calabrese A, Giannattasio C, Almuzara CM, D'auria S, Staiano M (2021) Emergent biosensing technologies based on fluorescence spectroscopy and surface plasmon resonance. Sensors 21(3):906. https://doi.org/10.3390/S21030906

Caporale A, Adorinni S, Lamba D, Saviano M (2021) Peptide–protein interactions: from drug design to supramolecular biomaterials. Molecules 26(5):1219. https://doi.org/10.3390/MOLECULES26051219

Carpenter AC, Paulsen IT, Williams TC (2018) Blueprints for biosensors: design, limitations, and applications. Genes 9(8):375. https://doi.org/10.3390/GENES9080375

Chai C, Xie Z, Grotewold E (2011) SELEX (systematic evolution of ligands by EXponential enrichment), as a powerful tool for deciphering the protein–DNA interaction space. Methods Mol Biol 754:249–258. https://doi.org/10.1007/978-1-61779-154-3_14

Chinchilla-Cárdenas DJ, Cruz-Méndez JS, Petano-Duque JM, García RO, Castro LR, Lobo-Castañón MJ, Cancino-Escalante GO (2024) Current developments of SELEX technologies and prospects in the aptamer selection with clinical applications. J Genet Eng Biotechnol 22(3):100400. https://doi.org/10.1016/J.JGEB.2024.100400

Cozzolino F, Iacobucci I, Monaco V, Monti M (2021) Protein-DNA/RNA interactions: an overview of investigation methods in the -omics era. J Proteome Res 20(6):3018–3030. https://doi.org/10.1021/ACS.JPROTEOME.1C00074/ASSET/IMAGES/LARGE/PR1C00074_0004.JPEG

Didarian R, Ozbek HK, Ozalp VC, Erel O, Yildirim-Tirgil N (2024) Enhanced SELEX platforms for Aptamer selection with improved characteristics: a review. Mol Biotechnol 1–16. https://doi.org/10.1007/S12033-024-01256-W/METRICS

Duguma AL, Bai X (2025) How the internet of things technology improves agricultural efficiency. Artif Intell Rev 58(2):1–26. https://doi.org/10.1007/S10462-024-11046-0/TABLES/5

Elbasi E, Zaki C, Topcu AE, Abdelbaki W, Zreikat AI, Cina E, Shdefat A, Saker L (2023) Crop prediction model using machine learning algorithms. Appl Sci 13(16):9288. https://doi.org/10.3390/APP13169288

Farooqui NA, Haleem M, Khan W, Ishrat M (2024) Precision agriculture and predictive analytics. In: Intelligent techniques for predictive data analytics. Wiley, pp 171–188. https://doi.org/10.1002/9781394227990.CH9

Fathi-Karkan S, Mirinejad S, Ulucan-Karnak F, Mukhtar M, Ghahramani Almanghadim H, Sargazi S, Rahdar A, Díez-Pascual AM (2023) Biomedical applications of aptamer-modified chitosan nanomaterials: an updated review. Int J Biol Macromol 238. https://doi.org/10.1016/J.IJBIOMAC.2023.124103

Fei A, Liu Q, Huan J, Qian J, Dong X, Qiu B, Mao H, Wang K (2015) Label-free impedimetric aptasensor for detection of femtomole level acetamiprid using gold nanoparticles decorated multiwalled carbon nanotube-reduced graphene oxide nanoribbon composites. Biosens Bioelectron 70:122–129. https://doi.org/10.1016/J.BIOS.2015.03.028

Fu Z, Xiang J (2020) Aptamers, the nucleic acid antibodies, in cancer therapy. Int J Mol Sci 21(8):2793. https://doi.org/10.3390/IJMS21082793

Fu Y, Liu T, Wang H, Wang Z, Hou L, Jiang J, Xu T (2024) Applications of nanomaterial technology in biosensing. J Sci Adv Mater Devices 9(2):100694. https://doi.org/10.1016/J.JSAMD.2024.100694

Getahun S, Kefale H, Gelaye Y (2024) Application of precision agriculture technologies for sustainable crop production and environmental sustainability: a systematic review. Sci World J 2024(1):2126734. https://doi.org/10.1155/2024/2126734

Ghobadi N (2013) Band gap determination using absorption spectrum fitting procedure. Int Nano Lett 3(1):1–4. https://doi.org/10.1186/2228-5326-3-2

Grieshaber D, MacKenzie R, Vörös J, Reimhult E (2008) Electrochemical biosensors—sensor principles and architectures. Sensors 8(3):1400–1458. https://doi.org/10.3390/S80314000

Guo X (2012) Surface plasmon resonance based biosensor technique: a review. J Biophotonics 5(7):483–501. https://doi.org/10.1002/JBIO.201200015

Gupta A, Bano A, Rai S, Pathak N, Sharma S (2020) New insights into application of nanoparticles for plant growth promotion: present and future prospects. In: Biogenic nanoparticles and their use in agro-ecosystems. Springer, Singapore, pp 259–279. https://doi.org/10.1007/978-981-15-2985-6_15

Hanif A, Farooq R, Rehman MU, Khan R, Majid S, Ganaie MA (2019) Aptamer based nanobiosensors: promising healthcare devices. Saudi Pharm J 27(3):312–319. https://doi.org/10.1016/J.JSPS.2018.11.013

Hayat A, Marty JL (2014) Aptamer based electrochemical sensors for emerging environmental pollutants. Front Chemistry 2(JUN). https://doi.org/10.3389/FCHEM.2014.00041

Holzinger M, Le Goff A, Cosnier S (2014) Nanomaterials for biosensing applications: a review. Front Chemistry 2(AUG):108707. https://doi.org/10.3389/FCHEM.2014.00063/PDF

Hou S, Zhang A, Su M (2016) Nanomaterials for biosensing applications. Nanomaterials 6(4):58. https://doi.org/10.3390/NANO6040058

Huang X, Zhu Y, Kianfar E (2021) Nano biosensors: properties, applications and electrochemical techniques. J Mater Res Technol 12:1649–1672. https://doi.org/10.1016/J.JMRT.2021.03.048

Huang Y, Qian Y, Wei H, Lu Y, Ling B, Qin Y (2023) A survey of deep learning-based object detection methods in crop counting. Comput Electron Agric 215:108425. https://doi.org/10.1016/J.COMPAG.2023.108425

Ingrao C, Strippoli R, Lagioia G, Huisingh D (2023) Water scarcity in agriculture: an overview of causes, impacts and approaches for reducing the risks. Heliyon 9(8). https://doi.org/10.1016/J.HELIYON.2023.E18507

Janghorban M, Aradanas I, Kazemi S, Ngaju P, Pandey R (2022) Recent advances, opportunities, and challenges in developing nucleic acid integrated wearable biosensors for expanding the capabilities of wearable technologies in health monitoring. Biosensors 12(11):986. https://doi.org/10.3390/BIOS12110986

Jaseena KU, Kovoor BC (2022) Deterministic weather forecasting models based on intelligent predictors: a survey. J King Saud Univ Comput Inform Sci 34(6):3393–3412. https://doi.org/10.1016/J.JKSUCI.2020.09.009

Javaid M, Haleem A, Singh RP, Suman R (2022) Enhancing smart farming through the applications of agriculture 4.0 technologies. Int J Intell Netw 3:150–164. https://doi.org/10.1016/J.IJIN.2022.09.004

Jha G, Debangshi U, Palla S, Nazrul F, Dey S, Dutta W, Bansal S (2025) Precision agriculture technologies for climate-resiliency and water resource. Management 351–381. https://doi.org/10.1007/978-3-031-76532-2_15

Jiang S, Liu M, Tantai W, Xu Q, Zou X, Ma F, Zhang CY (2021) Aptamer-mediated rolling circle amplification for label-free and sensitive detection of histone acetyltransferase activity. Chem Commun 57(16):2041–2044. https://doi.org/10.1039/D0CC07763A

Jo M, Ahn JY, Lee J, Lee S, Hong SW, Yoo JW, Kang J, Dua P, Lee DK, Hong S, Kim S (2011) Development of single-stranded DNA aptamers for specific bisphenol a detection. Oligonucleotides 21(2):85–91. https://doi.org/10.1089/OLI.2010.0267/ASSET/IMAGES/LARGE/FIGURE4.JPEG

Kah M, Tufenkji N, White JC (2019) Nano-enabled strategies to enhance crop nutrition and protection. Nat Nanotechnol 14(6):532–540. https://doi.org/10.1038/s41565-019-0439-5

Kangarshahi BM, Naghib SM (2024) Nanogenosensors based on aptamers and peptides for bioelectrochemical cancer detection: an overview of recent advances in emerging materials and technologies. Discov Appl Sci 6(2):1–26. https://doi.org/10.1007/S42452-024-05681-Z

Kanjana D (2017) Advancement of nanotechnology applications on plant nutrients management and soil improvement. In: Nanotechnology: food and environmental paradigm. Springer, Singapore, pp 209–234. https://doi.org/10.1007/978-981-10-4678-0_12

Kaphle A, Navya PN, Umapathi A, Daima HK (2018) Nanomaterials for agriculture, food and environment: applications, toxicity and regulation. Environ Chem Lett 16(1):43–58. https://doi.org/10.1007/S10311-017-0662-Y

Kara N, Ayoub N, Ilgu H, Fotiadis D, Ilgu M (2023) Aptamers targeting membrane proteins for sensor and diagnostic applications. Molecules 28(9):3728. https://doi.org/10.3390/MOLECULES28093728

Kashyap PL, Kumar S, Jasrotia P, Singh DP, Singh GP (2019) Nanosensors for plant disease diagnosis: current understanding and future perspectives. In: Nanoscience for sustainable agriculture. Springer, pp 189–205. https://doi.org/10.1007/978-3-319-97852-9_9

Kaushal M, Wani SP (2017) Nanosensors: frontiers in precision agriculture. In: Nanotechnology: an agricultural paradigm. Springer, Cham, pp 279–291. https://doi.org/10.1007/978-981-10-4573-8_13

Khaledian S, Nikkhah M, Shams-bakhsh M, Hoseinzadeh S (2017) A sensitive biosensor based on gold nanoparticles to detect Ralstonia solanacearum in soil. J Gen Plant Pathol 83(4):231–239. https://doi.org/10.1007/S10327-017-0721-Z

Khan NI, Song E (2020) Lab-on-a-chip systems for aptamer-based biosensing. Micromachines 11(2):220. https://doi.org/10.3390/MI11020220

Kim SK, Karadeniz F (2012) Biological importance and applications of squalene and Squalane. Adv Food Nutr Res 65:223–233. https://doi.org/10.1016/B978-0-12-416003-3.00014-7

Kou X, Zhang X, Shao X, Jiang C, Ning L (2020) Recent advances in optical aptasensor technology for amplification strategies in cancer diagnostics. Anal Bioanal Chem 412(25):6691–6705. https://doi.org/10.1007/S00216-020-02774-7

Kouadio L, El Jarroudi M, Belabess Z, Laasli SE, Roni MZK, Amine IDI, Mokhtari N, Mokrini F, Junk J, Lahlali R (2023) A review on UAV-based applications for plant disease detection and monitoring. Remote Sens 15(17):4273. https://doi.org/10.3390/RS15174273/S1

Kushwaha M, Singh S, Singh V, Dwivedi S (2024) Precision farming: a review of methods, technologies, and future prospects. Int J Environ Agric Biotechnol 9(2):242–253. https://doi.org/10.22161/IJEAB.92.27

Lee Y, Lee S, Jin CM, Kwon JA, Kang T, Choi I (2018) Facile fabrication of large-scale porous and flexible three-dimensional plasmonic networks. ACS Appl Mater Interfaces 10(33):28242–28249. https://doi.org/10.1021/ACSAMI.8B11055/SUPPL_FILE/AM8B11055_SI_001.PDF

Lyu Y, Chen G, Shangguan D, Zhang L, Wan S, Wu Y, Zhang H, Duan L, Liu C, You M, Wang J, Tan W (2016) Generating cell targeting aptamers for nanotheranostics using cell-SELEX. Theranostics 6(9):1440–1452. https://doi.org/10.7150/THNO.15666

Mahmoudian F, Ahmari A, Shabani S, Sadeghi B, Fahimirad S, Fattahi F (2024) Aptamers as an approach to targeted cancer therapy. Cancer Cell Int 24(1):1–22. https://doi.org/10.1186/S12935-024-03295-4

Mairal T, Cengiz Özalp V, Lozano Sánchez P, Mir M, Katakis I, O'Sullivan CK (2008) Aptamers: molecular tools for analytical applications. Anal Bioanal Chem 390(4):989–1007. https://doi.org/10.1007/S00216-007-1346-4/METRICS

Malhotra BD, Ali MA (2018) Nanomaterials in biosensors: fundamentals and applications. Nanomater Biosens 1–74. https://doi.org/10.1016/B978-0-323-44923-6.00001-7

Marinello F, Zou X, Liu Z, Zhu X, Zhang W, Qian Y, Li Y, Karunathilake EMBM, Le AT, Heo S, Chung YS, Mansoor S (2023) The path to smart farming: innovations and opportunities in precision agriculture. Agriculture 13(8):1593. https://doi.org/10.3390/AGRICULTURE13081593

Mgendi G (2024) Unlocking the potential of precision agriculture for sustainable farming. Discov Agric 2(1):1–24. https://doi.org/10.1007/S44279-024-00078-3

Min JH, Woo MK, Yoon HY, Jang JW, Wu JH, Lim CS, Kim YK (2014) Isolation of DNA using magnetic nanoparticles coated with dimercaptosuccinic acid. Anal Biochem 447(1):114–118. https://doi.org/10.1016/J.AB.2013.11.018

Morris KN, Jensen KB, Julin CM, Weil M, Gold L (1998) High affinity ligands from in vitro selection: complex targets. Proc Natl Acad Sci USA 95(6):2902–2907. https://doi.org/10.1073/PNAS.95.6.2902/ASSET/BC61725E-DCE8-4403-9F97-8B25E048D86F/ASSETS/GRAPHIC/PQ0484377005.JPEG

Naresh V, Lee N (2021) A review on biosensors and recent development of nanostructured materials-enabled biosensors. Sensors 21(4):1109. https://doi.org/10.3390/S21041109

Nath P, Mahtaba KR, Ray A (2023) Fluorescence-based portable assays for detection of biological and chemical analytes. Sensors 23(11):5053. https://doi.org/10.3390/S23115053

Ni X, Castanares M, Mukherjee A, Lupold SE (2011) Nucleic acid Aptamers: clinical applications and promising new horizons. Curr Med Chem 18(27):4206–4214. https://doi.org/10.2174/092986711797189600

Nimjee SM, Oney S, Volovyk Z, Bompiani KM, Long SB, Hoffman M, Sullenger BA (2009) Synergistic effect of aptamers that inhibit exosites 1 and 2 on thrombin. RNA 15(12):2105–2111. https://doi.org/10.1261/RNA.1240109

Nsibande SA, Forbes PBC (2016) Fluorescence detection of pesticides using quantum dot materials—a review. Anal Chim Acta 945:9–22. https://doi.org/10.1016/J.ACA.2016.10.002

Odeh F, Nsairat H, Alshaer W, Ismail MA, Esawi E, Qaqish B, Al Bawab A, Ismail SI (2019) Aptamers chemistry: chemical modifications and conjugation strategies. Molecules 25(1):3. https://doi.org/10.3390/MOLECULES25010003

Padhiary M, Saha D, Kumar R, Sethi LN, Kumar A (2024) Enhancing precision agriculture: a comprehensive review of machine learning and AI vision applications in all-terrain vehicle for farm automation. Smart Agric Technol 8:100483. https://doi.org/10.1016/J.ATECH.2024.100483

Pérez-de-Luque A (2017) Interaction of nanomaterials with plants: what do we need for real applications in agriculture? Front Environ Sci 5(APR):259755. https://doi.org/10.3389/FENVS.2017.00012/BIBTEX

Pola-López LA, Camas-Anzueto JL, Martínez-Antonio A, Luján-Hidalgo MC, Anzueto-Sánchez G, Ruíz-Valdiviezo VM, Grajales-Coutiño R, González JHC (2018) Novel arsenic biosensor "POLA" obtained by a genetically modified E. coli bioreporter cell. Sensors Actuators B Chem 254:1061–1068. https://doi.org/10.1016/J.SNB.2017.08.006

Pramod Pawase P, Madhukar Nalawade S, Ashok Walunj A, Balasaheb Bhanage G, Bhaskar Kadam P, Durgude AG, Patil MR (2024) Comprehensive study of on-the-go sensing and variable rate application of liquid nitrogenous fertilizer. Comput Electron Agric 216:108482. https://doi.org/10.1016/J.COMPAG.2023.108482

Prasad R, Bhattacharyya A, Nguyen QD (2017) Nanotechnology in sustainable agriculture: recent developments, challenges, and perspectives. Front Microbiol 8(JUN):264976. https://doi.org/10.3389/FMICB.2017.01014/PDF

Presnova G, Presnov D, Krupenin V, Grigorenko V, Trifonov A, Andreeva I, Ignatenko O, Egorov A, Rubtsova M (2017) Biosensor based on a silicon nanowire field-effect transistor functionalized by gold nanoparticles for the highly sensitive determination of prostate specific antigen. Biosens Bioelectron 88:283–289. https://doi.org/10.1016/J.BIOS.2016.08.054

Putzbach W, Ronkainen NJ (2013) Immobilization techniques in the fabrication of nanomaterial-based electrochemical biosensors: a review. Sensors 13(4):4811–4840. https://doi.org/10.3390/S130404811

Qi J, Nsoh B, Katimbo A, Guo H, Heeren DM, Nakabuye HN, Qiao X, Ge Y, Rudnick DR, Wanyama J, Bwambale E, Kiraga S (2024) Internet of things-based automated solutions utilizing machine learning for smart and real-time irrigation management: a review. Sensors 24(23):7480. https://doi.org/10.3390/S24237480

Rai M, Ribeiro C, Mattoso L, Duran N (2015) Nanotechnologies in food and agriculture. In: Nanotechnologies in food and agriculture. Springer, pp 1–347. https://doi.org/10.1007/978-3-319-14024-7/COVER

Rajak P, Ganguly A, Adhikary S, Bhattacharya S (2023) Internet of things and smart sensors in agriculture: scopes and challenges. J Agric Food Res 14:100776. https://doi.org/10.1016/J.JAFR.2023.100776

Reiss P, Protière M, Li L (2009) Core/shell semiconductor nanocrystals. Small 5(2):154–168. https://doi.org/10.1002/SMLL.200800841

Rejeb A, Abdollahi A, Rejeb K, Treiblmaier H (2022) Drones in agriculture: a review and bibliometric analysis. Comput Electron Agric 198:107017. https://doi.org/10.1016/J.COMPAG.2022.107017

Resch-Genger U, Grabolle M, Cavaliere-Jaricot S, Nitschke R, Nann T (2008) Quantum dots versus organic dyes as fluorescent labels. Nat Methods 5(9):763–775. https://doi.org/10.1038/nmeth.1248

Reverdatto S, Burz DS, Shekhtman A (2015) Peptide Aptamers: development and applications. Curr Top Med Chem 15(12):1082–1101. https://doi.org/10.2174/1568026615666150413153143

Ruigrok VJB, Levisson M, Hekelaar J, Smidt H, Dijkstra BW, van der Oost J (2012) Characterization of Aptamer-protein complexes by X-ray crystallography and alternative approaches. Int J Mol Sci 13(8):10537–10552. https://doi.org/10.3390/IJMS130810537

Sabir RM, Mehmood K, Sarwar A, Safdar M, Muhammad NE, Gul N, Athar F, Majeed MD, Sattar J, Khan Z, Fatima M, Hussain MA, Akram HMB (2024) Remote sensing and precision agriculture: a sustainable future. World Sustain Ser Part F3301:75–103. https://doi.org/10.1007/978-3-031-63430-7_4

Sajeevan A, Sukumaran RA, Panicker LR, Kotagiri YG (2025) Trends in ready-to-use portable electrochemical sensing devices for healthcare diagnosis. Microchim Acta 192(2):1–24. https://doi.org/10.1007/S00604-024-06916-X

Sekhon SS, Lee SH, Lee KA, Min J, Lee BT, Kim KW, Ahn JY, Kim YH (2017) Defining the copper binding aptamotif and aptamer integrated recovery platform (AIRP). Nanoscale 9(8):2883–2894. https://doi.org/10.1039/C6NR09408B

Sen RK, Prabhakar P, Bisht N, Patel M, Mishra S, Yadav AK, Venu DV, Gupta GK, Solanki PR, Ramakrishnan S, Mondal DP, Srivastava AK, Dwivedi N, Dhand C (2021) 2D materials-based Aptamer biosensors: present status and way forward. Curr Med Chem 29(37):5815–5849. https://doi.org/10.2174/0929867328666211213115723

Shangguan D, Li Y, Tang Z, Cao ZC, Chen HW, Mallikaratchy P, Sefah K, Yang CJ, Tan W (2006) Aptamers evolved from live cells as effective molecular probes for cancer study. Proc Natl Acad Sci USA 103(32):11838–11843. https://doi.org/10.1073/PNAS.0602615103/ASSET/C2A7B7C5-0C96-4B92-A051-39ECF368704D/ASSETS/GRAPHIC/ZPQ0300629630007.JPEG

Sharma TK (2014) Nucleic acid Aptamers as an emerging diagnostic tool for animal pathogens. Adv Anim Vet Sci 2(1). https://doi.org/10.14737/JOURNAL.AAVS/2014.2.1.50.55

Sharma K, Shivandu SK (2024) Integrating artificial intelligence and Internet of Things (IoT) for enhanced crop monitoring and management in precision agriculture. Sens Int 5:100292. https://doi.org/10.1016/J.SINTL.2024.100292

Shen L, Chen Z, Li Y, He S, Xie S, Xu X, Liang Z, Meng X, Li Q, Zhu Z, Li M, Le XC, Shao Y (2008) Electrochemical DNAzyme sensor for lead based on amplification of DNA−Au bio-Bar codes. Anal Chem 80(16):6323–6328. https://doi.org/10.1021/AC800601Y

Singh Sekhon B (2014) Nanotechnology in Agri-food production: an overview. Nanotechnol Sci Appl 7(2):31–53. https://doi.org/10.2147/NSA.S39406

Soule EE, Yu H, Olson L, Naqvi I, Kumar S, Krishnaswamy S, Sullenger BA (2022) Generation of an anticoagulant aptamer that targets factor V/Va and disrupts the FVa-membrane interaction in normal and COVID-19 patient samples. Cell Chem Biol 29(2):215–225.e5. https://doi.org/10.1016/J.CHEMBIOL.2022.01.009/ATTACHMENT/FC4B60E8-FC2A-49D0-A163-483B8D753DB0/MMC3.PDF

Soussi A, Zero E, Sacile R, Trinchero D, Fossa M (2024) Smart sensors and smart data for precision agriculture: a review. Sensors 24(8):2647. https://doi.org/10.3390/S24082647

Srivastava AK, Dev A, Karmakar S (2018) Nanosensors and nanobiosensors in food and agriculture. Environ Chem Lett 16(1):161–182. https://doi.org/10.1007/S10311-017-0674-7

Suhito IR, Koo KM, Kim TH (2020) Recent advances in electrochemical sensors for the detection of biomolecules and whole cells. Biomedicines 9(1):15. https://doi.org/10.3390/BIOMEDICINES9010015

Sun H, Zu Y, Miller AOA, Jacques J, Eynde V (2015) A highlight of recent advances in Aptamer technology and its application. Molecules 20(7):11959–11980. https://doi.org/10.3390/MOLECULES200711959

Talan A, Mishra A, Eremin SA, Narang J, Kumar A, Gandhi S (2018) Ultrasensitive electrochemical immuno-sensing platform based on gold nanoparticles triggering chlorpyrifos detection in fruits and vegetables. Biosens Bioelectron 105:14–21. https://doi.org/10.1016/J.BIOS.2018.01.013

Talari FF, Bozorg A, Faridbod F, Vossoughi M (2021) A novel sensitive aptamer-based nanosensor using rGQDs and MWCNTs for rapid detection of diazinon pesticide. J Environ Chem Eng 9(1):104878. https://doi.org/10.1016/J.JECE.2020.104878

Vats M, Mendiratta T, Singh H, Kumar A, Verma M, Behera K, Kumar S (2024) Application of magnetic nanomaterials as biosensors. Nanotechnol Life Sci 2024:65–88. https://doi.org/10.1007/978-3-031-72004-8_5

Vigneshvar S, Sudhakumari CC, Senthilkumaran B, Prakash H (2016) Recent advances in biosensor technology for potential applications—an overview. Front Bioeng Biotechnol 4(FEB):177235. https://doi.org/10.3389/FBIOE.2016.00011/BIBTEX

Wang J, Zhang Y, Chen Y, Hong S, Sun Y, Sun N, Pei R (2017) In vitro selection of DNA aptamers against renal cell carcinoma using living cell-SELEX. Talanta 175:235–242. https://doi.org/10.1016/J.TALANTA.2017.07.049

Wang M, Xu Y, Yang Y, Mu B, Nikitina MA, Xiao X (2022) Vis/NIR optical biosensors applications for fruit monitoring. Biosens Bioelectron X:11. https://doi.org/10.1016/J.BIOSX.2022.100197

Weerathunge P, Ramanathan R, Shukla R, Sharma TK, Bansal V (2014) Aptamer-controlled reversible inhibition of gold nanozyme activity for pesticide sensing. Anal Chem 86(24):11937–11941. https://doi.org/10.1021/AC5028726/SUPPL_FILE/AC5028726_SI_001.PDF

Xiao Y, Pan T, Da W, Liu Y, Chen S, Chen D, Liu K, Zheng Y, Xie D, Gao Y, Xu H, Sun Y, Tan W (2024) Aptamer-drug conjugates-loaded bacteria for pancreatic cancer synergistic therapy. Signal Transduct Target Ther 9(1):1–17. https://doi.org/10.1038/s41392-024-01973-3

Xu J, Cui Y, Zhang S, Zhang M (2024) The evolution of precision agriculture and food safety: a bibliometric study. Front Sustain Food Syst 8:1475602. https://doi.org/10.3389/FSUFS.2024.1475602/PDF

Yan J, Xiong H, Cai S, Wen N, He Q, Liu Y, Peng D, Liu Z (2019) Advances in aptamer screening technologies. Talanta 200:124–144. https://doi.org/10.1016/J.TALANTA.2019.03.015

Yang L, Zhang X, Ye M, Jiang J, Yang R, Fu T, Chen Y, Wang K, Liu C, Tan W (2011) Aptamer-conjugated nanomaterials and their applications. Adv Drug Deliv Rev 63(14–15):1361–1370. https://doi.org/10.1016/J.ADDR.2011.10.002

Yang C, Jiang Y, Hao SH, Yan XY, Hong DF, Naranmandura H (2021) Aptamers: an emerging navigation tool of therapeutic agents for targeted cancer therapy. J Mater Chem B 10(1):20–33. https://doi.org/10.1039/D1TB02098F

Yang Y, Mali P, Arthur L, Molaei F, Atsyo S, Geng J, He L, Ghatrehsamani S (2025) Advanced technologies for precision tree fruit disease management: a review. Comput Electron Agric 229:109704. https://doi.org/10.1016/J.COMPAG.2024.109704

Ye W, Yu M, Wang F, Li Y, Wang C (2022) Multiplexed detection of heavy metal ions by single plasmonic nanosensors. Biosens Bioelectron 196:113688. https://doi.org/10.1016/J.BIOS.2021.113688

Yoon S, Rossi JJ (2018) Aptamers: uptake mechanisms and intracellular applications. Adv Drug Deliv Rev 134:22–35. https://doi.org/10.1016/J.ADDR.2018.07.003

Zhang Y, Lai BS, Juhas M (2019) Recent advances in Aptamer discovery and applications. Molecules 24(5):941. https://doi.org/10.3390/MOLECULES24050941

Zhang N, Chen Z, Liu D, Jiang H, Zhang Z-K, Lu A, Zhang B-T, Yu Y, Zhang G, Zhang N, Chen Z, Liu D, Jiang H, Zhang Z-K, Lu A, Zhang B-T, Yu Y, Zhang G, Hk Z-KZ et al (2021) Structural biology for the molecular insight between Aptamers and target proteins. Int J Mol Sci 22(8):4093. https://doi.org/10.3390/IJMS22084093

Zhao WW, Xu JJ, Chen HY (2017) Photoelectrochemical enzymatic biosensors. Biosens Bioelectron 92:294–304. https://doi.org/10.1016/J.BIOS.2016.11.009

Zhou J, Rossi J (2017) Aptamers as targeted therapeutics: current potential and challenges. Nat Rev Drug Discov 16(3):181–202. https://doi.org/10.1038/NRD.2016.199

Zhou L, Wang J, Li D, Li Y (2014) An electrochemical aptasensor based on gold nanoparticles dotted graphene modified glassy carbon electrode for label-free detection of bisphenol A in milk samples. Food Chem 162:34–40. https://doi.org/10.1016/J.FOODCHEM.2014.04.058

Zhuo Z, Yu Y, Wang M, Li J, Zhang Z, Liu J, Wu X, Lu A, Zhang G, Zhang B (2017) Recent advances in SELEX technology and Aptamer applications in biomedicine. Int J Mol Sci 18(10):2142. https://doi.org/10.3390/IJMS18102142

Chapter 7
Nanobiosensors for Precision Plant Disease Management

Mohsen Mohamed Elsharkawy

Abstract Global population growth and rising food consumption pose significant concerns for food safety. Crop loss is mostly caused by several diseases, including bacterial, fungal, and viral diseases. Diagnosis of pathogens that is accurate and quick is crucial for crop health monitoring in order to stop the spread of disease and support efficient management techniques. Implementing biosensors in agricultural settings can help tackle this problem. Biosensors can aid in sustainable agricultural activity by detecting pathogens and pesticides in crops. Farmers may enhance crop production by detecting diseases early with the use of these biosensors. Nanobiosensor-based technology offers nondestructive, minimally intrusive, cost-effective, and user-friendly solutions for detecting plant infections directly. This chapter discusses the way several biosensor types are used in agriculture to monitor a range of plant pathogens, considering the issues that need to be resolved. This chapter also discusses the uses and benefits of the kinds of biosensors, such as electrochemical, optical, and plant wearable technologies, in smart farming.

Keywords Nanoparticles · Nanobiosensors · Diagnosis · Pathogen · Disease management

7.1 Introduction

A major threat to preserving the world's food supply is the incidence of diseases and their spatial spread (Kashyap et al. 2017a). Under suitable conditions, some phytopathogen infections can result in full crop loss (Ray et al. 2017). Furthermore, the combined effects of globalization, climate change, increased human and pathogen

M. M. Elsharkawy (✉)
Agricultural Botany Department, Faculty of Agriculture, Kafrelsheikh University, Kafr Elsheikh, Egypt
e-mail: mohsen.abdelrahman@agr.kfs.edu.eg

J.-T. Chen (ed.), *Nanobiosensors for Crop Monitoring and Precision Agriculture*, Sustainability Sciences in Asia and Africa,
https://doi.org/10.1007/978-981-96-8335-2_7

mobilization, and the quick development of pathogen vectors contribute to the spread of dangerous plant diseases (Kashyap et al. 2018). For instance, *Tilletia indica* presents a significant issue to the grain sector due to quarantine requirements that may obstruct the worldwide transportation of afflicted wheat grain rather than a direct production loss (Kashyap et al. 2018). The pathogen quickly changes and adapts to various disease management strategies, even if resistant-crop cultivars and fungicides are used to strategically combat infections that arise in agricultural areas. Thus, the development of sensitive, genuine, and fast detection tests for plant pathogens that are present at minimal levels and even in plants that serve as symptomless carriers is crucial. Crop pathogen detection methods include bioassays, visual inspection, serology, microscopy, and molecular techniques, such as DNA fingerprinting, polymerase chain reaction, isothermal amplification assays, quantitative real-time polymerase chain reaction (PCR), reverse transcription-PCR, microarrays, next-generation sequencing, and nucleic acid blot assays (Singh et al. 2014; Kaur et al. 2019). Several reviews have focused on detecting and diagnosing agricultural diseases using molecular protocols (De Boer and López 2012; Nezhad 2014; Martinelli et al. 2015; Le and Vu 2017). The serological protocols such as radio-immunosorbent assay, immunodiffusion test, western blot analysis, enzyme-linked immunosorbent assay, dipstick immunoassay, dot immunobinding assay, and tissue blot immunoassay were utilized for pathogen diagnosis (Souiri et al. 2014; Sharma et al. 2017; Elsharkawy et al. 2022). Molecular assays may identify a single pathogen in one sample that may include several infections. Serological assays are ideal for simultaneously detecting many pathogens in a sample. Primers unique to species are created from the pathogen's DNA regions and used for DNA amplification. Numerous infections can be diagnosed and identified using amplicons made from nucleic acid sequences of a pathogen as a DNA barcode (Kashyap et al. 2017b). Although serological and molecular detection methods are quick and accurate, they cannot easily used in farmers' fields. The accurate and focused design of oligonucleotides and probes is crucial to their effectiveness. Furthermore, cross-contamination of chemicals and reagents during the PCR experiment might result in fabricated negative findings. The restricted use of PCR-based tests in field settings is another disadvantage (Martinelli et al. 2015). Sensor-based developments have been used to create portable instruments for crop pathogen field diagnostics in order to overcome these limitations (Kashyap et al. 2017a). For experts interested in identifying, characterizing, and tracking agricultural pathogens, these sensors provide a variety of accessories (Kashyap et al. 2016). Thermal, electrochemical, optical, and piezoelectric sensors are the main types of biosensors, which are grouped according to the principles of biorecognition and signal transduction mechanisms (Kwak et al. 2017). Nanobiosensors and biomarkers may be used together to quickly, precisely, and accurately collect data about early infection and disease progression in field settings (Ray et al. 2017). Nanosensors can be categorized according to their component materials, detection targets, and transmission signals. According to recent findings, field deployment of nanosensors is a great way to reduce the costly usage of agrochemicals for sustaining crop growth and health, making agricultural farming more resilient and secure (Jasrotia et al. 2018). This chapter emphasizes the significant technological advancements in plant pathogen detection and quantification, with a particular emphasis on nanobiosensors.

7.2 Types of Nanosensors

7.2.1 Fluorescence Resonance Energy Transfer (FRET) Nanosensor

An example of optical sensors that work on the basis of fluorescence resonance energy transmission and its quantity is the FRET nanosensor (Algar and Krull 2008). FRET theoretically uses two distinct types of fluorescent dyes, known as an acceptor (which absorbs excited energy) and a donor (which transfers energy following excitation by a light source), through dipole–dipole intermolecular interactions (Stanisavljevic et al. 2015). Energy transfer happens once the donor dye emission spectrum coincides with the acceptor dye excitation spectrum (Grecco and Verveer 2011). This happens if the gap between the dyes is 1–10 nm. Optimal FRET pairing relies on a substrate-specific binding domain flanked by acceptor and donor dyes at separate terminals (Ellinger and Voigt 2014). Substrate binding causes a shift in the orientation or distance of two fluorophores, which may be measured as energy transmission (Kwak et al. 2017).

7.2.2 Surface-Enhanced Raman Scattering (SERS) Nanosensor

SERS is a special kind of spectroscopic detection instrument. This kind of nanosensor is often made up of metal nanoparticle (NP) coatings that, once excited by a laser, enhance the characteristics of the attached molecule's Raman scattering pattern (Kahraman et al. 2017). The chemical effect and the electromagnetic effect are the two distinct processes that cause SERS effects (Sun et al. 2008). Enhancing the electromagnetic field close to the nanoparticle surface is accomplished via the SERS nanosensor's electromagnetic effect. Remarkably, this long-range increase in the electromagnetic impact takes place further from the surface of the nanoparticle. Conversely, chemical effects include carrying charges between guest molecules and nanoparticles at short distances. This chemical reaction happens when an analyte is absorbed on a nanoparticle-coated surface (Wei et al. 2015). Au and Ag nanoparticles are commonly utilized because of their powerful and adjustable plasmon resonance in the near-infrared (NIR) range. Several sensors have been created for disease detection in the agricultural field by utilizing SERS (Chen and Park 2016).

7.2.3 Electrochemical Nanosensor (ECN)

The electrochemical nanosensor is a powerful and novel method for quick crop disease detection because of its relative simplicity of use, affordability, high sensitivity, portability, and direct data-processing capabilities. Generally, three distinct electrode types—working electrode, counter electrode, and reference electrode—make up electrochemical nanosensors (Barry and O'Riordan 2015). Several tools, including voltammetry, conductometry, potentiometry, and amperometry, have been used to measure the change of any chemical signal or biological process into electrical signals (Qureshi et al. 2009). The greater active surfaces of electrochemical electrodes made with nanomaterials (NMs) have the benefit of increased sensitivity and a wider range of intensity recognition for redox and ion species associated with plants (Clausmeyer and Schuhmann 2016). These benefits allow ECN to be used for horticulture crops and field disease monitoring (Srivastava et al. 2018).

7.2.4 Piezoelectric Nanosensor (PZN)

The piezoelectric nanosensor (PZN) opens up new possibilities for the immediate detection of diseases in plants by converting mechanical energy, ultrasonic vibrations, and biofluid hydraulic energy of any biological process into electrical energy (Byrne et al. 2009). Piezoelectric sensors primarily operate on the principle of tracking and determining mass changes brought on by the biomolecular reaction of antibodies with their corresponding antigenic determinants (Mondal et al. 2022). For example, by computing the frequency variations of the quartz crystal microbalance (QCM), the alterations in the quartz crystal's resonance frequency caused by the changes in mass were detected.

7.3 Nanobiosensors

An independent, integrated device that uses a sensitive bioreceptor to detect biomolecules is called a biosensor (Virk et al. 2024). Therefore, recognition components are coupled with physical or chemical transducers to identify biological substances (Bhalla et al. 2016). The biological components engage in interaction alongside the analytes, while the transducer transforms the data gathered from the interaction into a measurable effect (Huang et al. 2021). The various forms of biosensors may be categorized based on their transducers and recognition components. Based on transducers, a biosensor can be electrochemical (impedimetric, potentiometric, amperometric, etc.), piezoelectric (quartz crystal microbalance, acoustic waves, etc.), calorimetric, or optical (refraction, reflection, absorption, fluorescence, transmission, waveguide, surface plasmon resonance, etc.) (Malhotra and Ali 2018).

Biosensors can be categorized according to their recognition components as organelle-based, whole-cell, aptamer-based, oligonucleotide-based, enzymatic, nucleic acid–based, and immunosensors (biomarker techniques or antigen–antibody interaction) (Malhotra and Ali 2018). Additionally, there are a number of flexible paper-based biosensors (such as lateral flow assays, dipstick tests, and microfluidic biosensors) that use optical detection, in addition to electrochemical, electrochemiluminescence, or chemiluminescence detection, to identify the biomarkers (Kuswandi and Ensafi 2019). In order to improve the sensitivity and functionality of the device, current research is concentrating on incorporating nanoparticles into the biosensor production process. Complex laboratory methods are effective and dependable for identifying contaminants in agricultural practices, but they are time-consuming, expensive, and require highly skilled operators. Thus, it is essential to create quick, dependable, easy-to-use, and on-site screening techniques. The development of nanotechnology demonstrated significant promise for resolving these issues (Neme et al. 2021). Nanobiosensors, which combine nanomaterials with biosensors, are sensitive, harmless, and developed to monitor various samples from the environment (Thakur et al. 2022). Nanobiosensors were employed to detect targets using metal oxides, metal nanoparticles, dendrimers, and magnetic nanoparticles (MNPs) (Singh 2021). A nanobiosensor is made up of three parts: a detector, a transducer, and a biological probe or recognition element (Naresh and Lee 2021). These portable, user-friendly devices have enormous potential in agriculture and have been used to monitor soil and plant physiological conditions on-site, including soil fertility, pH, moisture content, mineral concentrations, level of toxic chemicals, and early pest and pathogen detection (Nadporozhskaya et al. 2022). Nanobiosensors provide a wide range of applications in sustainable agricultural methods (Sharma et al. 2021).

7.3.1 Nanomaterial-Based Biosensors and Sustainable Agriculture

A biomaterial is produced whenever a single piece of apparatus generates good-quality analytical data using a specific biomolecule that is responsive to certain biomolecules and bioreactors. In this scenario, biological substances are detected by physical or chemical sensors that employ recognition components (Huang et al. 2021). Organic coatings are associated with analytics. Sensors can quickly identify modifications in organic coatings related to analytics. Several kinds of biosensors may be categorized using recognition components and sensors. Biosensors use a variety of transducers, including impedimetric, electromagnetic, pulse metric, piezoelectric, quartz crystal, plasmon waveguide, reflection absorption transmission, and fluorescence surface. Biosensor types can be enzyme-, aptamer-, oligonucleotide-, nucleic acid-, or immunosensor-based, according to the recognition component.

The limitations of conventional techniques may be solved through the creation of biosensors employing nanoparticles. Biological sensors based on polymer particles provide several benefits over conventional techniques, such as high specificity and sensitivity and quick and effective response periods (Chakroborty et al. 2023). Along with portable equipment, the techniques may also identify extremely low concentrations of analytes that could endanger domestic animals and vegetation. Nanoparticles can be employed in amperometric biosensors and multiplexing devices because of their versatility. The present focus of biosensing studies is on the utilization of biological probes along with nanoparticles, including captivating, coal nanotubes, coefficient dots, and metallurgic (Ahmad et al. 2022). In this field, fullerenes, silicon oxide nanoparticles, and titanium nanoparticles are occasionally employed. On the other hand, encapsulation may also be achieved using polymerases such as chitosans, zeolites, and polyacrylic acids (Malhotra and Ali 2018). Many NPs may be used to create nanobiosensors. The most common anion nanoparticles in the metals are the well-known AuNPs (gold nanoparticles), which are followed by CuNPs (copper nanoparticles)and Fe_3O_4, which are amazing nanoparticles. Numerous research studies frequently employ hybrid nanomaterials, such as cadmium tetrachloride quantum dots (QDs), magnetic nanoparticles, MNCs (magnetic nanochains)/AuNPs, number dots, and nano-sized magnetic beads. Monitor reliable agricultural methods in terms of food security, phytopathogen detection and the arbitrary use of microbicides were proven to be quite effective. Numerous pollutants, such as pesticides, heavy metals, endocrine disruptors, dioxin, and microbial infections, and biological oxygen demand may be detected using nanobiosensors at picomolar concentrations (Verma and Rani 2021). Zinc oxide and carbon nanomaterials have multiple applications in agriculture due to their great automatic optics, short virulence, synthetic properties, and biocompatibility. Nanomaterials having metal characteristics, including platinum and gold, offer unique sensing capabilities and are commonly employed (Yu et al. 2021). Biosensors made of nano-selenium and chitosan/gold nanoparticle composites have significant agricultural potential (Bao et al. 2015). Graphene can be chemically functionalized into metal composites because of its biocompatibility profile. Due to these special properties, graphene-based biosensors are outstanding choices for biosensors (Verma et al. 2022). Nanofabrication has several uses, such as sensing loamy soil, detecting pathogens in vegetation, regulating growth, detecting pesticides, and detecting metals (Ali et al. 2021). Developing sensitive nanobiosensors requires a deeper knowledge of biological interactions and efficient bio-conjugation techniques (Verma 2017).

Innovative technology, such as nano-based sensors, is essential for sustainable agriculture (Table 7.1). These offer affordable monitoring with great sensitivity. Nanostructures can enhance sensor performance by improving reactivity, accuracy, and multiphoton observations, as well as providing adaptability (Butnariu and Butu 2019). Nanomaterial-based electrochemical and optical sensors may be used to identify environmental pollutants and heavy metals in soil media. Engineered nanomaterials are perfectly adapted to transferring this alteration since they can be utilized as sensors in plants, allowing for extensive crop health monitoring at high spatial precision (Giraldo et al. 2019). Plant health can be positively or negatively

Table 7.1 Examples of nanosensors or nanobiosensors used in pathogen or pesticide detection

Nanosensor or nanobiosensor types	Application	Citation
SPR nanosensor	Triaminic Pesticide detection	Yılmaz et al. (2017)
SPR nanosensor	Cymbidium mosaic virus (CymMV) detection	Antonacci et al. (2017)
Artificial-nose nanobiosensor	Pathogen detection	Cui et al. (2018)
QD nanosensor	Pathogen detection	Esker et al. (2018)
AChE biosensor	*Phytophthora palmivora* detection	James et al. (2019)
Optical nanosensor	Detection of peptides associated with plant pathogens	Wu et al. (2020b)
Electrochemical nanosensor	Pesticide detection	Li et al. (2022)

impacted by nanoparticles and agrochemicals, depending on their physicochemical qualities and concentrations.

Agricultural methods have benefited from automation as it ensures effective crop control and ecological compatibility with the least amount of labor (Saravanadevi et al. 2022). Plant nanosensors can facilitate communication between farmers and plants, thereby guaranteeing plant sustainability and health (Khandelwal et al. 2019). Unique signaling compounds linked to stress and nutritional deficits can be detected by plant-integrated nanoparticles, which convert them into optical or electrical signals in detection devices. Among the nano-based tools utilized in the automated evaluation of the nutritional balance, stress levels, and health state of plants are nanobiosensors, international navigation systems using satellites, and global positioning systems. This has made it possible to oversee several agricultural projects with one person without the need for a physical inspection. Effective resource management is achieved by real-time plant health monitoring, which significantly improves growth and survival. It has cleared the path for sustainable agriculture by improving soil health and crop development through genetic engineering, resource management, and the strategic use of agrochemicals (Giraldo et al. 2019). Nanobiosensors play a crucial role in developing modern organic farming systems that rely on accurate sensing and nanobionic approaches. The broad reaction of a crop across an entire field may serve as the basis for nanobiosensors. The mathematical concept of nanobionics is used to create sensors that monitor the movement of various phytohormones and signals (Giraldo et al. 2019). Nanobiosensors often improve precision farming, which improves output while utilizing fewer pesticides and chemical fertilizers, because of their ability to sense, evaluate, and identify changes. The research of plant pathology is primarily concerned with lowering the possibility of devastating crop loss by controlling plant diseases directly or indirectly (Li et al. 2021). Plant diseases can be managed by using pesticides topically and in the soil. However, ecological issues may arise from exposure to the harmful substances. Long-term use of these pesticides may potentially cause the pathogens to develop resistance strains. An excellent method for identifying, monitoring, and combating pathogen infections in plants is the use of nanosensors (Table 7.1). The farmer can benefit from increased crop output through the appropriate use of nanosensors (Muthumalai et al. 2022). This is an achievement toward precision farming, which will transform conventional farming into intelligent, economical, and sustainable farming. The majority of plant sensors connected to digital devices that rely on nanotechnology have undergone regulated laboratory testing. To track H_2O_2, a crucial signaling molecule linked to the start of stress in plants, near-infrared fluorescent single-walled carbon nanotubes (SWCNTs) modified with DNA aptamer that interacts with hemin were developed and used to treat *Arabidopsis thaliana* leaves (Sharma et al. 2021). These sensors are helpful for remote plant health monitoring and have been shown to detect initial indications of stress in plants (Wu et al. 2020a). MnNPs (magnetic nanoparticles) were studied as a nano-priming agent to reduce salt stress in *Capsicum annuum* during germination when administered with 100 mM of NaCl. Energy dispersive spectroscopy (EDS) and scanning electron microscopy (SEM) demonstrated that MnNPs penetrated the seed coating to form NP complexes with biological molecules. Likewise, dynamic light scattering demonstrated that MnNPs' hydrodynamic size was greater in NaCl than in Milli-Q water. The SEM picture showed that MnNPs formed clusters once they penetrated the seed coating while priming. Since the most crucial elements in proteins are N, C, O, and S, elemental mapping demonstrates the way proteins and MnNPs interact (Ye et al. 2020). Recently, plasmonic nanoparticles and smartphone fingerprinting have been used to noninvasively identify plant diseases in greenhouses and laboratories. The conversion into sensing in plants led to a notable enhancement in the sustainability of agriculture (Johnson et al. 2021). Despite the widespread use and growing interest in using smart nanosensors to monitor crop and soil health in contemporary agriculture, the precision and usefulness of these products in the field have not yet been maximized (Pandey 2018). Such devices have not yet been verified for use in phenotyping facilities and natural agricultural settings, since they are susceptible to variations in weather, development patterns, plant growth, etc. (Kumar et al. 2021). Therefore, extensive research is required to use nanosensors to support sustainable agriculture.

7.3.2 Sensing Technology with Nanoparticles for Pathogen Detection

To reduce the harm to the environment, care must be taken while using pesticides, fertilizers, and heavy metals in agricultural activities. Emerging technology must be used to regularly assess the health of the soil and the potential appearance of any diseases. To monitor vast field areas in real time, accurate sensing devices that are portable and easy to use must be manufactured. Nanotechnology might improve any biosensor's performance to match its actual use in agriculture (Table 7.1). Sensitive analytical methods and remarkable advancements in nanofabrication technology allow for the production of responsive sensors. Numerous fields, including crop cultivation and harvesting, disease detection, and soil characteristics, might benefit from the use of nanosensors. NPs are appropriate for improved sensing systems due to their distinct surface chemistry, electrical and thermal properties, increased sensitivities, and improved detection limits (Yao et al. 2009). Upgraded detection limits are provided by nanoparticles as a form of sensor for the diagnosis of many viral, fungal, and bacterial diseases (Yao et al. 2009). Typically, diagnostic instruments that either directly detect pathogens or identify instructive substances implicated in the progression of disease are made using nanoparticles. Gold nanorods functionalized with antibodies were used to quickly identify the infections in ornamental crops caused by Cymbidium mosaic virus (CymMV) and Odontoglossum ringspot virus (ORSV) (Lin et al. 2014). The SPR (surface plasmon resonance) was utilized to develop an immunosensor based on gold nanoparticles for the verification of *Tilletia indica* infection in wheat (Singh et al. 2010). Fluorescent-silica nanoparticles were utilized specifically in conjunction with *Xanthomonas axonopodis* pv. *vesicatoria* antigen to validate bacterial spot disease in solanaceous crops (Yao et al. 2009). Wang et al. (2010) developed an electrochemical sensor that detects *Sclerotinia sclerotiorum* using a modified gold electrode and Cu nanoparticles. The study found that adding copper nanoparticles to gold electrodes can enhance the electrochemical performance of salicylic acid sensors. Fernández-Baldo et al. (2010) used microfluidics to create a carbon-based screen-printed electrode for measuring *Botrytis cinerea* infection levels in apples, grapes, and pears. They utilized a competitive immunoassay approach to detect *B. cinerea* antigens immobilized on 3-aminopropyl-modified micro-magnetic beads. James (2013) used a lithographically patterned nanowire electrodeposition (LPNE) approach with a polypyrrole (PPy) nanoribbon–adapted chemiresistive sensor to fabricate a biosensor capable of detecting Cucumber mosaic virus (CMV) at a detection threshold of 10 ng ml^{-1}. A sensor for *Aspergillus niger* food detection was developed based on nanoparticles and a copper oxide nanolayer (Etefagh et al. 2013). It was created using sol-gel and spray pyrolysis procedures, and the biosensing qualities of the created nanostructure were measured using the electrical resistance concept. Zhang et al. (2013) used the idea of integrating immunomagnetic separation with upconversion nanoparticles (UCNPs) as biolabeling to detect plant viruses quickly and accurately. They used several antibodies to coat 100-nm-sized magnetic nanoparticles (MNPs) and used them as signal probes to detect Arabis mosaic virus (ArMV), Bean pod mottle virus (BPMV), and Tomato ringspot virus (ToRSV). The main advantages of this assay are its ability to detect many independent organic fluorophores simultaneously and eliminate nonspecific viral absorption, its low fluorescence background and high sensitivity, and its ability to be selective in tracing target viruses. In addition, the proposed assay is simple to use and can quickly—within 2.5 h—diagnose plant samples. Furthermore, Rettcher et al. (2015) revealed the way MNPs may be used to create an easy-to-use magnetic immunoassay for the detection of Grapevine fan leaf virus (GFLV). Monoclonal antibodies that recognize the GFLV capsid protein were immobilized onto immunofiltration columns for this experiment, and MNPs were then linked. To measure the titer of GFLV in leaf sap, immunofiltration with magnetic labeling in a double-antibody sandwich structure was used. A two-frequency magnetic activation field was used to create frequency signals in the resonant detecting coil that, mechanistically, matched the virus concentrations in the immunofiltration columns using a magnetic-frequency-mixing approach. Nanoparticles formed from semi-conductive metal oxide were synthesized and used to identify volatile organic compounds (VOCs). For example, the effectiveness of TiO_2 and SnO_2 nanoparticles was demonstrated in detecting the chemical p-ethylguaiacol produced by *Phytophthora cactorum* (Fang et al. 2014). In addition to being used to detect disease-marking VOCs, nanoparticles have the potential to be used to identify molecules that are pathogen-originating (Fang and Ramasamy 2015).

7.4 Smart Nanobiosensors for Crop Health Monitoring

As an exciting new area in the diagnosis of plant diseases and hazardous chemicals, smart nanobiosensors—like wearable sensors—can open the door to better plant biology research and agricultural practices. The sensors used in a wearable detecting system will be placed directly on the surfaces of various plant components, including stems and leaves. Research has shown a great deal of excitement in wearable sensors because of their affordability, adaptability, portability, and biocompatibility. In order to measure the development of plants and the temperature and humidity levels on the crop's leaf surface, a flexible strain sensor was designed (Nassar et al. 2018). Additional sophisticated leaf sensors, such as tape-based graphene sensors, are available to track the plant's water-related activity in immediate detail (Oren et al. 2017).

A wearable biosensor (based on a laser-induced graphene strategy) was developed for the electrochemical sensing of pesticides like methyl parathion (Zhao et al. 2020). AuNPs may be used to improve this detection method, and smartphones can wirelessly receive the detection results. Furthermore, a graphene-based method for detecting plant volatile organic compounds (VOCs) under tomato late blight infection and mechanical damage was developed (Li et al. 2021). This detection device can identify the emission of VOCs as a result of stress within four days of tomato infection and 1 h following mechanical injury. A polymer tattoo was created to detect the oxidative damage that prolonged exposure to ozone causes to fruits (Kim et al. 2020). Wearable sensors, as well as e-monitoring devices like e-tongues, e-noses, and e-eyes, have gained popularity in identifying and controlling plant diseases (Mohammad-Razdari et al. 2022).

7.5 Nanomaterial-Based Biosensor for Pesticide Residue Detection

Chemicals or biological substances used to combat pests and weeds are known as pesticides (Mfarrej and Rara 2019). The ongoing use of pesticides for crop production causes contamination of the food chain and natural habitat and other related risks. Conventional approaches that are frequently used to identify pesticides, such as GC/MS (Gas Chromatography-Mass Spectrometry), are very accurate but have low detection limits and need additional work as well as well-equipped labs. Additional benefits of using NMs include increased sensitivity, reduced sample size, and quicker detection. The binding of a specific molecule to a biosensor can be evaluated using many methods, including electric potential, color changes, and fluorescence. Array-based analysis involves fixing biomolecules to a substrate and measuring many analytes at once (Ghormade et al. 2011). On-site pesticides present in water from agricultural fields and other sources can be identified using AuNPs (Lisha and Pradeep 2009). Immunochromatographic test strips for the identification of several pesticides indicate AuNP function. Most of the pesticides are inhibitors of the enzyme acetylcholinesterase (AChE), which is crucial to human health. The enzymatic reaction may be used to create biosensors, such as a localized surface plasmon resonance fiber-optic biosensor that measures pesticide levels by monitoring AChE inhibition (Guo et al. 2009). In order to quantify the enzyme, it was adsorbed on the surface of AuNPs and the light changed upon blockage with a pesticide. Carbon-shell silver nanoparticles have been used to create a comparable non-enzymatic nanosensor for the colorimetric assessment of organophosphorus pesticides in agricultural goods (Bilal et al. 2022). Dimethoate, an organophosphate pesticide that suppresses AChE enzyme activity, can be hazardous to the environment and kill organisms. Therefore, developing sensitive and accurate pesticide residue monitoring systems is crucial (Liu et al. 2022). A nanosensor based on AgNPs (gold nanoparticles) combined with peroxidase-like activity was developed for dimethoate detection in plants (Hsu et al. 2017).

7.6 Toxin Detection by Nanosensors

Toxins can be detected early using traditional procedures. Traditional procedures are time-consuming and costly and have limited sensitivity. To address these limitations, biosensors have been developed (Girigoswami et al. 2021). Colorimetric sensors using AuNPs and plasma resonance were used to detect bacterial toxins in different products (Marin et al. 2021). A competitive immunoassay using gold nanoparticles was developed to detect organochlorine in samples (Lisa et al. 2009). Among the toxins created by inappropriate handling and processing conditions are bacterial toxins such as enterotoxins, endotoxins, and emetic toxins and mycotoxins like aflatoxins, fumonisins, zearalenone, and ochratoxin A (Fayyaz et al. 2022). Toxins generated by bacterial and fungal sources can cause a variety of foodborne diseases, painful gastrointestinal conditions, and cardiovascular abnormalities. A biosensor for ricin and staphylococcal enterotoxin B has been developed (Jia et al. 2022). Recently, viroids, prions, bacteria, protozoa, fungi, microbial toxins, and antibiotics from microbial sources have all been detected using biosensor techniques based on graphene and its derivatives, such as graphene QDs, graphene oxide, and graphene-based nanocomposites (Pourmadadi et al. 2023). *Aspergillus* sp. produces the mycotoxin B1, a substance with carcinogenic and mutagenic characteristics (Jallow and Govender 2021). Aptamers can detect aflatoxin B1, which contaminates agricultural goods.

Temperature, humidity, gas composition, and microbial agents are just a few of the many factors that nanobiosensors can measure to identify microbial contamination (Wlodkowic and Karpiński 2021). The functionality of nanobiosensors for detecting microbial contamination depends on the specific sensor type. However, the basic concept remains the same: the sensors identify alterations in their environment caused by microbial populations. Certain bioactive compounds or antibodies that possess a particular affinity for particular microorganisms are used by some of these sensors to enable specific detection. Others depend on nanoparticles, which undergo optical or electrical changes in response to the presence of microorganisms, producing a signal that can be detected (Nile et al. 2020).

7.7 Potential Applications of Nanosensors in Plant Disease Diagnosis

Plant diseases reduce agricultural output and threat food security for people and animals. To feed the global population, food production is expected to expand by 35–50% between 2012 and 2050 (FAO 2019). Plant diseases are responsible for around 16% of agricultural output losses worldwide (Ficke et al. 2018). Plant pathology research attempts to decrease crop loss through direct or indirect interventions (Li et al. 2020).

Plant disease diagnosis methods gained an innovative dimension due to nanobiosensor-based technologies by providing a cost-effective, easy-to-use, nondestructive, and minimally intrusive system with improved specificity and sensitivity. Nanomaterials can serve as a quick diagnostic tool for plant pathogens. There are two types of biorecognition components found in electrochemical biosensors: biocomplexing and biocatalytic. Although enzymes are frequently utilized as biorecognition components in biochemical-sensing programs, they typically function as labels in pathogen detection via a subsequent binding step (Choi et al. 2018). This is the process by which analytes react with molecules or how directed molecular components cause a biosensor to respond. A number of variables, including biosensor performance, assay design, material properties, volume, the makeup of the pathogen-containing sample, and the particular application, can affect this matter. Each of these variables influences the selection of the best measurement format (Jeerapan et al. 2020). A variety of electrochemical methods for pathogen detection can use functionalized electrodes. Among these approaches are impedimetric, potentiometric, conductimetric, ion charge/field effect, and amperometric techniques. The sort of electrochemical method used for pathogen detection is frequently classified by the measured signal, with signals delivered either continuously or occasionally (Saxena et al. 2018).

Biosensors can be employed directly or indirectly to identify pathogens or as indicators for particular diseases. *Tilletia indica* was detected in wheat using AuNP-based immunosensors (Singh et al. 2010). *Xanthomonas axonopodis* was identified in *Solanaceae* using fluorescent SiNPs (silicon nanoparticles) connected with Ab (Yao et al. 2009). Although the discipline is still in its infancy, it has a lot of promise that must be fulfilled in practical field applications. Using a nano-chip, a sort of microarray with a fluorescent oligo probe, is a key approach for detecting hybridization (López et al. 2009). *Phytoplasma aurantifolia* was detected using QD fluorescence resonance energy transfer–based immunosensors, which demonstrated great sensitivity (100%) (Rad et al. 2012). Plants develop distinct volatile signatures whenever they become infected with any pathogen, and the identification of these signature molecules may aid in confirming the existence of infection. Despite the fact that plants emit p-ethylguaiacol when infected with the fungus *Phytophthora cactorum*, SnO_2 and TiO_2 used an electrochemical sensing method to detect it on screen-printed carbon electrodes (Fang et al. 2014).

Recently, Huang et al. (2014) developed a QCM-based biosensor for the quick detection of Maize chlorotic mottling virus (MCMV). After coating the gold surfaces of QCM crystals, MCMV-specific antibodies were cross-linked to the surface to enable precise and accurate viral inoculum detection. The enhanced sensitivity, selectivity, and repeatability of the MCMV biosensor suggest that the piezoelectricity-driven QCM sensor would be a preferable option for accurate, dependable, and quick diagnoses. Safarpour et al. (2012) showed the way fluorescent nanoparticles might be used to create a diagnostic system that could detect *Polymyxa betae*. They used a monoclonal antibody that demonstrated specificity for *P. betae* to biofunctionalize quantum dots (QDs) in their work. The electrostatic interaction idea was used to dip an antibody that corresponded to the glutathione-S-transferase protein. Mutual kinship between the antigen and antibody caused resonance dipole–dipole coupling, which started the transmission of fluorescence resonance energy. It is important to note that this nanobiosensor analysis only needs 20 μl of biosample, which means it may be helpful in urgent circumstances where there is a shortage of samples. Bakhori et al. (2012) created a FRET-based DNA biosensor that uses QD to detect *Ganoderma boninense* in oil palm. Sandwich hybridization enabled the discovery of similar gene sequences by FRET signals in *G. boninense*. The developed DNA-based nanobiosensor is accurate and sufficiently sensitive for the detection of *G. boninense*. Rad et al. (2012) developed a QD-based nanobiosensor to detect *Candidatus Phytoplasma aurantifolia* in lime trees with high accuracy. This diagnostic technology uses an immunodominant membrane protein from phytoplasma to generate specific binding antibodies.

Carbon nanoparticles are an ideal material for developing fluorometry-driven nanosensors due to their unique properties (Zuo et al. 2015). A fluorometric nanosensor for Citrus tristeza virus (CTV) detection employing carbon nanoparticles was released (Shojaei et al. 2016). The researchers coupled cadmium-telluride QDs with an antibody produced against CTV. Additionally, attachment of the coat protein of CTV onto the surface of carbon NPs was conducted to establish an immunocomplex for effective quenching interaction. It was shown that once the target proteins were present in the plant, free coat proteins competed with the displacement of the carbon NP–tagged coat protein, resulting in the production of fluorescence and QD restoration. Compared to ELISA (Enzyme-Linked Immunosorbent Assay), the described fluorometry nanosensor was found to be much more sensitive and accurate. The aforementioned study findings suggest that FRET in combination with quantum dots has special qualities that make it easier to build and construct new, higher-quality sensors for use in the field.

A species-specific, label-free diagnostic approach for *Phytophthora ramorum* was developed (Yuksel et al. 2015). They achieved nucleic acid–based *P. ramorum* detection by using LATE-PCR (Linear-After-The-Exponential PCR) to amplify the Ypt1 (GTP-binding protein) region of *P. ramorum* and then hybridizing the adsorbed capture probes with the pathogen-originating Ypt1 gene sequence. They realized from their experimental results that a SERS-based nanosensor may function in conjunction with adenine-free capture probes and has the capacity to detect adenine-containing DNA targets from the infected leaves of *Rhododendron*. Additionally, a new isothermal amplification technique was used to show that the suggested assay is feasible in field settings. The findings showed that SERS-based nanosensors, which are primarily reliant on DNA–DNA hybrid formations, might be used in compact handheld Raman devices that enable quick pathogen detection in field settings. A quick, sensitive, and specific method was developed to identify several pathogens (*Pseudomonas syringae*, *Botrytis cinerea*, and *Fusarium oxysporum*) that infect *Arabidopsis thaliana* and tomato using SERS-labeled nanotags (Lau et al. 2016). The created nanosensor was discovered to be more sensitive than traditional PCR and capable of identifying pathogenic DNA in just two copies.

SnO_2 and TiO_2 NPs detection capacities were investigated in order to create an electrochemical sensor that would identify strawberry fruits infected with *Phytophthora cactorum* (Muthumalai et al. 2022). P-ethylguaiacol, a volatile compound generated by fungus in infected strawberry fruits, was detected electrochemically using screen-printed carbon electrodes improved with nanoparticles (SnO_2 and TiO_2). The results of the experiment demonstrated that an electrode encapsulated with nanoparticles had improved sensitivity. Furthermore, the p-ethylguaiacol signal's amperometric identification is unaffected by other plant volatiles (Fang et al. 2014). The study also showed that metal oxide nanoparticles offer a viable and affordable substitute for expensive electrode materials like platinum and gold used in amperometric sensor construction. An electrochemical nanosensor was recently created to identify *P. syringae* infections in plant samples prior to the onset of symptoms (Yuksel et al. 2015). The created sensor operated on the basis of isothermal nucleic acid amplification of the pathogen by recombinase polymerase. The procedure is more accurate than conventional PCR-driven diagnostic tests, according to the inventors' confirmation.

7.8 Challenges

The widespread implementation of agritechnologies has been impeded by a number of potential obstacles. The agricultural industry is a low-profit business that is difficult to capture, so the use of new technologies is very important to increase the profits. This chapter has explained the potential use of nanosensors and their uses in the detection and management of plant diseases. In order to guarantee high-quality food production, nanosensors include attaching or reacting with target species or strains and eventually transforming into detectable signals. This allows for the rapid, accurate, and early identification of disease-inducing substances. They also provide the advantages of accuracy, precision, mobility, and easy-to-use pathogen detection in field settings. Minimizing the expenses related to the production of nanobiosensors is still a difficulty. To increase the durability and versatility of nanobiosensors, alternative, less costly biological components like cells or enzymes, new stabilizing matrices, and chitosan for immobilizing nanoparticles may be developed (Alvarado et al. 2019). The effective application of nano-silica linked to validamycin is useful for the controlled pesticides that are dissolved in water. Recent developments in biosensor nanotechnology have helped solve issues in agriculture. Plant growers encounter a number of difficulties, including decreased stress, effective resource usage, and crop loss from environmental and disease circumstances. New tools are needed to monitor plant physiological and developmental responses in immediate detail. Nanomaterials in these emerging technologies transform plant chemical responses into digital data, which may be analyzed using biosensors and other electronic instruments (Giraldo et al. 2019). There are currently relatively few reports on marketed nanobiosensors used in the food and agricultural sectors. To fully utilize NPs as nanosensors in agriculture and related fields, fundamental research is still needed to improve current sensor technology, sensing strategies, and analytical tools and procedures for early disease identification.

7.9 Future Nanobiosensor Development

The world's yearly food production has to be properly handled, preserved, or converted into items with additional value in order to minimize food waste. The development of biosensors has gone through numerous stages. Originally distinct in earlier generations, the biocatalyst and transducers were gradually so tightly combined that the absence of one affected the function of the other. Modern biosensors eliminate the need for a mediator. Their electrode's surface exhibits a direct reduction in the enzyme (Zeinhom et al. 2018). Biosensors enable smart and precise agriculture for end users, farmers, and researchers. Agricultural biosensors are classified by the type of biorecognition system used. Substrate–enzyme–coenzyme, antigen–antibody, and nucleic acids' complementary sequences are examples of commonly used biorecognition systems. Additionally, cells from plants, animals, microbes, and tissues can all be used as parts of biorecognition. Crops are treated with a variety of fungicides in powder or liquid form (Dork et al. 2017). However, these medications also endanger human health as they inhibit AChE, which can lead to cognitive and intellectual issues as well as, perhaps, death (Chen et al. 2016). Computational biology, particularly deep learning and machine learning, will be essential to realizing the enormous promise of the next generation of nanobiosensor evolution for agriculture (Dork et al. 2017). More efficient sensors can result from the prediction and optimization of these interactions with the aid of computational modeling (Swierczewska et al. 2012). The enormous data produced by field-deployed nanobiosensor networks may be analyzed using machine learning methods. This can facilitate preemptive actions by enabling continuous tracking of pest epidemics (Buja et al. 2021). Models utilizing deep learning may be used to forecast agricultural yields, optimize fertilization and irrigation, and offer early disease epidemic prediction by combining sensor data with environmental and meteorological data (Elbasi et al. 2023).

7.10 Conclusion

The widespread application of nanobiosensors in the relevant agri-food sectors is steadily increasing. Interestingly, these fast and accurate sensors were used to identify plant diseases and residual pesticides, which helps anticipate and reduce yield losses in agriculture. Contemporary methods for fabricating nanosensors encompass both bottom-up and top-down methodologies. While the main fabrication methods in the top-down strategy are reportedly electron beam lithography, fiber pulling, electrodeposition, and chemical etching, in the bottom-up strategy, nanosensors can be made from smaller parts that are preassembled in specific designs to form a completed sensor. It has been demonstrated that standard fabrication procedures work well since they can be carried out in-house and generate accurate, uniformly sized particles more quickly and cheaply. However, a variety of targets, such as gases, ions, and biomolecules, are simply detectable. Performance restrictions are caused by the size, shape, and distributional dispersion of nanoparticles, which restricts current production processes. However, the creation of wide-ranging nanosensors and more affordable substitutes can increase their commercialization and eliminate the limitations on their accuracy and precision. It is necessary to create new analytical techniques to identify, assess, and control the effects of every nanomaterial in the natural environment. A comprehensive analysis of nanomaterials' size, dose, exposure duration, structures, surface chemistry, immune response, retention duration, accumulation, and other characteristics is necessary for determining their environmental effect. Considering their adaptability, durability, and dynamic use in agroecosystems, the multifaceted potential of these fast and specific identification systems deserves more investigation. The manipulation and global dissemination of such knowledge necessitates the development of a comprehensive databank and international cooperation in strategy, law, and policymaking.

Furthermore, the relevant authorities should offer precise guidelines and criteria for reducing the risks related to the use of nanotechnological products. Our hypothesis is that governments and academics can use nanotechnology to improve agriculture through focused studies and advancements on sustainable practices.

References

Ahmad T, Iqbal A, Halim SA, Uddin J, Khan A, El Deeb S et al (2022) Recent advances in electrochemical sensing of hydrogen peroxide (H_2O_2) released from cancer cells. Nano 12(9):1475. https://doi.org/10.3390/nano12091475

Algar WR, Krull UJ (2008) Quantum dots as donors in fluorescence resonance energy transfer for the bioanalysis of nucleic acids, proteins, and other biological molecules. Anal Bioanal Chem 391:1609–1618

Ali HS, Fakhri MA, Khalifa Z (2021) Optical and structural properties of the gold nanoparticles ablated by laser ablation in ethanol for biosensors. J Phys Conf Ser 1795(1):012065. https://doi.org/10.1088/1742-6596/1795/1/012065

Alvarado K, Bolaños M, Camacho C, Quesada E, Vega-Baudrit J (2019) Nanobiotechnology in agricultural sector: overview and novel applications. J Biomater Nanobiotechnol 10(02):120–141. https://doi.org/10.4236/jbnb.2019.102007

Antonacci A, Arduini F, Moscone D, Palleschi G, Scognamiglio V (2017) Nanostructured (bio) sensors for smart agriculture. TrAC Trends Anal Chem 98:95–103. https://doi.org/10.1016/j.trac.2017.10.022

Bakhori NM, Yusof NA, Abdullah AH, Hussein MZ (2012) Development of fluorescence-based DNA biosensor utilizing quantum dot for early detection of *Ganoderma boninense*. IPCBEE 48(26):138–142. https://doi.org/10.7763/IPCBEE.2012.V48.26

Bao Y, De Keersmaecker H, Corneillie S, Yu F, Mizuno H, Zhang G et al (2015) Tunable ratiometric fluorescence sensing of intracellular pH by aggregation induced emission-active hyperbranched polymer nanoparticles. Chem Mater 27(9):3450–3455. https://doi.org/10.1021/acs.chemmater.5b00858

Barry S, O'Riordan A (2015) Electrochemical nanosensors: advances and applications. Rep Electrochem 2016(6):1–14

Bhalla N, Jolly P, Formisano N, Estrela P (2016) Introduction to biosensors. Essays Biochem 60:1–8

Bilal S, Saleem S, Akhtar N, Sami AJ, Gokce G, Hayat A (2022) A novel cellulosic non-enzymatic nanosensor based on carbon shell silver (Ag@C) nanoparticles for colorimetric detection of Chlorpyrifos in agricultural products. JSFA Rep 2(10):511–523

Buja I, Sabella E, Monteduro AG, Chiriacò MS, De Bellis L, Luvisi A et al (2021) Advances in plant disease detection and monitoring: from traditional assays to in-field diagnostics. Sensors 21(6):2129. https://doi.org/10.3390/s21062129

Butnariu M, Butu A (2019) Plant nanobionics: application of nanobiosensors in plant biology. In: Plant nanobionics: approaches in nanoparticles, biosynthesis, and toxicity. Springer, Cham, pp 337–376. https://doi.org/10.1007/978-3-030-16379-2_12

Byrne B, Stack E, Gilmartin N, O'Kennedy R (2009) Antibody-based sensors: principles, problems and potential for detection of pathogens and associated toxins. Sensors 9:4407–4445. https://doi.org/10.3390/s90604407

Chakroborty S, Pal K, Nath N, Singh V, Barik A, Soren S et al (2023) Sustainable synthesis of multifunctional nanomaterials from rice wastes: a comprehensive review. Environ Sci Pollut Res 30(42):95039–95053. https://doi.org/10.1007/s11356-023-29235-9

Chen J, Park B (2016) Recent advancements in nanobioassays and nanobiosensors for foodborne pathogenic bacteria detection. J Food Prot 79:1055–1069

Chen M, Hou C, Huo D, Bao J, Fa H, Shen C (2016) An electrochemical DNA biosensor based on nitrogen-doped graphene/Au nanoparticles for human multidrug resistance gene detection. Biosens Bioelectron 85:684–691. https://doi.org/10.1016/j.bios.2016.05.051

Choi S, Han SI, Jung D, Hwang HJ, Lim C, Bae S et al (2018) Highly conductive, stretchable and biocompatible Ag–Au core–sheath nanowire composite for wearable and implantable bioelectronics. Nat Nanotechnol 13(11):1048–1056. https://doi.org/10.1038/s41565-018-0226-8

Clausmeyer J, Schuhmann W (2016) Nanoelectrodes: applications in electrocatalysis, single-cell analysis and high-resolution electrochemical imaging. TrAC Trends Anal Chem 79:46–59

Cui HF, Wu WW, Li MM, Song X, Lv Y, Zhang TT (2018) A highly stable acetylcholinesterase biosensor based on chitosan-TiO2-graphene nanocomposites for detection of organophosphate pesticides. Biosens Bioelectron 99:223–229. https://doi.org/10.1016/j.bios.2017.07.068

De Boer SH, López MM (2012) New grower-friendly methods for plant pathogen monitoring. Annu Rev Phytopathol 50(1):197–218

Dork D, Anastassiades M, Spitzke M, Dork D (2017) Development of a QuEChERS-based method for the simultaneous determination of acidic pesticides, their esters, and conjugates following alkaline hydrolysis. J Agric Food Chem 65(6):1296–1305. https://doi.org/10.1021/acs.jafc.6b05407

Elbasi E, Zaki C, Topcu AE, Abdelbaki W, Zreikat AI, Cina E et al (2023) Crop prediction model using machine learning algorithms. Appl Sci 13(16):9288. https://doi.org/10.3390/app13169288

Ellinger D, Voigt CA (2014) The use of nanoscale fluorescence microscopic to decipher cell wall modifications during fungal penetration. Front Plant Sci 5:270. https://doi.org/10.3389/fpls.2014.00270

Elsharkawy MM, Alotibi FO, Al-Askar AA, Adnan M, Kamran M, Abdelkhalek A, Behiry SI, Saleem MH, Ahmad AA, Khedr AA (2022) Systemic resistance induction of potato and tobacco plants against potato virus Y by *Klebsiella oxytoca*. Life 12(10):1521. https://doi.org/10.3390/life12101521

Esker PD, Sparks A, Amorim L, Bergamin Filho A, Caffi T, Castilla N et al (2018) Concepts, approaches, and avenues for modelling crop health and crop losses. Eur J Agron 100:4–18. https://doi.org/10.1016/j.eja.2018.04.003

Etefagh R, Azhir E, Shahtahmasebi N (2013) Synthesis of CuO nanoparticles and fabrication of nanostructural layer biosensors for detecting *Aspergillus niger* fungi. Sci Iran 20:1055–1058

Fang Y, Ramasamy RP (2015) Current and prospective methods for plant disease detection. Biosensors 5(3):537–561. https://doi.org/10.3390/bios5030537

Fang Y, Umasankar Y, Ramasamy RP (2014) Electrochemical detection of p-ethylguaiacol, a fungi infected fruit volatile using metal oxide nanoparticles. Analyst 139:3804–3810

FAO (2019) Moving forward on food loss and waste reduction. The State of Food and Agriculture 2019. https://www.fao.org/3/ca6030en/CA6030EN.pdf

Fayyaz K, Nawaz A, Olaimat AN, Akram K, Farooq U, Fatima M et al (2022) Microbial toxins in fermented foods: health implications and analytical techniques for detection. J Food Drug Anal 30(4):523–537. https://doi.org/10.38212/2224-6614.3431

Fernández-Baldo MA, Messina GA, Sanz MI, Raba J (2010) Microfluidic immunosensor with micromagnetic beads coupled to carbon-based screen-printed electrodes (SPCEs) for determination of *Botrytis cinerea* in tissue of fruits. J Agric Food Chem 10:11201–11206

Ficke A, Cowger C, Bergstrom G, Brodal G (2018) Understanding yield loss and pathogen biology to improve disease management: *Septoria nodorum* blotch-a case study in wheat. Plant Dis 102:696–707

Ghormade V, Deshpande MV, Paknikar KM (2011) Perspectives for nano-biotechnology enabled protection and nutrition of plants. Biotechnol Adv 29:792–803. https://doi.org/10.1016/j.biotechadv.2011.06.007

Giraldo JP, Wu H, Newkirk GM, Kruss S (2019) Nanobiotechnology approaches for engineering smart plant sensors. Nat Nanotechnol 14:541–553. https://doi.org/10.1038/S41565-019-0470-6

Girigoswami A, Ghosh MM, Pallavi P, Ramesh S, Girigoswami K (2021) Nanotechnology in detection of food toxins–focus on the dairy products. Biointerf Res Appl Chem 11(6):14155–14172. https://doi.org/10.33263/BRIAC116.1415514172

Grecco HE, Verveer PJ (2011) FRET in cell biology: still shining in the age of super-resolution? ChemPhysChem 12:484–490. https://doi.org/10.1002/cphc.201000795

Guo YR, Liu SY, Gui WJ, Zhu GN (2009) Gold immunochromatographic assay for simultaneous detection of carbofuran and triazophos in water samples. Anal Biochem 389:32–39. https://doi.org/10.1016/j.ab.2009.03.020

Hsu CW, Lin ZY, Chan TY, Chiu TC, Hu CC (2017) Oxidized multiwalled carbon nanotubes decorated with silver nanoparticles for fluorometric detection of dimethoate. Food Chem 224:353–358. https://doi.org/10.1016/j.foodchem.2016.12.095

Huang X, Xu JM, Ji HF, Li GF, Chen HJ (2014) Quartz crystal microbalance-based biosensor for rapid and sensitive detection of maize chlorotic mottle virus. Anal Methods 6:4530–4536

Huang X, Zhu Y, Kianfar E (2021) Nano biosensors: properties, applications and electrochemical techniques. J Mater Res Technol 12:1649–1672

Jallow S, Govender NP (2021) Ibrexafungerp: a first-in-class oral triterpenoid glucan synthase inhibitor. J Fungi 7(3):163. https://doi.org/10.3390/jof7030163

James C (2013) Polypyrrole nanoribbon based chemiresistive immunosensors for viral plant pathogen detection. Anal Methods 5:3497–3502

James A, Franco F, Merca EM, Rodriguez S, Johny F, Balidon-Veronica P et al (2019) DNA-based electrochemical nanobiosensor for the detection of *Phytophthora palmivora* (Butler) Butler, causing black pod rot in cacao (*Theobroma cacao* L.) pods. Physiol Mol Plant Pathol 107:14–20. https://doi.org/10.1016/j.pmpp.2019.04.004

Jasrotia P, Kashyap PL, Bhardwaj AK, Kumar S, Singh GP (2018) Scope and applications of nanotechnology for wheat production: a review of recent advances. Wheat Barley Res 10(1):1–14. https://doi.org/10.25174/2249-4065/2018/76672

Jeerapan I, Sonsa-Ard T, Nacapricha D (2020) Applying nanomaterials to modern biomedical electrochemical detection of metabolites, electrolytes, and pathogens. Chemosensors 8(3):71. https://doi.org/10.3390/chemosensors8030071

Jia X, Wang K, Li X, Liu Z, Liu Y, Xiao R et al (2022) Highly sensitive detection of three protein toxins via SERS-lateral flow immunoassay based on SiO_2@ Au nanoparticles. Nanomed Nanotechnol Biol Med 41:102522. https://doi.org/10.1016/j.nano.2022.102522

Johnson MS, Sajeev S, Nair RS (2021) Role of nanosensors in agriculture. In: Proceedings of 2nd IEEE International Conference on Computational Intelligence and Knowledge Economy ICCIKE 2021, pp 58–63. https://doi.org/10.1109/ICCIKE51210.2021.9410709

Kahraman M, Mullen ER, Korkmaz A, Wachsmann-Hogiu S (2017) Fundamentals and applications of SERS-based bioanalytical sensing. Nano. https://doi.org/10.1515/nanoph-2016-0174

Kashyap PL, Rai P, Sharma S, Chakdar H, Kumar S, Pandiyan K, Srivastava AK (2016) Nanotechnology for the detection and diagnosis of plant pathogens. In: Ranjan S et al (eds) Nanoscience in food and agriculture 2, sustainable agriculture reviews 21. Springer, Basel, pp 253–276. https://doi.org/10.1007/978-3-319-39306-3_8

Kashyap PL, Kumar S, Srivastava AK (2017a) Nanodiagnostics for plant pathogens. Environ Chem Lett 15:7–13. https://doi.org/10.1007/s10311-016-0580-4

Kashyap PL, Rai P, Srivastava AK, Kumar S (2017b) Trichoderma for climate resilient agriculture. World J Microbiol Biotechnol 33:155. https://doi.org/10.1007/s11274-017-2319-1

Kashyap PL, Srivastava AK, Tiwari SP, Kumar S (2018) Microbes for climate resilient agriculture. Wiley, Hoboken, p 376

Kaur SI, Kashyap PL, Kang SS, Sharma A (2019) Detection and diagnosis of seed-borne viruses and virus-like pathogens. In: Kumar R, Gupta A (eds) Seed borne diseases of agricultural crops: detection, diagnosis & management. Springer. https://doi.org/10.1007/978-981-32-9046-4_7

Khandelwal A, Joshi R, Mukherjee P, Singh SD, Shrivastava M (2019) Use of bio-based nanoparticles in agriculture. Nanotechnol Agric Adv Sustain Agric 2019:89–100. https://doi.org/10.1007/978-981-32-9370-0_6

Kim JJ, Fan R, Allison LK, Andrew TL (2020) On-site identification of ozone damage in fruiting plants using vapor-deposited conducting polymer tattoos. Sci Adv 6:eabc3296

Kumar A, Sharma PK, Singh S, Verma JP (2021) Impact of engineered nanoparticles on microbial communities, soil health and plants. Adv Sci Technol Innov 2021:201–215. https://doi.org/10.1007/978-3-030-66956-0_14

Kuswandi B, Ensafi AA (2019) Perspective-paper-based biosensors: trending topic in clinical diagnostics developments and commercialization. J Electrochem Soc 167:037509

Lau HY, Wang Y, Wee EJH, Botella JR, Trau M (2016) Field demonstration of a multiplexed point-of-care diagnostic platform for plant pathogens. Anal Chem 88(16):8074–8081. https://doi.org/10.1021/acs.analchem.6b01551

Le DT, Vu NT (2017) Progress of loop-mediated isothermal amplification technique in molecular diagnosis of plant diseases. Appl Biol Chem 60:169–180. https://doi.org/10.1007/s13765-017-0267-y

Li Z, Yu T, Paul R, Fan J, Yang Y, Wei Q (2020) Agricultural nanodiagnostics for plant diseases: recent advances and challenges. Nanosc Adv 2:3083–3094

Li Z, Zhou J, Dong T, Xu Y, Shang Y (2021) Application of electrochemical methods for the detection of abiotic stress biomarkers in plants. Biosens Bioelectron 182:113105

Li H, Qi H, Chang J, Gai P, Li F (2022) Recent progress in homogeneous electrochemical sensors and their designs and applications. TrAC Trends Anal Chem 156:116712. https://doi.org/10.1016/j.trac.2022.116712

Lin H-Y, Huang C-H, Lu S-H, Kuo I-T, Chau L-K (2014) Direct detection of orchid viruses using nanorod-based fiber optic particle plasmon resonance immunosensor. Biosens Bioelectron 51:371–378

Lisa M, Chouhan RS, Vinayaka AC, Manonmani HK, Thakur MS (2009) Gold nanoparticles based dipstick immunoassay for the rapid detection of dichlorodiphenyltrichloroethane: an organochlorine pesticide. Biosens Bioelectron 25(1):224–227. https://doi.org/10.1016/j.bios.2009.05.006

Lisha KP, Pradeep T (2009) Enhanced visual detection of pesticides using gold nanoparticles. J Environ Sci Health Part B 44:697–705. https://doi.org/10.1080/03601230903163814

Liu P, Zhao M, Zhu H, Zhang M, Li X, Wang M et al (2022) Dual-mode fluorescence and colorimetric detection of pesticides realized by integrating stimulus responsive luminescence with oxidase-mimetic activity into cerium-based coordination polymer nanoparticles. J Hazard Mater 423:127077. https://doi.org/10.1016/j.jhazmat.2021.127077

López MM, Llop P, Olmos A, Marco-Noales E, Cambra M, Bertolini E (2009) Are molecular tools solving the challenges posed by detection of plant pathogenic bacteria and viruses? Mol Biol 11:13–46

Malhotra BD, Ali MA (2018) Nanomaterials in biosensors: fundamentals and applications. In: Nanomaterials for biosensors. Elsevier, pp 1–74

Marin M, Nikolic MV, Vidic J (2021) Rapid point-of-need detection of bacteria and their toxins in food using gold nanoparticles. Compr Rev Food Sci Food Saf 20(6):5880–5900. https://doi.org/10.1111/1541-4337.12839

Martinelli F, Scalenghe R, Davino S, Panno S, Scuderi G, Ruisi P, Villa P, Stroppiana D, Boschetti M, Goulart LR, Davis CE, Dandekar AM (2015) Advanced methods of plant disease detection: a review. Agron Sustain Dev 35:1–25. https://doi.org/10.1007/s13593-014-0246-1

Mfarrej MFB, Rara FM (2019) Competitive, sustainable natural pesticides. Acta Ecol Sin 39(2):145–151. https://doi.org/10.1016/j.chnaes.2018.08.005

Mohammad-Razdari A, Rousseau D, Bakhshipour A, Taylor S, Poveda J, Kiani H (2022) Recent advances in E-monitoring of plant diseases. Biosens Bioelectron:113953

Mondal R, Dam P, Chakraborty J, Paret ML, Katı A, Altuntas S, Sarkar R, Ghorai S, Gangopadhyay D, Mandal AK, Husen A (2022) Potential of nanobiosensor in sustainable agriculture: the state-of-art. Heliyon 8(12):e12207. https://doi.org/10.1016/j.heliyon.2022.e12207. PMID: 36578430; PMCID: PMC9791828

Muthumalai K, Gokila N, Haldorai Y, Rajendra Kumar RT (2022) Nanosensors for crop protection: design and fabrication. Nanosens Smart Agric 2022:403–422. https://doi.org/10.1016/B978-0-12-824554-5.00041-0

Nadporozhskaya M, Kovsh N, Paolesse R, Lvova L (2022) Recent advances in chemical sensors for soil analysis: a review. Chemosensors 10:35

Naresh V, Lee N (2021) A review on biosensors and recent development of nanostructured materials-enabled biosensors. Sensors 21:1109

Nassar JM, Khan SM, Villalva DR, Nour MM, Almuslem AS, Hussain MM (2018) Compliant plant wearables for localized microclimate and plant growth monitoring. Npj Flex Electron 2:1–12

Neme K, Nafady A, Uddin S, Tola YB (2021) Application of nanotechnology in agriculture, postharvest loss reduction and food processing: food security implication and challenges. Heliyon 7:e08539

Nezhad AS (2014) Future of portable devices for plant pathogen diagnosis. Lab Chip 14:2887–2904

Nile SH, Baskar V, Selvaraj D, Nile A, Xiao J, Kai G (2020) Nanotechnologies in food science: applications, recent trends, and future perspectives. Nano-micro Lett 12:45–34. https://doi.org/10.1007/s40820-020-0383-9

Oren S, Ceylan H, Schnable PS, Dong L (2017) High-resolution patterning and transferring of graphene-based nanomaterials onto tape toward roll-to-roll production of tape-based wearable sensors. Adv Mater Technol 2:1700223

Pandey G (2018) Challenges and future prospects of agri-nanotechnology for sustainable agriculture in India. Environ Technol Innov 11:299–307. https://doi.org/10.1016/J.ETI.2018.06.012

Pourmadadi M, Shamsabadipour A, Aslani A, Eshaghi MM, Rahdar A, Pandey S (2023) Development of Polyvinylpyrrolidone-based nanomaterials for biosensors applications: a review. Inorg Chem Commun 152:110714. https://doi.org/10.1016/j.inoche.2023.110714

Qureshi A, Kang WP, Davidson JL, Gurbuz Y (2009) Review on carbon-derived, solid-state, micro and nano sensors for electrochemical sensing applications. Diam Relat Mater 18:1401–1420

Rad F, Mohsenifar A, Tabatabaei M, Safarnejad MR, Shahryari F, Safarpour H, Foroutan A, Mardi M, Davoudi D, Fotokian M (2012) Detection of candidatus *Phytoplasma aurantifolia* with a quantum dots FRET-based biosensor. J Plant Pathol 94(3):525–534

Ray M, Ray A, Dash S, Mishra A, Gopinath Achary K, Nayak S, Singh S (2017) Fungal disease detection in plants: traditional assays, novel diagnostic techniques and biosensors. Biosens Bioelectron 87:708–723

Rettcher S, Jungk F, Kühn C, Krause H-J, Nölke G, Commandeur U, Fischer R, Schillberg S, Schröper F (2015) Simple and portable magnetic immunoassay for rapid detection and sensitive quantification of plant viruses. Appl Environ Microbiol 81:3039–3048. https://doi.org/10.1128/AEM.03667-14

Safarpour H, Safarnejad M, Tabatabaei M, Mohsenifar A, Rad F, Basirat M, Shahryari F, Hasanzadeh F (2012) Development of a quantum dots FRET-based biosensor for efficient detection of *Polymyxa betae*. Can J Plant Pathol 34(4):507–515. https://doi.org/10.1080/07060661.2012.709885

Saravanadevi K, Renuga Devi N, Dorothy R, Joany RM, Rajendran S, Nguyen TA (2022) Nanotechnology for agriculture: an introduction. Nanosens Smart Agric 3–23. https://doi.org/10.1016/B978-0-12-824554-5.00013-6

Saxena KK, Qian J, Reynaerts D (2018) A review on process capabilities of electrochemical micromachining and its hybrid variants. Int J Mach Tools Manuf 127:28–56. https://doi.org/10.1016/j.ijmachtools.2018.01.004

Sharma S, Rai P, Rai S, Srivastava M, Kashyap PL, Sharma A, Kumar S (2017) Genomic revolution in crop disease diagnosis: a review. In: Singh SS (ed) Plants and microbes in an ever-changing environment. Nova Science Publishers, Inc, Hauppauge, pp 257–293

Sharma P, Pandey V, Sharma MMM, Patra A, Singh B, Mehta S, Husen A (2021) A review on biosensors and nanosensors application in agroecosystems. Nanoscale Res Lett 16:1–24

Shojaei TR, Salleh MAM, Sijam K, Rahim RA, Mohsenifar A, Safarnejad R, Tabatabaei M (2016) Fluorometric immunoassay for detecting the plant virus Citrus tristeza using carbon nanoparticles acting as quenchers and antibodies labeled with CdTe quantum dots. Microchim Acta 183:2277. https://doi.org/10.1007/s00604-016-1867-7

Singh RP (2021) Recent trends, prospects, and challenges of nanobiosensors in agriculture. In: Biosensors in agriculture: recent trends and future perspectives. Concepts and strategies in plant sciences. Springer, Cham

Singh S, Singh M, Agrawal VV, Kumar A (2010) An attempt to develop surface plasmon resonance based immunosensor for Karnal bunt (*Tilletia indica*) diagnosis based on the experience of nano-gold based lateral flow immune-dipstick test. Thin Solid Films 519:1156–1159

Singh R, Kumar S, Kashyap PL, Srivastava AK, Mishra S, Sharma AK (2014) Identification and characterization of microsatellite from *Alternaria brassicicola* to assess cross-species transferability and utility as a diagnostic marker. Mol Biotechnol 56:1049–1059

Souiri A, Zemzami M, Amzazi S, Ennaji MM (2014) Polyclonal and monoclonal antibody-based methods for detection of plant viruses. Eur J Sci Res 123(3):281–295

Srivastava AK, Dev A, Karmakar S (2018) Nanosensors and nanobiosensors in food and agriculture. Environ Chem Lett 16(1):161–182. https://doi.org/10.1007/s10311-017-0674-7

Stanisavljevic M, Krizkova S, Vaculovicova M, Kizek R, Adam V (2015) Quantum dots-fluorescence resonance energy transfer-based nanosensors and their application. Biosens Bioelectron 74(2015):562–574

Sun LL, Song YH, Wang L, Guo CL, Sun YJ, Liu Z, Li Z (2008) Ethanol-induced formation of silver nanoparticle aggregates for highly active SERS substrates and application in DNA detection. J Phys Chem C 112:1415–1422

Swierczewska M, Liu G, Lee S, Chen X (2012) High-sensitivity nanosensors for biomarker detection. Chem Soc Rev 41(7):2641–2655. https://doi.org/10.1039/c1cs15238f

Thakur M, Wang B, Verma ML (2022) Development and applications of nanobiosensors for sustainable agricultural and food industries: recent developments, challenges and perspectives. Environ Technol Innov:102371

Verma ML (2017) Nanobiotechnology advances in enzymatic biosensors for the Agri-food industry. Environ Chem Lett 15(4):555–560. https://doi.org/10.1007/s10311-017-0640-4

Verma ML, Rani V (2021) Biosensors for toxic metals, polychlorinated biphenyls, biological oxygen demand, endocrine disruptors, hormones, dioxin, phenolic and organophosphorus compounds: a review. Environ Chem Lett 19(2):1657–1666. https://doi.org/10.1007/s10311-020-01116-4

Verma D, Ranjan KR, Mukherjee MD, Solanki PR (2022) Bioinspired synthesis of hematite nanoparticles-reduced graphene oxide composite for application in bisphenol a detection: a new in-sight. Biosens Bioelectron 11:100217. https://doi.org/10.1016/j.biosx.2022.100217

Virk V, Deepak H, Taneja K, Srivastava R, Giri S (2024) Amelioration in nanobiosensors for the control of plant diseases: current status and future challenges. Front Nanotechnol 6:1310165. https://doi.org/10.3389/fnano.2024.1310165

Wang Z, Wei F, Liu SY, Xu Q, Huang JY, Dong XY, Yu JH, Yang Q, Zhao YD, Chen H (2010) Electrocatalytic oxidation of phytohormone salicylic acid at copper nanoparticles-modified gold electrode and its detection in oilseed rape infected with fungal pathogen *Sclerotinia sclerotiorum*. Talanta 80:1277–1281

Wei H, Abtahi SMH, Vikesland PJ (2015) Plasmonic colorimetric and SERS sensors for environmental analysis. Environ Sci Nano 2:120–135. https://doi.org/10.1039/c4en00211c

Wlodkowic D, Karpiński TM (2021) Live-cell systems in real-time biomonitoring of water pollution: practical considerations and future perspectives. Sensors 21(21):7028. https://doi.org/10.3390/s21217028

Kwak S-Y, Wong MH, Lew TTS, Bisker G, Lee MA, Kaplan A, Dong J, Liu AT, Koman VB, Sinclair R, Hamann C, Strano MS (2017) Nanosensor technology applied to living plant systems. Ann Rev Anal Chem 10:113–140

Wu H, Nißler R, Morris V, Herrmann N, Hu P, Jeon SJ, Kruss S, Giraldo JP (2020a) Monitoring plant health with near-infrared fluorescent H_2O_2 nanosensors. Nano Lett 20(2020):2432–2442. https://doi.org/10.1021/acs.nanolett.9b05159

Wu H, Nißler R, Morris V, Herrmann N, Hu P, Jeon SJ et al (2020b) Monitoring plant health with near-infrared fluorescent H_2O_2 nanosensors. Nano Lett 20(4):2432–2442. https://doi.org/10.1021/acs.nanolett.9b05159

Yao KS, Li SJ, Tzeng KC, Cheng TC, Chang CY, Chiu CY, Liao CY, Hsu JJ, Lin ZP (2009) Fluorescence silica nanoprobe as a biomarker for rapid detection of plant pathogens. Adv Mater Res 79–82:513–516

Ye Y, Cota-Ruiz K, Hernández-Viezcas JA, Valdés C, Medina-Velo IA, Turley RS, Peralta-Videa JR, Gardea-Torresdey JL (2020) Manganese nanoparticles control salinity-modulated molecular responses in *Capsicum annuum* L. Through priming: a sustainable approach for agriculture. ACS Sustain Chem Eng 8:1427–1436. https://doi.org/10.1021/acssuschemeng.9b05615

Yılmaz F, Saylan Y, Akgonullu S, Çimen D, Derazshamshir A, Bereli N et al (2017) Surface plasmon resonance based nanosensors for detection of triazinic pesticides in agricultural foods. In: New pesticides and soil sensors. Academic, pp 679–718

Yu J, Wang D, Geetha N, Khawar KM, Jogaiah S, Mujtaba M (2021) Current trends and challenges in the synthesis and applications of chitosan-based nanocomposites for plants: a review. Carbohydr Polym 261:117904. https://doi.org/10.1016/j.carbpol.2021.117904

Yuksel S, Schwenkbier L, Pollok S, Weber K, Cialla-May D, Popp J (2015) Label-free detection of *Phytophthora ramorum* using surface-enhanced Raman spectroscopy. Analyst 140:7254–7262

Zeinhom MMA, Wang Y, Song Y, Zhu MJ, Lin Y, Du D (2018) A portable smart-phone device for rapid and sensitive detection of *E. coli* O157: H7 in yoghurt and egg. Biosens Bioelectron 99:479–485. https://doi.org/10.1016/j.bios.2017.08.002

Zhang M, Chen W, Chen X, Zhang Y, Lin X, Wu Z, Li M (2013) Multiplex immunoassays of plant viruses based on functionalized upconversion nanoparticles coupled with immunomagnetic separation. J Nanomater. https://doi.org/10.1155/2013/317437

Zhao F, He J, Li X, Bai Y, Ying Y, Ping J (2020) Smart plant-wearable biosensor for in-situ pesticide analysis. Biosens Bioelectron 170:112636

Zuo P, Lu X, Sun Z et al (2015) A review on syntheses, properties, characterization and bioanalytical applications of fluorescent carbon dots. Microchim Acta 183:519–542. https://doi.org/10.1007/s00604-015-1705-3

Chapter 8
Nanobiosensors in Plant Environmental Stress Management: Methods and Current Achievements

Vinod Goyal, Sarita, and Shweta Mehrotra

Abstract Nanobiosensors are emerging as crucial tools in agriculture, revolutionizing stress management and resource optimization. These sensors enable early detection of plant stress at the molecular level, such as identifying biomarkers that signal drought or disease. This early detection allows for timely interventions, such as adjusting irrigation or applying fertilizers, leading to more efficient resource use and reduced crop losses. Nanobiosensors also contribute to precision agriculture by providing continuous real-time monitoring of environmental factors like soil moisture, temperature, and nutrient levels, allowing for targeted applications of water, fertilizers, and pesticides. The integration of nanobiosensors with Internet of Things (IoT) technology further enhances their effectiveness by enabling remote monitoring and data collection, which aids in decision-making. This integration facilitates real-time tracking of plant health and soil conditions, helping farmers optimize agricultural practices and improve yields while minimizing environmental impact. Furthermore, nanobiosensors play a significant role in research, helping scientists understand plant responses to various stresses and develop stress-resistant crop varieties. Nanobiosensors are particularly valuable in combating climate change impacts on agriculture. They help monitor drought stress, detect pathogens, and assess soil health, which is vital for sustainable farming practices. By enabling precise, data-driven interventions, these sensors can reduce input costs, enhance productivity, and reduce the environmental footprint of agriculture. As research continues to advance, nanobiosensors hold immense potential to improve agricultural resilience, ensuring food security in a changing climate and fostering sustainable agricultural practices for the future.

V. Goyal · Sarita
Department of Botany and Plant Physiology, CCS Haryana Agricultural University, Hisar, Haryana, India

S. Mehrotra (✉)
Department of Biotechnology, Guru Jambheshwar University of Science & Technology, Hisar, India
e-mail: shwetamehrotragoyal@gmail.com

J.-T. Chen (ed.), *Nanobiosensors for Crop Monitoring and Precision Agriculture*, Sustainability Sciences in Asia and Africa,
https://doi.org/10.1007/978-981-96-8335-2_8

Keywords Agriculture · Biomarkers · Environmental stress · Internet of Things · Nanobiosensors · Pathogen · Real-time monitoring · Soil quality

8.1 Introduction

Nanobiosensors represent cutting-edge analytical instruments that merge biological recognition components, including enzymes, antibodies, or nucleic acids, with transducers at the nanoscale. These advanced devices are engineered to identify specific biological or chemical indicators and convert them into measurable electronic outputs. Integrating nanotechnology in these sensors significantly enhances their sensitivity, precision, and portability, facilitating ongoing, instantaneous monitoring of multiple environmental variables.

Nanobiosensors are rapidly emerging as vital tools in plant environmental stress management, providing innovative methods for real-time monitoring and assessment. Nanobiosensors are revolutionizing agriculture by offering innovative solutions for precision farming and sustainable practices. These devices enable real-time, accurate detection of critical factors affecting plant health, such as nutrient availability, soil conditions, and biotic or abiotic stresses, as shown in Fig. 8.1. Advanced nanosensors offer data-driven analysis of soil nutrients, moisture levels, and environmental factors, empowering farmers to maximize resource efficiency in fertilizer and water usage. These microscopic devices can detect subtle shifts in soil composition, such as nitrogen or phosphorus concentrations, enabling precise application methods that prevent overuse and mitigate the risk of environmental degradation (Zain et al. 2023). Environmental stressors like drought, high salinity, or pathogen

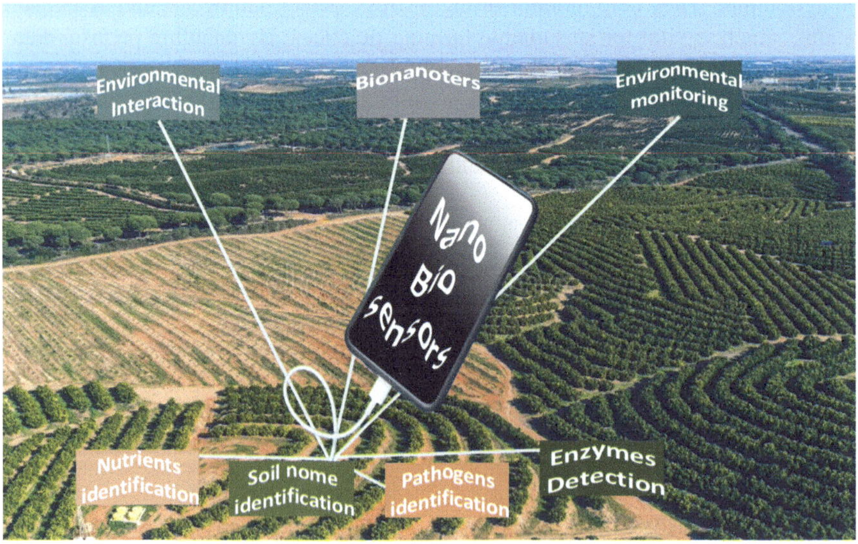

Fig. 8.1 Representation of nanobiosensors and their application in agriculture

infection trigger molecular-level changes in plants, including fluctuations in metabolite content and increased oxidative stress. The rapid identification of these biomarkers by nanobiosensors enables early detection and timely intervention. This advanced technology contributes to minimizing crop losses and enhancing the resilience of plants against adverse conditions (Miguel-Rojas and Pérez-de-Luque 2023). Researchers have developed various biosensors for detecting plant stress markers. Hu et al. (2020) created an electrochemical biosensor that detects IAA and SA in soybean seedlings experiencing saline stress, with detection limits of 1.99 μM and 3.30 μM, respectively. This system's sensitivity was improved by incorporating multiwall carbon nanotubes (MWNT) and carbon black. In a separate study, Wang et al. (2019a, b) engineered an electrochemical-based immune biosensor to identify vitronectin-like proteins (VN), which are biomarkers present on plant cell surfaces during heavy metal stress. Their system achieved detection limits of 18.5 nmol/L for cadmium and 25.6 nmol/L for lead. Additional electrochemical biosensors capable of detecting phytohormones under various stress conditions have been reported by other researchers (He et al. 2020; Wang et al. 2020; Cao et al. 2020). Moreover, a surface plasmon resonance-based technique for abscisic acid (ABA) detection was developed, employing AuNPs-aptamer conjugates and reaching a detection limit of 0.33 μM (Wang et al. 2017).

Transducers form the basis for classifying biosensors into several categories. These include electrochemical biosensors, which encompass amperometric, impedimetric, and potentiometric types; calorimetric biosensors; piezoelectric biosensors, such as those utilizing quartz crystal microbalance or acoustic wave technology; and optical biosensors, which employ various principles like reflection, refraction, absorption, transmission, fluorescence, surface plasmon resonance, and waveguide (Malhotra and Ali 2018). The classification of biosensors encompasses various categories based on their recognition elements, including aptamer-based, enzymatic, nucleic acid-based, and oligonucleotide-based systems. Additionally, biosensors can be categorized as immunosensors, which utilize antigen-antibody interactions or biomarker approaches, as well as whole-cell and organelle-based sensors, among other types (Malhotra and Ali 2018).

8.2 Role of Nanobiosensors

8.2.1 Early Detection of Stress

In agriculture, nanobiosensors serve a crucial function in detecting molecular-level stress indicators, allowing for prompt and accurate interventions. These advanced tools can identify subtle changes in plant physiology, including the generation of reactive oxygen species (ROS), considered early warning signs of stress. By enabling farmers to recognize potential problems before visible damage manifests, these sensors help maintain crop productivity and quality. Nanobiosensors play a crucial role in identifying stress indicators at the molecular level, facilitating early interventions that can prevent crop losses.

8.2.1.1 Molecular-Level Monitoring

Nanobiosensors can detect specific biomarkers such as proteins, nucleic acids, or metabolites that indicate stress conditions (Sharma et al. 2020). By utilizing nanomaterials with high sensitivity and specificity, these sensors can monitor changes in plant physiology before visible symptoms manifest, allowing for the detection of stress due to drought, salinity, or disease. Nanobiosensors provide ultra-sensitive monitoring of plant stress biomarkers, such as ROS and specific metabolites. For example, ROS production increases under abiotic stresses like drought, salinity, and heat, acting as both a damaging agent and a signaling molecule (Hasanuzzaman et al. 2020), as shown in Fig. 8.2. These sensors can detect biotic stresses by identifying pathogen-associated molecular patterns (PAMPs) and other indicators of infections, ensuring accurate and rapid diagnosis (Aina et al. 2024).

8.2.1.2 Timely Intervention and Real-Time Monitoring

Early detection enables farmers to implement timely strategies to mitigate stress effects. The real-time data provided by nanobiosensors enables immediate intervention, such as adjusting irrigation or applying biostimulants, to counteract stress impacts. For instance, salinity-induced oxidative stress can be mitigated by employing measures informed by ROS levels detected by sensors (Das et al. 2024). This

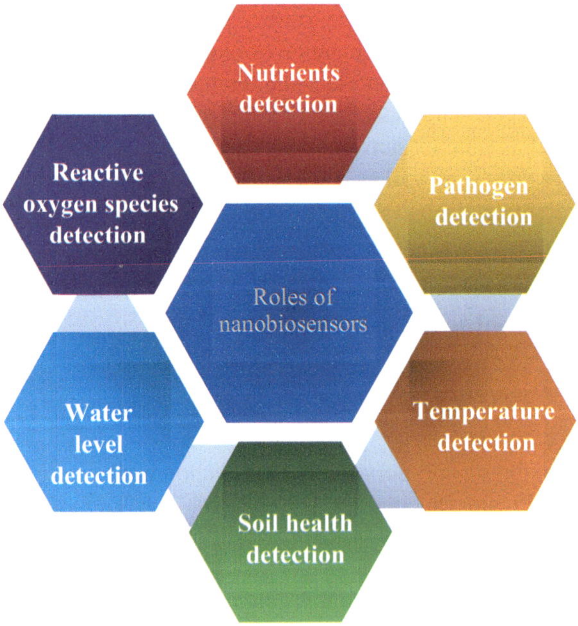

Fig. 8.2 Role of nanobiosensors in agriculture

proactive approach minimizes the physiological and economic losses associated with delayed responses. If a nanobiosensor detects increasing levels of abscisic acid (ABA), a stress hormone, farmers can adjust irrigation schedules to alleviate drought stress (Zhao et al. 2021).

Nanobiosensors provide real-time data, revolutionizing how plant health and environmental conditions are monitored. These sensors continuously track physiological and environmental parameters, enabling dynamic and responsive management strategies to optimize crop health and productivity. Continuous assessment of plant health and continuous monitoring of factors such as soil moisture, nutrient levels, and temperature enables a comprehensive understanding of plant health dynamics (Wang et al. 2019a, b). This real-time data helps farmers react promptly to unfavorable conditions. By integrating nanobiosensors with data analytics, farmers can create feedback loops that optimize resource management (Cai et al. 2020). For instance, if sensors indicate low nutrient levels, farmers can apply fertilizers precisely when needed, minimizing waste and enhancing crop yield.

Real-time monitoring offered by nanobiosensors plays a pivotal role in continuously tracking plant responses to environmental changes. These sensors can measure fluctuations in crucial parameters such as temperature, humidity, soil moisture, and nutrient levels, which are essential for maintaining optimal growth conditions (Virk et al. 2024; Mathivanan 2021). Moreover, nanobiosensors can detect stress markers like reactive oxygen species (ROS) or nutrient deficiencies, providing early warnings about potential threats and enabling timely intervention to mitigate damage (Miguel-Rojas and Pérez-de-Luque 2023; Voke et al. 2021).

The data collected by nanobiosensors is crucial for creating adaptive feedback loops in agriculture. Nanobiosensors can detect soil nutrient depletion, enabling precise and timely fertilizer applications to prevent wastage and over-fertilization (Virk et al. 2024; Antonacci et al. 2018). Sensors that monitor soil moisture levels can help optimize irrigation schedules, promoting water conservation and reducing unnecessary water usage (Miguel-Rojas and Pérez-de-Luque 2023; Rana et al. 2024). Nanobiosensors capable of detecting pest-related biomarkers allow for targeted pesticide use, minimizing environmental impact and reducing chemical waste (Rai et al. 2012; Mondal et al. 2022).

8.2.2 Precision Agriculture

Precision agriculture leverages advanced technologies to enhance farming efficiency, and detailed data contributions from modern tools amplify its effectiveness. Precision agriculture focuses on the judicious use of resources, such as water, fertilizers, and pesticides, to enhance farming efficiency and sustainability. By employing advanced technologies like Variable Rate Technology (VRT), farmers can tailor input applications to match the specific needs of different areas within a field, ensuring optimal resource allocation (Ahmad et al. 2018). This targeted approach minimizes input waste while maximizing crop health and productivity. For example,

VRT utilizes prescription maps or real-time sensor data to adjust input rates, addressing variability in soil conditions and crop requirements (Getahun et al. 2024). This not only reduces the environmental impact but also enhances economic returns by preventing over-application and wastage of inputs like fertilizers and water. Through the integration of precision farming technologies, sustainable agricultural practices are becoming increasingly effective, addressing both productivity and environmental sustainability challenges (Monteiro et al. 2021).

Real-time data collection through sensors and mapping tools is transforming crop management by providing farmers with actionable insights. Precision agriculture systems analyze critical parameters like soil health, crop performance, and weather patterns to ensure optimal use of resources. Sensors monitor soil moisture and nutrient content, enabling tailored irrigation and fertilizer application strategies. Mapping tools, often integrated with GPS and satellite imagery, allow farmers to identify spatial variability within their fields. This precise data helps allocate resources effectively, minimizing waste and improving yield. Furthermore, weather data supports the adjustment of farming practices to mitigate the impacts of environmental changes. By combining these technologies, farmers can make informed decisions, optimize crop health, and contribute to sustainable agricultural practices. Nanobiosensors significantly contribute to precision agriculture by optimizing resource usage based on real-time data. These sensors enable targeted application of water, fertilizers, and pesticides, leading to reduced input costs and environmental impact (Schmidhalter and Kafkafi 2019). By analyzing sensor data, farmers can apply resources only where and when they are needed. Nanobiosensors generate vast amounts of data that can inform agronomic decisions. By employing data analytics and machine learning, farmers can develop tailored strategies that enhance productivity while minimizing environmental degradation (Huang et al. 2021).

Nanobiosensors are pivotal in research aimed at understanding plant responses to various stressors, facilitating the development of stress-resistant varieties. Through the application of nanobiosensors, researchers can gain insights into the molecular mechanisms behind plant stress responses, such as the signaling pathways involved in drought and salinity tolerance (Tuteja 2020). This knowledge is vital for improving crop resilience. Nanobiosensors are designed to detect specific signaling molecules, such as hydrogen peroxide (H_2O_2), involved in stress responses. By monitoring these molecules in real time, researchers can study how plants respond to biotic and abiotic stresses, such as drought, salinity, and pathogens (Virk et al. 2024). These sensors also provide insights into oxidative stress mechanisms by measuring reactive oxygen species (ROS) levels, facilitating the identification of stress-tolerant traits (Al-Khayri et al. 2023). By identifying and monitoring stress-related biomarkers, nanobiosensors enable targeted breeding or genetic engineering of crops with enhanced stress tolerance. Additionally, the integration of nanotechnology in precision agriculture allows for testing new crop varieties under controlled stress conditions, ensuring their effectiveness in mitigating environmental challenges (Miguel-Rojas and Pérez-de-Luque 2023). The insights gained from nanobiosensor research can guide breeding programs aimed at developing stress-resistant plant varieties. For instance, identifying genetic markers associated with stress resistance allows for marker-assisted selection (MARS) in breeding (González et al. 2022).

8.3 Methods

8.3.1 Types of Nanoparticles

Nanomaterials, particularly nanoparticles such as gold, silver, and carbon nanotubes (CNTs), are widely used in the development of advanced biosensors for detecting biomarkers associated with plant stress. These materials provide several advantages due to their unique properties. The gold nanoparticles (AuNPs) are widely used due to their ease of functionalization, biocompatibility, and strong surface plasmon resonance, making them ideal for binding to specific biomolecules (Vera et al. 2021) and allowing for effective detection of biomarkers in plant stress studies (Lai et al. 2024). The AgNPs exhibit strong antibacterial properties, silver nanoparticles also improve the sensitivity of biosensors for detecting plant stress markers (Ibrahim et al. 2021) and can enhance the optical and electrochemical signals in sensors (Deng et al. 2020). Carbon nanotubes (CNTs) materials have high conductivity and surface area, which facilitate improved charge transfer and sensor sensitivity (Chen et al. 2019) and enhance the performance of sensors used to monitor plant stress responses (Sihombing et al. 2024). Nanoparticles can amplify signals from the target analytes, leading to lower detection limits. The surface plasmon resonance exhibited by gold nanoparticles can significantly enhance the sensitivity of optical sensors (Linke et al. 2022). The unique physical and chemical properties of these nanoparticles allow for increased sensitivity and specificity in detecting plant stress biomarkers. For example, gold nanoparticles improve the colorimetric detection of biomarkers, while carbon nanotubes enhance signal transduction, leading to more accurate stress detection (Malik et al. 2023). These enhancements enable real-time monitoring and precise diagnostics, crucial for developing stress-resistant crop varieties.

8.3.2 Detection Techniques

Detection techniques, such as fluorescent sensors and electrochemical sensors, are crucial for monitoring plant stress by detecting specific metabolites or ions released during stress responses. These sensors provide real-time, quantitative data that can help in assessing plant health.

Fluorescent-based nanosensors are designed to detect various biological and chemical agents, making them highly sensitive for identifying plant stress markers. These sensors work by emitting fluorescence upon interaction with specific analytes, allowing for precise detection of metabolites associated with oxidative stress or other physiological changes (Sargazi et al. 2022). Fluorescent sensors utilize fluorophores that emit light upon excitation, allowing for sensitive detection of stress-related metabolites (Liu et al. 2020). These sensors can provide real-time monitoring of plant health by measuring changes in fluorescence intensity.

Electrochemical sensors, including those used for detecting reactive oxygen species (ROS) or environmental toxicants, offer high sensitivity and selectivity. They function by measuring the electrical signals generated during electrochemical reactions with target analytes. They offer high sensitivity and rapid response times. These sensors are particularly effective for detecting changes in the ion concentrations like calcium or magnesium associated with stress responses (Zhang et al. 2021) or metabolic shifts that occur during plant stress, such as the presence of hydrogen peroxide or nitrate (Duanghathaipornsuk et al. 2021; Hara and Singh 2021).

8.3.3 Microfluidics

Microfluidics has emerged as a powerful tool in plant stress research, enabling the analysis of small sample volumes with high precision. Integrated microfluidic systems offer the advantage of handling minuscule amounts of biological material, allowing for detailed study of plant responses under various stress conditions. This approach facilitates the high-throughput analysis of metabolites, ions, and other biomarkers released during stress, providing deeper insights into plant health and adaptation mechanisms. Microfluidic devices enable the use of very small sample volumes and can handle picoliter to microliter volumes, allowing for efficient use of precious biological samples and enabling high-throughput screening (Huang et al. 2018). This capability reduces reagent costs and minimizes sample waste, making it an ideal choice for applications in plant stress research. Recent advancements have focused on optimizing microfluidic channels for precise fluid handling and efficient detection of plant metabolites under stress conditions (Malik et al. 2023). Microfluidic systems are increasingly being integrated into plant stress response studies to explore how plants react to drought, salinity, or pathogen attacks. These systems allow researchers to simulate stress environments at a microscale and monitor real-time physiological changes, enabling researchers to observe plant responses at a cellular level (Rogers et al. 2019). For example, recent research has used microfluidic platforms to analyze how reactive oxygen species (ROS) accumulate in plants under oxidative stress, which is a key marker of stress responses (Sihombing et al. 2024).

8.3.4 Molecular Imprinting

Molecular imprinting is a technique used to create synthetic receptors that can specifically bind to target molecules, such as stress-related compounds in plants, improving detection accuracy. This method involves creating a polymer matrix with specific binding sites for target molecules, allowing for high selectivity (Khan et al. 2020). For example, sensors can be designed to detect specific plant hormones associated with stress, such as jasmonic acid. The "molecularly imprinted polymers" (MIPs) are created, which are tailored to recognize and interact with specific biomolecules associated with plant stress, such as reactive oxygen species (ROS) or specific hormones. The ability of these sensors to selectively bind to stress markers enhances the sensitivity and specificity of detection, enabling them to distinguish between similar compounds, leading to more accurate readings of plant stress indicators (Nishida et al. 2021), making them highly effective for monitoring plant health in real time. The key advantage of molecular imprinting is its ability to create highly selective binding sites for specific compounds, mimicking the natural binding of antibodies to antigens. This selectivity ensures that the sensor detects only the stress-related markers of interest, even in complex biological environments. Recent studies highlight the use of MIPs for detecting specific plant stress biomarkers such as antioxidants and metabolic by-products (BelBruno 2018). By providing highly specific recognition sites, molecularly imprinted sensors offer superior detection accuracy compared to conventional sensors. The resulting sensors can identify and quantify low concentrations of stress markers with minimal interference, improving the reliability of stress diagnostics in plants (Li et al. 2024).

8.3.5 Field Deployments

Field deployments of agricultural monitoring systems benefit significantly from portable and user-friendly designs of nanobiosensors, enabling on-site, real-time monitoring in various agricultural settings. Advances in miniaturization and integration of sensors into handheld devices enable farmers to monitor crops in real time, making it easier to assess plant health directly in the field (Gao et al. 2022). These systems are designed to be lightweight and easy to transport, making them ideal for farmers who need to move between different fields and crops. The portability ensures that the monitoring equipment can be quickly deployed without the need for extensive setup, facilitating timely data collection for better decision-making. Additionally, user-friendly interfaces are essential in ensuring that farmers, regardless of their technological expertise, can easily interpret the data and take necessary actions. This accessibility is crucial for widespread adoption in agriculture (Patel et al. 2020). Modern apps and smart farming tools, such as KisanVishwa, incorporate intuitive designs, making it easier for farmers to use sensors and access critical farm information (Patokar and Gohokar 2024). This enhances the overall effectiveness of agricultural monitoring systems, providing real-time insights on factors like crop health and environmental conditions.

8.4 Current Achievements

8.4.1 Drought Stress Monitoring

Drought stress monitoring is significantly enhanced by the use of nanobiosensor systems that enable sensitive detection of key metabolites associated with stress responses, such as abscisic acid (ABA), allowing for timely interventions (Rai et al.

2020). These sensors utilize various nanomaterials to achieve high specificity and sensitivity, providing crucial information about plant stress levels. ABA is a plant hormone that plays a critical role in the plant's response to drought, and its levels increase when plants experience water scarcity. Several nanobiosensors have been designed to detect these stress-related metabolites, providing real-time data on plant stress status. This allows farmers to monitor crop health more efficiently and make data-driven decisions. By detecting changes in ABA levels, farmers can adjust irrigation practices to better manage water use, ensuring that crops receive optimal water supply during drought conditions, thus reducing water wastage and improving crop resilience (Rajak et al. 2023). By monitoring ABA levels, farmers can optimize irrigation schedules, applying water only when necessary, which enhances water use efficiency (Li et al. 2021). This capability is particularly valuable in arid regions where water resources are limited.

8.4.2 Pathogen Detection

Nanobiosensors have been effectively utilized in plant pathogen detection, enabling rapid diagnosis of specific pathogens and improving disease management in agriculture. These sensors are capable of identifying pathogen-specific biomarkers at early stages, which allows for swift intervention and targeted treatment. This timely detection of pathogens like fungi and bacteria within hours (Shen et al. 2021) helps reduce crop losses by preventing the spread of diseases and optimizing the use of pesticides. Furthermore, the enhanced sensitivity and accuracy of nanobiosensors provide significant improvements over traditional diagnostic methods, offering farmers the ability to act quickly and efficiently. The ability to detect pathogens early in their lifecycle is crucial for minimizing crop damage and ensuring better crop yields (Shivashakarappa et al. 2022; Virk et al. 2024). By enabling timely interventions, such as targeted fungicide applications, these sensors help reduce crop losses significantly (Zhang et al. 2020). Early detection is crucial in managing plant health and ensuring food security.

8.4.3 Soil Quality Assessment

Nanobiosensors are increasingly being used to assess soil health, including key factors such as nutrient status, pH, microbial activity, and moisture content. These sensors provide real-time, precise data that allows for continuous monitoring of soil conditions, ensuring that farmers can quickly detect imbalances or deficiencies (Xue et al. 2021). By accurately tracking soil health, nanobiosensors enable more informed decisions regarding irrigation, fertilization, and other management practices. By providing real-time data on soil quality, farmers can make informed decisions regarding soil amendments and fertilization.

This leads to enhanced management strategies that optimize soil productivity and plant resilience, ensuring that crops thrive under varying environmental conditions. Improved understanding of soil health contributes to better management strategies, such as crop rotation and cover cropping, which enhance soil structure and fertility (Kumar et al. 2020).

With their ability to monitor soil quality in a timely and cost-effective manner, nanobiosensors are transforming sustainable agricultural practices (Mondal et al. 2022; Miguel-Rojas and Pérez-de-Luque 2023).

8.4.4 Integration with IoT

The integration of nanobiosensors with Internet of Things (IoT) technology has significantly enhanced the ability to monitor various parameters remotely. This integration enables real-time data collection and continuous monitoring of critical factors such as environmental conditions, plant health, pathogen presence, and soil conditions remotely (Singh et al. 2022). Data can be accessed in real time via mobile applications, facilitating timely decisions. By linking nanobiosensors with IoT systems, data can be transmitted to cloud-based platforms for analysis, allowing for timely and informed decision-making in agriculture and healthcare. This remote monitoring capability, combined with advanced data analysis, improves the efficiency of detecting anomalies and enables predictive actions, thus optimizing resource management and crop protection (Kulkarni et al. 2022).

Nanobiosensors are playing a pivotal role in advancing sustainable agriculture by enabling precision interventions. These sensors provide real-time data on plant health, soil conditions, and environmental factors, allowing farmers to apply inputs such as water, nutrients, and pesticides only when and where they are needed. This precision helps optimize resource use and minimize waste, contributing to more sustainable practices in agriculture (Virk et al. 2024). Additionally, by reducing the overuse of chemicals and water, nanobiosensors help decrease the environmental impact of farming, promoting an eco-friendlier approach to crop production (Miguel-Rojas and Pérez-de-Luque 2023). IoT-enabled nanobiosensors collect vast amounts of data that can be analyzed for trends and patterns, helping farmers optimize agricultural practices based on evidence rather than intuition (Patel et al. 2021).

8.4.5 Contributions to Sustainable Practices

Nanobiosensors significantly contribute to sustainable agricultural practices by enabling precise interventions based on real-time data. The ability to detect stress at early stages allows for targeted applications of water, fertilizers, and pesticides, reducing resource waste and enhancing crop yield (Gao et al. 2023). By optimizing resource use and minimizing chemical applications, nanobiosensors help decrease the environmental footprint of agriculture, promoting sustainability (Sharma et al. 2022).

8.5 Conclusion and Future Prospects in Agricultural Resilience

As climate change poses increasing threats to agricultural productivity, nanobiosensors play a vital role in ensuring global food security. Nanobiosensors have revolutionized agriculture by offering precise, real-time monitoring of plant health, soil quality, and environmental conditions, enabling resource-efficient practices and reducing environmental impact. These advancements bolster agricultural resilience by mitigating biotic and abiotic stresses, crucial in addressing challenges posed by climate change. Ongoing advancements in nanomaterials and detection techniques will likely lead to even more sensitive and specific sensors, capable of monitoring a wider range of plant stressors. By enabling precise and timely interventions, these technologies can help mitigate the adverse effects of climate-induced stresses such as drought, disease outbreaks, and soil degradation (Patel et al. 2021). The ability to monitor and respond to these challenges will be critical in maintaining food supply chains and ensuring that agricultural systems can adapt to changing climatic conditions (Xue et al. 2021). Additionally, integrating these sensors with IoT technology, artificial intelligence, and machine learning, farmers can achieve real-time monitoring and data collection, enabling predictive analytics and more effective interventions leading to enhanced crop yields, reduced input costs, and minimized environmental impact (Gao et al. 2023; Zhang et al. 2020) due to informed decision-making and optimized resource use with minimized environmental impact. By optimizing inputs and improving crop productivity, nanobiosensors play a vital role in ensuring global food security, supporting sustainable farming systems to meet the demands of a growing population in a rapidly changing climate. Nanobiosensors represent a critical innovation in agriculture, addressing immediate challenges while paving the way for a more resilient and sustainable future in food production. Their role in enhancing agricultural practices will be indispensable as we navigate the complexities of a changing climate.

References

Ahmad L, Mahdi SS, Ahmad L, Mahdi SS (2018) Variable rate technology and variable rate application. In: Satellite farming: an information and technology based agriculture. Springer, pp 67–80

Aina O, Bakare OO, Fadaka AO, Keyster M, Klein A (2024) Plant biomarkers as early detection tools in stress management in food crops: a review. Planta 259(3):60

Al-Khayri JM, Rashmi R, Surya Ulhas R, Sudheer WN, Banadka A, Nagella P et al (2023) The role of nanoparticles in response of plants to abiotic stress at physiological, biochemical, and molecular levels. Plan Theory 12(2):292

Antonacci A, Arduini F, Moscone D, Palleschi G, Scognamiglio V (2018) Nanostructured (bio) sensors for smart agriculture. TrAC Trends Anal Chem 98:95–103

BelBruno JJ (2018) Molecularly imprinted polymers. Chem Rev 119(1):94–119

Cai H et al (2020) Utilizing nanotechnology in agriculture for sustainability. Sustain For 12(9):3817

Cao JT, Dong YX, Ma Y, Wang B, Ma SH, Liu YM (2020) A ternary CdS@ Au-g-C3N4 heterojunction-based photoelectrochemical immunosensor for prostate specific antigen detection using graphene oxide-CuS as tags for signal amplification. Anal Chim Acta 1106:183–190
Chen H et al (2019) Carbon nanotube-based electrochemical sensors. Biosens Bioelectron 130:368–381
Das A, Pal S, Chakraborty N, Hasanuzzaman M, Adak MK (2024) Regulation of reactive oxygen species metabolism and oxidative stress signaling by abscisic acid pretreatment in rice (Oryza sativa L.) seedlings through sub1A QTL under salinity. *Plant*. Stress 11:100422
Deng Y et al (2020) The application of silver nanoparticles in biosensors. Sensors Actuators B Chem 306:127564
Duanghathaipornsuk S, Farrell EJ, Alba-Rubio AC, Zelenay P, Kim DS (2021) Detection technologies for reactive oxygen species: fluorescence and electrochemical methods and their applications. Biosensors 11(2):30
Gao Y et al (2022) Portable nanobiosensors for agricultural applications: current status and future perspectives. Trends Biotechnol 40(3):299–313
Gao Y et al (2023) Sustainable agricultural practices enabled by nanobiosensors. Agron Sustain Dev 43:30
Getahun S, Kefale H, Gelaye Y (2024) Application of precision agriculture technologies for sustainable crop production and environmental sustainability: a systematic review. Sci World J 2024(1):2126734
González MT et al (2022) Utilizing molecular markers for the development of stress-resistant crop varieties. Mol Breed 42:10
Hara TO, Singh B (2021) Electrochemical biosensors for detection of pesticides and heavy metal toxicants in water: recent trends and progress. ACS ES&T Water 1(3):462–478
Hasanuzzaman M, Bhuyan MB, Zulfiqar F, Raza A, Mohsin SM, Mahmud JA et al (2020) Reactive oxygen species and antioxidant defense in plants under abiotic stress: revisiting the crucial role of a universal defense regulator. Antioxidants 9(8):681
He KC, Wang HR, Yang H, Sun LJ, Liu W, Bao N (2020) Digitalized pencil trace modified electrodes for real time evaluation of salicylic acid in detached Arabidopsis thaliana leaves during regeneration. Anal Chim Acta 1120:59–66
Hu Y, Wang X, Wang C, Hou P, Dong H, Luo B, Li A (2020) A multifunctional ratiometric electrochemical sensor for combined determination of indole-3-acetic acid and salicylic acid. RSC Adv 10(6):3115–3121
Huang X et al (2018) Microfluidics for plant research: applications and future prospects. Plant Cell Rep 37:1135–1149
Huang G et al (2021) Data-driven precision agriculture: integration of sensors and machine learning. Comput Electron Agric 183:106064
Ibrahim N, Jamaluddin ND, Tan LL, Mohd Yusof NY (2021) A review on the development of gold and silver nanoparticles-based biosensor as a detection strategy of emerging and pathogenic RNA virus. Sensors 21(15):5114
Khan Y et al (2020) Molecularly imprinted polymers: a powerful tool for biosensing. Trends Anal Chem 124:115779
Kulkarni MB, Ayachit NH, Aminabhavi TM (2022) Recent advancements in nanobiosensors: current trends, challenges, applications, and future scope. Biosensors 12(10):892
Kumar A et al (2020) Integrated soil health management for sustainable agriculture. Sustain For 12(9):3845
Lai H, Huang R, Weng X, Huang B, Yao J, Pian Y (2024) Classification and applications of nanomaterials in vitro diagnosis. Heliyon 10(11):e32314
Li Y et al (2021) Smart irrigation management using nanobiosensors for drought stress detection. Agric Water Manag 242:106392
Li Y, Luo L, Kong Y, Li Y, Wang Q, Wang M et al (2024) Recent advances in molecularly imprinted polymer-based electrochemical sensors. Biosens Bioelectron 116018
Linke D et al (2022) Nanomaterials for enhanced sensor performance. Mater Today 47:135–145

Liu Y et al (2020) Fluorescent biosensors for plant stress monitoring. Plant Methods 16:67
Malhotra BD, Ali MA (2018) Nanomaterials for biosensors. In: Nanomaterials in biosensors: fundamentals and applications. Elsevier, pp 1–74
Malik S, Singh J, Goyat R, Saharan Y, Chaudhry V, Umar A, Ibrahim AA, Akbar S, Ameen S, Baskoutas S (2023) Nanomaterials-based biosensor and their applications: a review. Heliyon 9(9):e19929
Mathivanan S (2021) Perspectives of nano-materials and nanobiosensors in food safety and agriculture. In: Novel nanomaterials. Intechopen, p 197
Miguel-Rojas C, Pérez-de-Luque A (2023) Nanobiosensors and nanoformulations in agriculture: new advances and challenges for sustainable agriculture. Emerg Top Life Sci 7(2):229–238
Mondal R, Dam P, Chakraborty J, Paret ML, Katı A, Altuntas S et al (2022) Potential of nanobiosensor in sustainable agriculture: the state-of-art. Heliyon 8(12)
Monteiro A, Santos S, Gonçalves P (2021) Precision agriculture for crop and livestock farming—brief review. Animals 11(8):2345
Nishida Y et al (2021) Advancements in molecularly imprinted polymers for biosensing applications. Sensors Actuators B Chem 329:129113
Patel V et al (2020) User-friendly nanobiosensor interfaces for agricultural monitoring. Sensors 20(15):4294
Patel S et al (2021) Data analytics in smart agriculture: the role of IoT and nanobiosensors. Sustain Comput Inform Syst 30:100503
Patokar AM, Gohokar VV (2024) Design and development of an intuitive android application for smart farming. Curr Agric Res J 12(1)
Rai V, Acharya S, Dey N (2012) Implications of nanobiosensors in agriculture. Sci Res 3(2)
Rai P et al (2020) Nanobiosensors for the detection of abscisic acid in plants: a review. Biosens Bioelectron 169:112578
Rajak P, Ganguly A, Adhikary S, Bhattacharya S (2023) Internet of things and smart sensors in agriculture: scopes and challenges. J Agric Food Res 14:100776
Rana L, Kumar M, Rajput J, Kumar N, Sow S, Kumar S et al (2024) Nexus between nanotechnology and agricultural production systems: challenges and future prospects. Disc Appl Sci 6(11):1–19
Rogers JA et al (2019) Microfluidic devices for studying plant stress responses. Nat Rev Mol Cell Biol 20:306–321
Sargazi S, Fatima I, Kiani MH, Mohammadzadeh V, Arshad R, Bilal M et al (2022) Fluorescent-based nanosensors for selective detection of a wide range of biological macromolecules: a comprehensive review. Int J Biol Macromol 206:115–147
Schmidhalter U, Kafkafi U (2019) Precision agriculture and sustainability: the role of sensors. Agronomy 9(9):482
Sharma S et al (2020) Nanobiosensors for agricultural applications: a review. J Nanobiotechnol 18(1):1–16
Sharma R et al (2022) The impact of nanobiosensors on sustainable agriculture. Environ Sci Pollut Res 29:10245–10258
Shen Y et al (2021) Nanobiosensors for rapid pathogen detection in agriculture: a review. Front Microbiol 12:611497
Shivashakarappa K, Reddy V, Tupakula VK, Farnian A, Vuppula A, Gunnaiah R (2022) Nanotechnology for the detection of plant pathogens. Plant Nano Biol 2:100018
Sihombing YA, Sari RI, Hermanto BR, Handayani M, Kusumocahyo SP, Ulum MF et al (2024) Enhanced uric acid detection using functionalized multi-walled carbon nanotube/AgNi nanocomposites: a comparative study on screen-printed carbon electrode (SPCE) and fabric-based biosensors. Sensors Actuators Rep 8:100223
Singh A et al (2022) Integrating nanobiosensors with IoT for smart agriculture. Comput Electron Agric 191:106568
Tuteja N (2020) Plant responses to abiotic stress: mechanisms and adaptations. Plant Biol 22(2):193–203

Vera CP et al (2021) Gold nanoparticles in biosensing: a review. Anal Chim Acta 1141:157–170
Virk V, Deepak H, Taneja K, Srivastava R, Giri S (2024) Amelioration in nanobiosensors for the control of plant diseases: current status and future challenges. Front Nanotechnol 6:1310165
Voke E, Pinals RL, Goh NS, Landry MP (2021) In planta nanosensors: understanding biocorona formation for functional design. ACS Sensors 6(8):2802–2814
Wang S, Li W, Chang K, Liu J, Guo Q, Sun H et al (2017) Localized surface plasmon resonance-based abscisic acid biosensor using aptamer-functionalized gold nanoparticles. PLoS One 12(9):e0185530
Wang X, Cheng M, Yang Q, Wei H, Xia A, Wang L et al (2019a) A living plant cell-based biosensor for real-time monitoring invisible damage of plant cells under heavy metal stress. Sci Total Environ 697:134097
Wang Y et al (2019b) Smart sensors for real-time monitoring of soil and crop status. Sensors 19(16):3448
Wang Q, Li X, Tang L, Fei Y, Pan Y, Sun L (2020) Based electroanalytical devices for in situ determination of free 3-indoleacetic acid and salicylic acid in living Pyropia haitanensis thallus under various environmental stresses. J Appl Phycol 32:485–497
Xue X et al (2021) Nanobiosensors for soil health assessment: current status and future prospects. Soil Biol Biochem 156:108184
Zain M, Ma H, Nuruzzaman M, Chaudhary S, Nadeem M, Shakoor N et al (2023) Nanotechnology based precision agriculture for alleviating biotic and abiotic stress in plants. Plant Stress 10:100239
Zhang Z et al (2020) The role of nanobiosensors in reducing crop losses due to pathogens. Plant Dis 104(1):89–95
Zhang Y et al (2021) Electrochemical biosensors for plant stress detection: principles and applications. Biosensors 11(10):327
Zhao Y et al (2021) Plant stress signaling: molecular mechanisms and biotechnological applications. Front Plant Sci 12:675184

Chapter 9
Nanobiosensors: Advanced Detection and Monitoring of Plant Hormones

Sadhana Giri and Verinder Virk

Abstract Plant hormones, known as phytohormones, are crucial in regulating plant growth, development, and responses to environmental stress. Traditional methods for detecting and monitoring these molecules are often slow, complex, and unsuitable for real-time applications. Nanobiosensors provide a game-changing solution, offering enhanced sensitivity, specificity, and the ability to perform real-time, on-site monitoring of phytohormones. This chapter focuses on the fusion of nanotechnology and biosensing to detect key plant hormones, including auxins, gibberellins, cytokinins, abscisic acid, and ethylene. It delves into the design strategies, materials, and functionalization methods used in developing nanobiosensors, explaining how these sensors interact with specific hormones. Key advances in nanomaterials, such as carbon nanotubes, quantum dots, and nanoparticles, are highlighted for their role in boosting detection accuracy and signal transmission. The chapter also explores the practical applications of nanobiosensors in agriculture, including precision farming, stress monitoring, and crop enhancement. Challenges such as sensor stability, selectivity in complex environments, and scalability are also addressed. Finally, the chapter offers a forward-looking view on improving nanobiosensor technology and its potential to promote sustainable agriculture and ensure food security.

Keywords Nanobiotechnology · Nanobiosensors · Phytohormones · Plant improvement

S. Giri · V. Virk (✉)
Department of Microbiology, Kanya Gurukula Campus, Gurukula Kangri (Deemed to be University), Haridwar, Uttarakhand, India
e-mail: virender.wahla@gkv.ac.in

J.-T. Chen (ed.), *Nanobiosensors for Crop Monitoring and Precision Agriculture*, Sustainability Sciences in Asia and Africa,
https://doi.org/10.1007/978-981-96-8335-2_9

9.1 Introduction

Plants are highly dynamic organisms, continuously adapting to their environment through a series of biochemical and physiological processes. Central to these processes are plant hormones, or phytohormones, which are naturally occurring chemicals that regulate growth, development, and responses to environmental stimuli. These hormones play pivotal roles in key plant functions such as seed germination, flowering, fruit ripening, and stress responses. Phytohormones, including auxins, gibberellins, cytokinins, abscisic acid (ABA), and ethylene, control the timing and coordination of these processes. The ability to monitor and quantify these hormones accurately is essential for improving crop yields, optimizing growth conditions, and ensuring food security, particularly in the face of challenges like climate change and rapid population growth (Mochida and Shinozaki 2023a, b). Traditionally, detecting and quantifying phytohormones has been accomplished using advanced techniques such as high-performance liquid chromatography (HPLC), gas chromatography-mass spectrometry (GC-MS), and enzyme-linked immunosorbent assays (ELISA) (Xu et al. 2021a, b). While these methods are accurate and sensitive, they are often time-consuming, require expensive equipment, and are not suitable for real-time applications. In the context of agriculture, the lack of fast and reliable detection methods limits the ability to monitor plant growth and stress responses continuously, preventing immediate intervention. This is particularly problematic in precision agriculture, where timely and data-driven decisions are critical for improving crop productivity and sustainability (Huang et al. 2023). Recent advances in nanotechnology have introduced new possibilities for real-time, on-site monitoring of plant hormones through the development of nanobiosensors. These sensors leverage the unique properties of nanomaterials to detect small amounts of molecules with high sensitivity and specificity, enabling real-time applications in agricultural settings (Zhang et al. 2022a, b). Nanobiosensors, which combine biological recognition elements with nanomaterials, offer significant advantages over traditional methods, including enhanced sensitivity, faster response times, and the ability to operate in complex environments such as soil or plant tissues (Tang et al. 2022). The design and functionality of nanobiosensors are shaped by the properties of the nanomaterials used, such as carbon nanotubes (CNTs), gold nanoparticles (AuNPs), quantum dots (QDs), and graphene. These materials offer high surface area, ease of functionalization, and efficient signal transduction capabilities. For example, CNTs have been used in electrochemical biosensors for plant hormone detection, enabling sensitive and rapid detection of hormones like auxins and gibberellins (Zhou et al. 2023a, b). Graphene-based sensors are also emerging as promising candidates due to their excellent conductivity, biocompatibility, and high surface area, making them suitable for monitoring plant stress responses and hormone signaling (Xu et al. 2021a, b). Functionalization of nanomaterials with biological recognition elements, such as antibodies, aptamers, enzymes, or molecularly imprinted polymers (MIPs), is crucial for ensuring the specificity of these sensors. The biological recognition element selectively binds to the target phytohormone, causing a detectable change in the sensor's properties, which can then be measured (Yang et al. 2023a, b). This approach allows for the detection of hormones even at trace levels, which is especially important in agricultural applications where hormone concentrations can vary widely depending on environmental conditions. Additionally, combining multiple nanomaterials and recognition elements allows for the development of multiplexed sensors capable of detecting multiple hormones simultaneously, further enhancing their utility in complex agricultural systems (Zhang et al. 2022a, b). The ability to monitor plant hormones in real time is crucial for precision farming. By incorporating nanobiosensors into agricultural systems, farmers can gain continuous insight into plant hormone levels, enabling more informed decisions related to irrigation, fertilization, and pest control. For instance, real-time monitoring of ABA levels can provide valuable information regarding plant water stress, allowing farmers to optimize irrigation schedules and reduce water usage (Wang et al. 2023a, b). Similarly, ethylene detection can be used to monitor the ripening process of fruits, enabling farmers to harvest crops at the optimal time and improve postharvest quality (Mochida and Shinozaki 2023a, b). Beyond their agricultural applications, nanobiosensors hold great promise for improving sustainability by reducing the use of chemical fertilizers and pesticides. By providing more precise information on plant stress and growth, these sensors enable targeted interventions that minimize waste and environmental impact (Tang et al. 2022). For example, by monitoring the levels of cytokinin, a hormone associated with cell division, nanobiosensors can help optimize the timing of plant growth regulators, reducing the need for excessive chemical inputs (Yang et al. 2023a, b). These technologies, if widely adopted, could contribute to more sustainable agricultural practices, which are essential in light of growing concerns about food security and environmental degradation. However, despite the promising potential of nanobiosensors, several challenges remain to be addressed. One of the primary concerns is sensor stability under field conditions. The environmental factors that affect plant growth, such as temperature fluctuations, humidity, and soil composition, can influence the performance and durability of nanobiosensors (Wang et al. 2023a, b). Ensuring long-term sensor stability and reliability is crucial for their practical application in agriculture. Additionally, the selectivity of these sensors in complex environments such as soil or plant tissues can be compromised by the presence of interfering substances, which may result in false positives or inaccurate measurements (Zhou et al. 2023a, b). Ongoing research is focused on improving sensor specificity through better functionalization strategies and optimizing the design of the recognition elements. Another challenge lies in scaling up the production of nanobiosensors for widespread use in agriculture. While laboratory-scale prototypes have demonstrated excellent performance, producing large quantities of these sensors at a cost-effective price remains a significant hurdle (Xu et al. 2021a, b). Moreover, the integration of these sensors into existing agricultural systems requires the development of robust, wireless communication platforms for real-time data transmission and analysis. This integration is critical for ensuring that the data collected by the sensors can be utilized effectively by farmers and agricultural managers to make informed decisions. Looking forward, the future of nanobiosensors in agriculture holds immense promise. As the global population continues to grow, the need for sustainable and efficient agricultural practices will only increase. Nanobiosensors can play a pivotal role in meeting this demand by providing farmers with accurate, real-time data that enables precise control over plant growth, irrigation, pest management, and nutrient optimization (Tang et al. 2022). The integration of nanobiosensors with emerging technologies such as artificial intelligence (AI), machine learning (ML), and remote sensing could create intelligent systems capable of autonomous monitoring and managing crop production sustainably and efficiently (Zhang et al. 2022a, b). Enhancing agricultural productivity and sustainability has been made possible by developments in sensor technologies, data analytics, and precision agriculture. Biosensors are widely considered one of the most effective tools for addressing the world's issues, which include climate change and a growing global population that will increase demand for food security and safety (Wang et al. 2022). Toxins and nutrients have accumulated in surface and ground waters as a result of the overuse of fertilizers and agrochemicals in agricultural fields. Fisheries are suffering recreational activities are dwindling and water purification expenses are rising as a result of these dangerous concentrations (Rashmi et al. 2020). Precision agriculture includes wireless field networking and nanosensors to monitor and control farming practices. It oversees pre- and postharvest activities and site-specific crops to ensure compliance (Lal 2015). The remarkable properties of the functional material used to create nanobiosensors under precision agriculture. This may help in the accurate identification of soil, moisture, nutrients, phytopathogens, and humidity (Autonacci et al. 2018). The receptors, transducers, and bio-recognition chemicals that make biosensors identify particular target molecules within the medium. Biosensors that rely on electrochemistry. High specificity and sensitivity on biochemical interaction provide easy-to-use, affordable, and mass-producible electrodes. They are anticipated to be utilized to create tiny portable instruments for carrying out multiple compound analyses at once (Malhotra and Chaubey 2014). Devices that incorporate nanomaterials can improve sample analysis precision and economy (Verma and Kapoor 2015); (Bilal et al. 2021). Over the past 10 years, nanomaterials used in electrochemical biosensors have significantly increased, attaining better selectivity, improved performance, and sensitivity (Acimovic et al. 2014). In order to detect food pesticides and mycotoxins, this review outlines recent developments in electrochemical biosensors in conjunction with nanomaterials.

The detection of nanomaterial in sample analysis is examined, and potential issues and solutions are discussed and compared. More sophisticated nanomaterial-based sensing technologies will be created in the upcoming years, improving public health and food safety. As organic matter breaks down, these biosensors have proven effective in assessing the flow of nutrients in the rhizosphere and identifying nutrient availability in a variety of soils (Yeomans et al. 1999). The release of root exudates in response to nitrate ($NO3^-$) and nitrogen (N) supply was monitored using a luminescent cellular biosensor, *Pseudomonas fluorescens* 10,586 Pucd607. The biosensor's response to glucose concentration remains unaffected by the varying nitrate concentrations in the plant. The study equivocally demonstrated that *Hordeum vulgare* significantly increases its rhizodeposition at the root apex when nitrate concentrations in the rhizosphere are reduced (Standing et al. 2003). Biosensors can provide continuous monitoring or early detection of diseases, contributing significantly to sustainable agriculture. They are vital in identifying pathogens that cause diseases, detecting pesticides, herbicides, and fertilizers, checking for food adulteration, heavy metal ion detection, measuring soil conditions, and identifying illegal additives (Hazarika et al. 2022). Historically, various laboratory techniques such as HPLC and gas chromatography (GC) have been commonly employed. However, these traditional methods have their drawbacks, including high costs, lengthy processing times, and the need for skilled personnel and specialized equipment to conduct experiments. A biosensor is an essential tool for assessing nutrient availability in various soil types throughout the decomposition of organic matter and for examining nutrient movement within the rhizosphere. Developed by the Nicolas Plumere group, this nitrate biosensor is a game changer for farmers. It requires just a single drop of plant juice for measurement and is designed with user-friendliness in mind, making it accessible even for those without specialized training. With this technology, farmers can easily monitor nitrate levels in their crops, empowering them to make informed decisions about their soil health and nutrient management.

9.2 Development of Biosensors in "Advancing Precision Agriculture: Wireless Field Networking and Nanosensors for Optimized Farming Practices"

According to a recent study reported by Kamat et al. (2023a, b), a unique molecularly imprinted microfluidic chip was developed that can sense nutrients including phosphate and nitrate in media. They have used a microcontroller-controlled potentiostat for data recording and acquisition, enabling the storage and transmission of data via a long-range radio module (LoRA). The germination rate of the legume crimson red and recording the nitrate and phosphate levels in media for 7 days have been utilized to verify the function of the microfluidic devices. In the future, researchers could leverage this affordable and efficient device for various real-time field applications. Potential uses include monitoring the soil microenvironment, selecting drought-resistant plant varieties, identifying root systems and pathogens, and enhancing nutrient uptake efficiency. This technology advances our understanding of plant health and soil dynamics, ultimately contributing to more sustainable agricultural practices. Another highly effective method for detecting plant diseases is the electronic nose (E-nose), which has been extensively employed in this field (Cellini et al. 2017; Cui et al. 2018; A.D. Wilson 2018). The E-nose is a powerful analytical tool designed for gas analysis, effectively simulating the biological aspect of the factory system. It integrates an advanced data-processing technique with a versatile multisensory array. To function optimally, the E-nose is explicitly engineered to be cross-sensitive, allowing it to detect multiple components within the sample being analyzed. Plant-wearable sensors have gained popularity for monitoring plant health. These devices can assist in detecting diseases or abnormalities in plants by tracking their state of health in real time (Qu et al. 2021a, b). Satellite or drone image sensing has been utilized to predict agricultural yields and growth, as well as to assess crop contamination caused by pests or diseases (Mogili and Deepak 2018). This technological advancement enables farmers to make informed decisions and optimize their crop management strategies. The sensor continuously provides temperature and humidity data to the smartphone, and the inbuilt GPS sensor in the smartphone gathers GPS data, including longitude and latitude. The smartphone's front camera is used to capture both data and images. Through the use of a Wi-Fi connection, the smartphone serves as an information gateway, collecting data from the sensor and sending it back to the distant server. The sensor data is received by the web service, which then stores it in MongoDB (Alfian et al. 2017). Plant growth status changes can be effectively tracked using a multimodal flexible sensor system that incorporates stacked ZnIn2S4 (ZIS) nanosheets as the sensing medium. A flexible sensor utilizing a zinc oxide thin semiconductor (ZIS) is employed to monitor light and humidity levels precisely near the leaves and in areas critical to plant transcription. The dehydration status of plants is accurately assessed by observing the opening and closing of plant stomata, which reliably indicate relative humidity levels (Lu et al. 2020).

Additionally, a wide range of materials has been utilized to develop electrochemical sensors specifically for detecting glucose in plants. These techniques play a vital role in monitoring fruit yield and sugar content, as they are directly correlated with the amount of glucose produced in chloroplasts (Naikoo et al. 2021). In the research study, a technique has been developed to monitor the expression of β-glucuronidase (GUS) in real time in transgenic tobacco plants by utilizing biomarker activity. This method employs the chronoamperometric technique and involves creating a microchip electrode configuration. These electrodes are then used to translate the activity of GUS enzymes in MSK8 tomato and BY-2 tobacco cells into measurable electric current signals. In the case of tomato cells expressing GUS, the detection sensitivity was recorded at 0.076 mA/mM cm^2, with a detection limit of 0.1 mM (Pandey et al. 2018). Biosensor application to plant cells can trigger stress involves a complex signaling network that reprograms gene activity and metabolism. Plant hormones and signaling molecules are essential for translating pathogen recognition into effective defense responses (Uslu and Grossman 2016; Mandadi and Scholthof 2013). Early signaling events involving changes in intracellular Ca2+ and reactive oxygen species are significantly influenced by plant hormones SA, JA, and ET (Bari and Jones 2009).

9.3 Types of Nanobiosensors for Real-Time Nutrient Tracking and Sustainable Farming

Plant biosensor development has recently improved our understanding of plant signaling, particularly its dynamics in development (Uslu and Grossman 2016). (Muniandy et al. 2019) cover different types of biosensors, including enzyme-based, electrochemical, DNA-based, whole-cell, ion-selective electrochemical, optical, nanomaterial-based, piezoelectric, microbial, and plant-wearable biosensors. Utilizing nanoparticles with a physical transducer probe, nanobiosensors offer an affordable, highly sensitive, portable, minimally invasive, and rapid method for real-time biological detection (Jain 2005). The different types of biosensors can be used for the nutrient detection in plants. Furthermore, fluorescence-based optical biosensors can assess the overall nutrient absorption of various crops and soils in agriculture. This method employs fluorescent dyes or molecules. When these molecules attach to a specific nutrient, they produce fluorescence. This indicates that fluorescence levels are directly proportional to nutrient concentration (Narsaiah et al. 2012). A wearable biosensor can analyze pesticides containing organophosphorus in real time on crop surfaces. It transmits the collected data to a smartphone device, enabling in-situ electrochemical analysis of pesticides on agricultural products (Zhao et al. 2020). Utilizing intelligent nanosensors for crop-specific nutrient allocation is key to achieving precision agriculture (Mittal et al. 2020). Nanosensors, which are portable, highly sensitive, and precise, are replacing traditional sensors. Datasets allow AI to analyze the environmental consequences of different fertilizers and their application rates, identifying the most efficient dosages to maximize production and minimize costs. Farming's move toward environmental sustainability will be aided by these (Chen and Yu 2022; Mkrttchian 2021). In the same vein, technological progress has increased crop output and modernized real-time harvesting, processing, marking, and monitoring. AI in agriculture assists farmers with daily tasks and precision farming, optimizing resource use for improved yields and quality. To improve farm productivity and reduce farmers' workload, smart farming utilizes modern technologies such as drones, IoT, and satellite imaging. Machine learning in IoT devices enables continuous monitoring and transmission of vital soil data (composition, N, P, K levels, temperature, humidity, and moisture) to a central cloud. Data analysis via ML algorithms produces actionable insights in this context (Islam et al. 2023) (Table 9.1).

9.4 Principles of Nanobiosensors for Plant Hormone Detection

Nanobiosensors are devices that integrate nanomaterials with biological recognition elements to detect specific target molecules, in this case, plant hormones. The general principle of these sensors involves a biological recognition element (e.g., an

Table 9.1 Components and functions of nanobiosensors

Component	Function	Examples
Biological recognition element	Selectively binds to the target molecule (plant hormones)	Antibodies, aptamers, enzymes
Target molecule	Enrollment in local colleges, 2005 *Source:* Fictitious data, for illustration purposes only Specific substances to be detected	Plant hormones (e.g., auxins, cytokinins, gibberellins)
Nanomaterials	Enhance the sensitivity and performance of the biosensor	Quantum dots, carbon nanotubes, gold nanoparticles
Signal transduction mechanism	Converts the biological interaction into a measurable signal	Fluorescence spectroscopy, electrical measurements, visual colorimetric analysis
Signal detection	Quantifies the measurable change caused by the interaction	Fluorescence spectroscopy, electrical measurements, visual colorimetric analysis
Advantages of nanomaterials	High surface area, unique optical/electrical properties	Enhanced sensitivity, improved detection limits

antibody, aptamer, or enzyme) that selectively binds to a target molecule, causing a measurable change in the sensor's physical or chemical properties. This change can manifest as a shift in electrical conductivity, a fluorescence signal, or a color change, which can be easily quantified. The high surface area and unique properties of nanomaterials, such as quantum dots, carbon nanotubes, and gold nanoparticles, significantly enhance the sensitivity and performance of these biosensors (Yang et al. 2022). Quantum dots, known for their exceptional optical properties, allow for highly sensitive fluorescence-based detection (Fig. 9.1) (Xu et al. 2021a, b). Carbon nanotubes, with their remarkable electrical conductivity and chemical stability, are often employed for the electrochemical sensing of plant hormones (Zhou et al. 2023a, b). Gold nanoparticles are widely used due to their tunable optical properties and biocompatibility, which are particularly effective in colorimetric sensing platforms (Zhang et al. 2022a, b). Aptamer-based biosensors have shown exceptional promise in hormone detection due to their high specificity and stability (Yang et al. 2023a, b). Furthermore, graphene-based biosensors have been utilized for detecting abscisic acid (ABA) and other stress-related hormones, enabling real-time monitoring of plant stress responses (Wang et al. 2023a, b). Microfluidic chips integrated with nanobiosensors allow for high-throughput hormone analysis with minimal sample volumes (Kamat et al. 2023a, b). In addition, wearable biosensors have emerged as an innovative approach for monitoring plant health, offering real-time insights into hormonal changes during growth and stress (Qu et al. 2021a, b). The durability and environmental adaptability of these sensors are critical in agricultural settings, where environmental variables can impact performance (Gupta and Kaur 2022). Recent advancements in electrochemical sensors have further reduced the detection limits for plant hormones, making them applicable to diverse agricultural

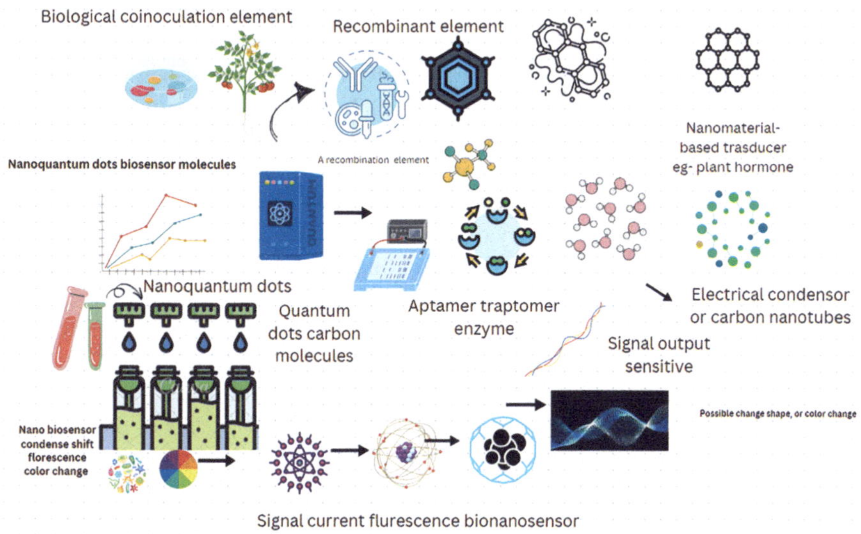

Fig. 9.1 Nano-quantum dots biosensor molecules

scenarios (Naikoo et al. 2021). These advancements collectively highlight the transformative potential of nanobiosensors in precision agriculture, enabling more efficient monitoring and management of plant physiological processes (Mittal et al. 2020).

9.4.1 *Nanomaterials for Biosensing*

Nanomaterials are crucial for enhancing the sensitivity and specificity of nanobiosensors. The use of nanomaterials allows sensors to detect hormones at lower concentrations than traditional techniques. Some of the most commonly used nanomaterials in plant hormone detection are:

- *Carbon Nanotubes (CNTs)*: Carbon nanotubes (CNTs) are cylindrical molecules composed entirely of carbon atoms, which exhibit unique electrical properties, a high surface area, and remarkable mechanical strength. These characteristics make CNTs ideal for electrochemical biosensing applications, especially in detecting hormonal changes in plant tissues. The high conductivity of CNTs enhances the sensitivity and specificity of biosensors, enabling the detection of trace amounts of hormones. CNT-based biosensors have demonstrated remarkable sensitivity in identifying plant hormones such as auxins, gibberellins, and cytokinins (Zhou et al. 2023a, b). Recent studies have further explored these applications. Zhang et al. (2022a, b) reported significant advancements in CNT-based electrochemical sensors, focusing on their use in detecting plant hormones with enhanced sensitivity. Similarly, Liu et al. (2021) highlighted the potential of CNTs and other nanomaterials in creating innovative sensors for plant hormone analysis, paving the way for real-time monitoring techniques. Chen et al. (2023) specifically noted that CNT-functionalized electrodes exhibit improved auxin detection capabilities due to their exceptional electron transport properties. In addition to auxins, the detection of cytokinins has also benefited from CNT-modified sensors, as demonstrated by Ramesh et al. (2022). These sensors leverage the unique surface chemistry of CNTs to achieve precise and rapid cytokinin analysis. Patel et al. (2020) provided a comprehensive review of how carbon nanomaterials, including CNTs, contribute to biosensor development in agricultural research. Their findings suggest that CNT-based biosensors hold promise for advancing precision agriculture practices. Further innovations were detailed by Ali et al. (2020), who discussed the functionalization of CNTs to enhance their biosensing capabilities. emphasized the versatility of CNTs in hormone detection systems, illustrating their potential for large-scale applications in plant biology. Moreover, Gao et al. (2022) demonstrated how CNTs can amplify electrochemical signals, significantly improving the detection limits for plant hormones. CNT-based sensors have also shown promise in detecting gibberellins, as reported by Chowdhury et al. (2023). Their research highlighted both the progress and challenges in using CNTs for gibberellin analysis, underscoring the need for continued innovation. Lastly, Singh and Kaur (2021) detailed the integration of CNT composites in biosensor technologies, showcasing their efficiency in monitoring hormonal changes across various plant systems.

- *Quantum Dots (QDs)*: Quantum dots are semiconductor nanoparticles with size-tunable optical properties, including fluorescence. When exposed to light, quantum dots emit fluorescence that is dependent on their size and composition. This allows them to be used in fluorescence-based biosensors for detecting hormones like ethylene and ABA. The ability of QDs to provide high brightness and photostability enables the detection of hormones at very low concentrations in complex biological matrices (Xu et al. 2021a, b).
- *Gold Nanoparticles (AuNPs)*: Gold nanoparticles are widely used in surface plasmon resonance (SPR) and colorimetric biosensors. AuNPs exhibit strong light-scattering and absorption properties that can amplify signals, making them ideal for detecting hormones in plant tissues. They can be functionalized with biological recognition elements, such as antibodies or aptamers, to selectively bind to specific plant hormones. AuNP-based sensors have been used to monitor hormones like auxins and gibberellins (Zhang et al. 2022a, b).
- *Graphene and Graphene Oxide (GO)*: Graphene is a single layer of carbon atoms arranged in a hexagonal lattice, known for its exceptional electrical conductivity, mechanical properties, and biocompatibility. When combined with graphene oxide (GO), which contains hydroxyl and carboxyl groups, these materials can be used to design sensors with high surface areas that facilitate efficient hormone detection. Graphene-based biosensors have shown great promise in detecting stress hormones such as ABA, which plays a significant role in drought and osmotic stress (Wang et al. 2023a, b) (Table 9.2).

Table 9.2 A comprehensive compilation of aptamers reported in the review, specifically designed for precisely detecting infectious pathogens, highlights their significant potential in diagnostic applications

Target	Specific nanomaterials	Binding affinity	Selex method	References
Chikungunya virus	The subsequent introduction of a silver reagent and its deposition on the AuNP surface created a gray contrast in the testing zone	50 pg/ml	Cell selex	
Candida albicans	AU1 and AD1 aptamers exhibited exceptional potential in detecting (1—3) β-d—glucans, a key component of *C. albicans*'s fungal cell wall. These aptamers demonstrated remarkable specificity and high binding affinity, underscoring their effectiveness in targeted fungal detection	79.76 nm ± 103.7 ± nm	Cell selex	Borghei et al. (2016)
Influenza virus (H3N2) and (H1N1)	A8 and A20 DNA aptamers and their optimized truncated sequences proved highly effective in detecting type A influenza viruses (H3N2 and H1N2). These aptamers exhibited exceptional binding and selectivity, established their utility in precise and reliable influenza diagnostics	5.56 nm ± 5.84 nm	Substantive selex	Bai et al. (2018)
Ebola virus and Ebola Sudan virus	The RNA aptamer 39SGP1A, enhanced with 2'fluropyrimidine (2' FY), demonstrated exceptional efficiency in detecting Ebola virus (EBOV) and Ebola Sudan virus (SUDV), ensuring high sensitivity and reliability in viral diagnostics	30 nm and 250 nm	Cell selex	
Escherichia coli ATCC8739	The FRET-based aptasensor achieved precise *Escherichia coli* ATCC 8739 detection, demonstrating a remarkable detection limit (LOD) of just 3 CFU/mL, ensuring high sensitivity and specificity in bacterial identification	–	Cell selex	–

9.4.2 *Functionalization for Specificity*

For nanobiosensors to be effective, they must selectively detect their target plant hormones in the presence of complex plant matrices. This specificity is achieved by functionalizing nanomaterials with biological recognition elements, which can interact with the target molecule. Common recognition elements include:

- *Antibodies*: Immunosensors, which use antibodies to bind specifically to the target hormone, are among the most common types of biosensors. The high affinity between antibodies and antigens ensures that these sensors are highly specific for the target hormones, making them suitable for detecting a variety of plant hormones, including auxins, gibberellins, and cytokinins.
- *Aptamers*: Aptamers are short DNA or RNA sequences that can fold into specific three-dimensional shapes, allowing them to bind to their target molecules with high affinity. Aptamers offer several advantages over antibodies, including ease of synthesis, stability under harsh conditions, and resistance to temperature fluctuations. Aptamer-based sensors have been used to detect hormones such as ABA (Yang et al. 2023a, b).
- *Molecularly Imprinted Polymers (MIPs)*: MIPs are synthetic polymers designed with specific binding sites that complement the shape and structure of a target molecule. These polymers are cost-effective and highly stable, making them a promising alternative to antibodies and aptamers for plant hormone detection. MIP-based sensors are capable of detecting a wide range of hormones with high specificity (Yang et al. 2022).

9.5 Applications of Nanobiosensors in Agriculture

The integration of nanobiosensors into agriculture presents significant opportunities for improving crop management, resource optimization, and sustainable farming practices. By providing real-time, on-site monitoring of plant hormone levels, these sensors can help farmers make informed decisions about irrigation, fertilization, pest control, and crop harvesting.

9.5.1 *Precision Agriculture and Crop Management*

In precision agriculture, nanobiosensors can provide real-time data on plant health and growth, enabling farmers to optimize resource usage. For instance:

- *Water Stress Monitoring*: The levels of ABA are known to increase in response to drought stress, making it a key marker for water deficit conditions. By using nanobiosensors to monitor ABA levels in plants, farmers can optimize irrigation schedules, improving water use efficiency and reducing waste (Wang et al. 2023a, b).

- *Growth Regulation*: Cytokinins and auxins play vital roles in regulating plant growth, particularly in cell division and root development. By monitoring these hormones in real time, nanobiosensors can help farmers optimize growth conditions, including the timing of fertilization and the application of growth regulators (Yang et al. 2023a, b).

9.5.2 *Crop Quality and Postharvest Management*

Nanobiosensors are also valuable in optimizing crop quality and postharvest processes. For example:

- *Ethylene Detection for Ripening Control*: Ethylene is a key hormone involved in fruit ripening and senescence. Nanobiosensors that detect ethylene levels can help farmers determine the optimal time for harvesting fruits and vegetables, ensuring that they are harvested at peak ripeness and quality (Mochida and Shinozaki 2023a, b).
- *Gibberellin Detection for Seed Germination*: Gibberellins are involved in seed germination and early growth. By detecting gibberellin levels in seeds, nanobiosensors can help ensure uniform germination and early plant development, leading to improved crop yields.

9.6 Challenges and Limitations

Despite the promising potential of nanobiosensors in transforming agricultural practices, several significant challenges and limitations must be overcome for their widespread implementation. One of the primary hurdles is the *selection of appropriate nanomaterials*, which are crucial for ensuring the sensor's accuracy and reliability in detecting specific agricultural parameters. Research by Zhang et al. (2023) highlighted the need for selecting nanomaterials with high specificity and stability for agricultural applications, as well as ensuring their scalability for mass production (Zhang et al. 2023). The diverse environmental conditions of agricultural fields, ranging from variable temperatures, humidity levels, and soil types, pose a challenge to the *stability and durability* of nanobiosensors. For instance, many nanomaterials suffer from degradation when exposed to harsh conditions, affecting their long-term performance (Gupta and Kaur 2022). Additionally, the *sensitivity* of these sensors can be compromised when dealing with the complex matrix of soil, water, or plant tissue, where interference from other substances can lead to false readings or reduced precision. A study by Choi et al. (2021) demonstrated that agricultural environments often introduce nonspecific interferences, diminishing sensor accuracy (Choi et al. 2021). The *integration of these sensors* into existing agricultural infrastructure also faces limitations, particularly in rural or remote areas where access to advanced technology is minimal. While nanobiosensors offer substantial advantages, the *high initial cost* of production and installation remains a significant economic barrier, especially for smallholder farmers. A cost-benefit analysis by Liu et al. (2022) showed that despite the effectiveness of nanobiosensors, the initial investment remains a substantial obstacle to their large-scale adoption in developing countries (Liu et al. 2022). Moreover, concerns regarding the *toxicity of nanomaterials* to nontarget organisms, including beneficial soil microbes or wildlife, raise safety issues that need to be carefully addressed through thorough risk assessments and regulatory frameworks. Studies such as those by Alivisatos et al. (2020) have highlighted the potential environmental risks of nanomaterials when released into agricultural ecosystems, prompting the need for stringent safety evaluations before their widespread use (Alivisatos et al. 2020). Despite these challenges, ongoing advancements in nanotechnology, coupled with cross-disciplinary collaborations, are pushing the development of more stable, affordable, and reliable nanobiosensors for agricultural applications. However, until these issues are effectively resolved, the widespread adoption of nanobiosensors will remain limited.

9.6.1 Stability and Durability

One of the primary challenges for nanobiosensors in agricultural applications is ensuring their stability and durability under field conditions. Environmental factors such as temperature fluctuations, humidity, and exposure to soil or plant tissues can significantly degrade sensor performance, reducing the reliability and lifespan of the devices. Agricultural environments are inherently dynamic, with frequent changes in temperature, moisture levels, and the presence of chemical substances in the soil or plants. These factors can cause physical degradation, chemical instability, or contamination of the nanomaterials used in the sensors. For instance, exposure to UV radiation, water, and oxidative stress can alter the properties of certain nanomaterials, leading to a loss of sensitivity or functionality (Gupta and Kaur 2022). Moreover, interactions with soil particles or plant fluids can result in biofouling, where biological materials accumulate on the sensor's surface, hindering its ability to accurately detect analytes. Studies have shown that maintaining long-term sensor performance in such environments is a significant challenge, especially when dealing with complex matrices like soil and plant tissues, where the high variability of conditions complicates sensor calibration and consistency (Choi et al. 2021). Nanomaterials such as gold nanoparticles or carbon nanotubes, often used in biosensors, can also suffer from agglomeration or surface modifications when exposed to environmental stressors, which compromises their efficiency (Zhang et al. 2023). Therefore, addressing the stability and durability of nanobiosensors is essential for their adoption in real-world agricultural settings. Researchers are exploring various strategies to enhance the resilience of these sensors, including the development of protective coatings, encapsulation techniques, and the use of more robust nanomaterials that can withstand harsh field conditions (Alivisatos et al. 2020).

9.7 Conclusion

Nanobiosensors represent a groundbreaking advancement in agricultural technology, offering significant potential for enhancing crop management, precision farming, and sustainability. By leveraging the unique properties of nanomaterials such as carbon nanotubes, quantum dots, gold nanoparticles, and graphene, these sensors can achieve unprecedented sensitivity and specificity in detecting plant hormones at very low concentrations. This capability is crucial for monitoring vital plant growth processes, including stress responses, growth regulation, and hormone-induced development. Furthermore, the integration of functionalization strategies with biological recognition elements, such as antibodies, aptamers, and molecularly imprinted polymers, enhances the specificity of nanobiosensors, making them effective even in complex biological matrices. The applications of nanobiosensors in agriculture are vast and transformative. From optimizing irrigation schedules based on real-time monitoring of stress hormones like abscisic acid (ABA) to regulating growth and ripening through the detection of cytokinins, auxins, and ethylene, these sensors enable farmers to make informed, data-driven decisions. This results in more efficient resource usage, improved crop yields, and enhanced quality control in postharvest processes. However, despite their immense potential, the widespread adoption of nanobiosensors faces challenges such as the need for greater stability and durability under field conditions, cost constraints, and regulatory concerns about the environmental impact of nanomaterials. As research and development continue to address these challenges, the future of nanobiosensors in agriculture looks promising, with the potential to revolutionize the way we monitor and manage crops for a more sustainable and productive agricultural system. Conclusively, nanobiosensors offer a powerful tool to bridge the gap between advanced technology and agriculture, providing real-time, on-site, and highly sensitive monitoring solutions for plant hormone detection. As these technologies evolve and become more cost-effective, they hold the promise to enhance precision agriculture practices, reduce waste, and contribute to global food security in the face of growing environmental challenges. Continued advancements in nanotechnology, material science, and sensor integration will be essential in unlocking the full potential of these devices, paving the way for a more sustainable and efficient agricultural future.

References

Acimovic J, Sanz V, Ferrer J (2014) Nanomaterials in electrochemical biosensors: from laboratory-scale applications to field detection. Sensors Actuators B Chem 195:107–115

Alfian G, Hossen SM, Kuo YH (2017) A smartphone-based remote sensing system for precision agriculture: real-time monitoring of plant growth and environment. Sensors 17(6):1384

Ali I, Wang J, Yao L (2020) Functionalization of carbon nanotubes to enhance their biosensing capabilities for precision agriculture. Environ Sci Technol 54(7):4195–4206

Alivisatos AP, Lee PK, McLean S (2020) Environmental risks of nanomaterials in agricultural ecosystems: a review and analysis of potential safety concerns. Environ Sci Technol 54(4):2051–2061

Autonacci M, Moro S, Palomba A (2018) Functional materials for precision agriculture: Nanobiosensors for soil and crop management. Agric Syst 165:178–189

Bai Y, Zhang Y, Lu Y (2018) DNA aptamers as sensors for influenza viruses H3N2 and H1N1: optimized sequences and diagnostic potential. J Mol Recognit 31(7):e2722

Bari R, Jones JDG (2009) Role of plant hormones in plant defense responses. Plant Mol Biol 69(4):479–488. https://doi.org/10.1007/s11103-008-9417-7

Bilal M, Iqbal HMN, Rasheed T (2021) Application of nanomaterials in electrochemical biosensors: recent advances, challenges, and future perspectives. Trends Food Sci Technol 108:235–247

Borghei M, Moradi M, Asghari S (2016) High binding affinity and specificity of AU1 and AD1 aptamers in detecting Candida albicans. Biosens Bioelectron 80:222–229

Cellini A, Bolognese S, Caputo G (2017) Electronic nose as a powerful tool for plant disease detection: an overview. Sensors Actuators B Chem 238:825–839

Chen W, Yu J (2022) Precision agriculture and the role of smart farming in sustainable development. Agric Syst 190:103131

Chen G, Liu H, Wang M (2023) CNT-functionalized electrodes for improved auxin detection: advances in electrochemical sensors for plant hormones. J Agric Food Chem 71(13):4592–4600

Choi J, Kwon H, Kim J (2021) Addressing non-specific interferences in Nanobiosensors for agricultural applications. Sensors Actuators B Chem 338:129948

Chowdhury S, Kumar R, Sharma P (2023) Advances in CNT-based sensors for gibberellin detection in plant systems. Sensors Actuators B Chem 336:129743

Cui L, Wang X, Zhang L (2018) A review of electronic nose applications for the detection of plant diseases. Sensors 18(8):2632

Gao L, Xu X, Zhang L (2022) Amplifying electrochemical signals using CNTs for the detection of plant hormones. Electroanalysis 34(5):1034–1042

Gupta S, Kaur H (2022) Durability and environmental adaptability of wearable Nanobiosensors in agricultural environments. Environ Monit Assess 194(9):524

Hazarika S, Saha S, Mondal A (2022) Biosensors in agriculture: a tool for detecting pathogens, pesticides, and food adulteration. Front Bioeng Biotechnol 10:774529

Huang Y, Li Q, Zhang F (2023) Advances in real-time monitoring of plant hormones for precision agriculture: challenges and opportunities. Environ Monit Assess 195(6):372

Islam MT, Kim HS, Lee SH (2023) Integration of machine learning and IoT for real-time soil monitoring in precision agriculture. J Agric Food Chem 71(1):12–24

Jain S (2005) Biosensors and their applications in agriculture and food industry. J Agric Food Chem 53(13):5009–5020

Kamat MS, Banu S, Nagaraj H (2023a) Development of a molecularly imprinted microfluidic chip for sensing nutrients in media. Sensors Actuators B Chem 378:132955

Kamat R, Yadav R, Sharma A (2023b) Microfluidic chips integrated with Nanobiosensors for real-time plant hormone analysis. Lab Chip 23(10):1872–1885

Lal R (2015) Precision agriculture: a key to global food security. Agric Syst 139:107–113

Liu H, Wei T, Zhang J (2021) Nanomaterials in plant hormone biosensing: applications and advancements in carbon nanotubes. Trends Plant Sci 26(3):245–256

Liu J, Zhang Z, Chen L (2022) Cost-benefit analysis of Nanobiosensors for smallholder farmers in developing countries. Agric Syst 191:103098

Lu D, Wang X, Zhang L (2020) A multimodal flexible sensor system for monitoring plant growth status using $ZnIn_2S_4$ nanosheets. Sensors Actuators B Chem 318:128267

Malhotra BD, Chaubey A (2014) Electrochemical biosensors: from research to commercialization. Sensors Actuators B Chem 198:118–129

Mandadi KK, Scholthof HB (2013) Plant immune responses against viruses: Defense mechanisms and resistance strategies. J Exp Bot 64(7):1494–1509
Mittal S, Soni P, Tiwari S (2020) Role of nano sensors in precision agriculture: an overview. Nano Struct Nano Objects 24:100224
Mkrttchian V (2021) Environmental sustainability in agriculture: the role of precision farming and nano sensors. Agric Sustain 5(2):101–113
Mochida K, Shinozaki K (2023a) Ethylene biosensors for fruit ripening control in precision agriculture. Front Plant Sci 14:906654
Mochida K, Shinozaki K (2023b) Real-time monitoring of ethylene levels in crops: implications for agriculture and post-harvest management. J Plant Growth Regul 42(2):475–488
Mogili U, Deepak M (2018) Satellite and drone-based crop sensing for predicting agricultural yield and disease assessment. Comput Electron Agric 153:63–76
Muniandy S, Ismail N, Ariffin K (2019) Advances in biosensor technology for agricultural applications: types and functions. Sensors Actuators B Chem 296:126598
Naikoo MI, Shah MA, Bashir S (2021) Advances in glucose biosensors for plant monitoring: from basic to applied techniques. Trends Plant Sci 26(8):815–828
Narsaiah K, Sharma R, Singh R (2012) Fluorescence-based optical biosensors for nutrient detection in agriculture. Food Rev Intl 28(4):365–377
Pandey S, Singh MK, Sharma A (2018) Real-time monitoring of β-glucuronidase (GUS) expression in transgenic tobacco plants using chronoamperometric biosensors. Sensors Actuators B Chem 256:148–156
Patel N, Singh G, Jindal S (2020) Carbon nanomaterials for the development of plant hormone biosensors: a review. Biosensors 10(3):23–34
Qu F, Zhang X, Li Y (2021a) Plant-wearable sensors for real-time monitoring of plant health and disease detection. Trends Plant Sci 26(5):444–457
Qu Z, Zhang M, Shi S (2021b) Wearable sensors for real-time monitoring of plant health: applications and challenges. Biosens Bioelectron 178:113014
Ramesh S, Kumar S, Suresh S (2022) Carbon nanotube-based biosensors for rapid detection of cytokinins. J Nanosci Nanotechnol 22(4):2841–2851
Rashmi A, Singh S, Jha R (2020) Environmental impacts of agrochemical overuse: implications for sustainability in agriculture. Environ Sci Pollut Res 27(4):3730–3743
Singh A, Kaur P (2021) Integration of CNT composites in biosensor technologies: efficiency in monitoring hormonal changes in plants. Sensors Actuators B Chem 342:129576
Standing K, Davies JM, McKenzie B (2003) Monitoring root exudate release in response to nitrate supply using a luminescent biosensor. Soil Biol Biochem 35(4):567–574
Tang Z, Zhang Y, Wu Q (2022) Nanobiosensors in precision farming: enabling sustainable agricultural practices through real-time monitoring of plant hormones and stress signals. Trends Food Sci Technol 111:121–134
Uslu A, Grossman R (2016) Plant hormones and signaling molecules in response to pathogen recognition and defense. Environ Exp Bot 130:1–14
Verma N, Kapoor D (2015) Nanomaterials for electrochemical biosensors: enhancing precision and sensitivity. Biosens Bioelectron 70:392–403
Wang Y, Zhang S, Li X (2022) Biosensors for precision agriculture: a review on nanomaterial-based sensors for sustainable farming. Sensors 22(15):5478
Wang H, Zhao X, Li L (2023a) Graphene-based biosensors for the detection of abscisic acid in response to plant stress. Sensors Actuators B Chem 372:132683
Wang X, Liu Y, Li L (2023b) Real-time monitoring of abscisic acid (ABA) levels in plants using Nanobiosensors for irrigation management. J Exp Bot 74(1):15–28
Wilson AD (2018) The electronic nose: a review of recent developments. Sensors 18(7):2164
Xu H, Song Q, Zhao Y (2021a) Quantum dot-based biosensors for plant hormone detection. Sensors Actuators B Chem 335:129681
Xu Z, Zhang J, Li H (2021b) Scaling up the production of Nanobiosensors: challenges and strategies for mass production and cost-effective implementation. Nano Today 38:100950

Yang X, Li J, Zhang L (2022) Molecularly imprinted polymers in Nanobiosensors for plant hormone detection. Biosens Bioelectron 82:58–66

Yang Y, Cheng L, Liu Y (2023a) Monitoring cytokinin levels in plants: applications of Nanobiosensors in regulating growth and reducing chemical inputs. Plant Sci 323:111557

Yang Z, Zhang X, Luo J (2023b) Aptamer-based biosensors for abscisic acid detection in agricultural applications. Anal Chim Acta 1234:68–76

Yeomans JC, Larkin RP, McGrath D (1999) The role of biosensors in monitoring nutrient dynamics and microbial activity in the rhizosphere. Soil Biol Biochem 31(6):873–878

Zhang L, Liu L, Chen C (2022a) Gold nanoparticles-based colorimetric sensors for the detection of plant hormones. Biosens Bioelectron 80:284–292

Zhang X, Wang R, Zhang Z (2022b) Integrating nanobiosensors with AI and machine learning: a step toward smart agriculture. IEEE Trans Artif Intell 3(4):342–354

Zhang L, Liu L, Chen C (2023) Advances in the stability and durability of nanomaterial-based biosensors for agricultural monitoring. Environ Sci Nano 10(7):2097–2107

Zhao Y, Zhang L, Huang S (2020) Real-time pesticide detection in agricultural crops using wearable electrochemical biosensors. Food Control 114:107248

Zhou X, Jiang Y, Wang Y (2023a) Carbon nanotubes-based electrochemical biosensors for detection of plant hormones. Electrochim Acta 416:140249

Zhou Y, Chen Z, Xie Q (2023b) Selectivity and reliability of nanobiosensors in complex agricultural environments: overcoming challenges for real-world applications. Biosens Bioelectron 212:114469

Chapter 10
Nanobiosensors for Detecting and Monitoring Soil Conditions

Samina Bibi, Siraj Uddin, Shiza Mukhtar, Javed Iqbal, Banzeer Ahsan Abbasi, and Umar Masood Quraishi

Abstract Real-time monitoring and assessment of soil health is critical for early detection of changes in soil physical, chemical, and biological conditions, enabling informed decision-making and implementation of suitable measures at the right time to improve soil quality and crop productivity. This chapter explores multiple factors that contribute to soil degradation. More specifically, it focuses on the detection of soil physical parameters, pathogens, and contaminants such as pesticides and heavy metals that deteriorate soil fertility. Additionally, human exposure to the toxic chemical pollutants present in soil poses potential health risks. Hence, effective assessment of soil health is critical to maintain ecosystem sustainability. Early detection of soil conditions also enables the development of precision farming and smart agricultural systems. Conventional analytical techniques are widely used for their high sensitivity and efficiency, but their high cost, complex sample preparation, and limited portability due to the large size of equipment limit their applicability for real-time monitoring of soil health. With technological progression, nanobiosensors have emerged as transformative tools for the precise monitoring of soil health in terms of pH, temperature, moisture, nutrients, pathogens, toxic gases, and soil contaminants. The integration of nanomaterials into biosensors enhances their selectivity, reproducibility, and sensitivity at the nanoscale. Nanoparticles (NPs) also improve the interaction of biosensors with target molecules, due to their high reactivity, small size, and large surface-to-volume ratio. Nanobiosensors are eco-friendly, biocompatible, highly sensitive, and portable with ease of use at different sites for real-time assessment. This chapter also reviews various types of

S. Bibi · S. Uddin · S. Mukhtar · U. M. Quraishi
Department of Plant Sciences, Quaid-i-Azam University, Islamabad, Pakistan
e-mail: saminabibi@bs.qau.edu.pk; umasood@qau.edu.pk

J. Iqbal
Department of Botany, Bacha Khan University, Charsadda, Khyber Pakhtunkhwa, Pakistan

B. A. Abbasi (✉)
Department of Botany, Rawalpindi Women University, Rawalpindi, Pakistan

J.-T. Chen (ed.), *Nanobiosensors for Crop Monitoring and Precision Agriculture*, Sustainability Sciences in Asia and Africa,
https://doi.org/10.1007/978-981-96-8335-2_10

nanobiosensors that have been reported in literature, highlighting their design and applications in the assessment of soil health.

Keywords Soil health · Nanobiosensor · Agricultural sustainability · Heavy metals · Pathogens · Pesticides

10.1 Introduction

Soil is a fundamental natural resource that supports plants, humans, and a variety of ecosystems. It comprises organic matter and/or minerals on earth's surface that support plants (Van Es 2017). Soil health refers to the ability of soil to sustain biological productivity and promote biodiversity. Various physical, chemical, and biological factors determine soil health, which is essential to promote agricultural sustainability (Arias et al. 2005; Maikhuri and Rao 2012). Soil rich in minerals, nutrients, and organic matter is not considered healthy if it causes damage to plants. Deterioration of soil health and quality caused by various anthropogenic activities and environmental factors has become a global concern (Fig. 10.1). Excessive fertilization, nutrient imbalance, soil contamination, salinization, acidification, and loss of soil structure are the critical factors that decline soil health (Jangra et al. 2024; Maikhuri and Rao 2012). To satisfy the increasing food demand for an ever-growing population, intensive agricultural practices further contribute to soil degradation and

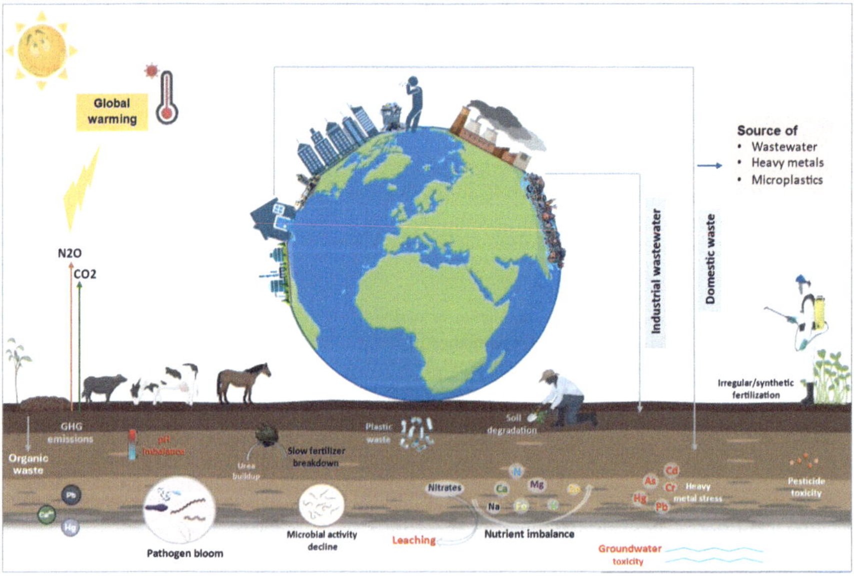

Fig. 10.1 Overview of soil pollution, key sources, and their effects on soil health

greenhouse gas (GHG) emissions (Ma et al. 2024). The increasing nitrogen fertilizer input triggers decomposition, nitrification, and denitrification processes in soil, leading to intensive GHG production, particularly nitrous oxide (N_2O) and carbon dioxide (CO_2) emissions (Khan et al. 2024).

According to an estimate, soil degradation leads to the loss of approximately 12 million hectares of agricultural land every year (Rickson et al. 2015). Additionally, human exposure to toxic chemical pollutants such as pesticides, pathogens, and heavy metals in soil, through dermal contact, inhalation, drinking water, and plant-based food consumption, poses potential risks to human health (Anas et al. 2024b; Sharma et al. 2021). In this context, effective assessment of soil health can help enhance crop productivity and environmental sustainability.

Conventional techniques employed for assessment of soil physical, chemical, and biological parameters require soil sampling, expensive equipment, laboratory experiments, and skilled technicians to operate the equipment. Chromatographic and spectroscopic techniques such as Gas chromatography (GC), mass spectroscopy (MS), high-performance liquid chromatography (HPLC), capillary electrophoresis, nuclear magnetic resonance (NMR) spectroscopy, ultra-visible (UV-Vis) spectroscopy, Fourier-transform infrared (FTIR) spectroscopy, and other techniques are currently used for measurement of pollutants in soil and environment due to their high sensitivity and efficiency but these techniques are laborious, expensive, time-consuming, and often inaccessible to farmers and small-scale stakeholders (Kundu et al. 2019). Moreover, these techniques are not suited for real-time or on-site measurements and instead require a long-time delay between sampling and measurements in lab. Addressing these challenges requires the development of cost-effective, real-time, and portable assessment methods. Although biosensors are already available for measurement of various soil parameters, nanotechnology offers enhanced specificity, sensitivity, and precise detection and quantification of analytes at nanoscale (Jangra et al. 2024). Nanobiosensors are highly effective for the assessment of soil quality and health in terms of pH, temperature, nutrients, pollutants, soil microbes, etc.

10.2 Nanobiosensors

A sensor is a device that is designed to detect changes in the properties of a specific physical, chemical, or biological analyte in a sample and translate them into a readable signal representing the detected variation. In biosensors, a biological component (e.g., DNA, enzyme, antibodies, microorganisms) is connected to a detector that measures changes in biological substances and transmits signals regarding the change observed or the presence of any physical, chemical, or biological component (Huang et al. 2021; Naresh and Lee 2021). Nanotechnology is an innovative technique that improves the performance of sensors and biosensors. Nanoparticles (NPs) have a small size and a large surface area-to-volume ratio, due to which they exhibit unique physical, chemical, biological, catalytic, and optical properties that

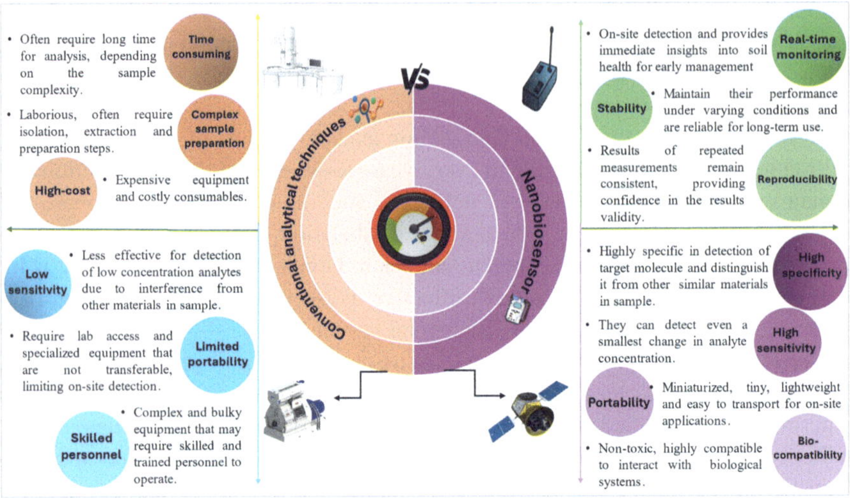

Fig. 10.2 Comparison of nanobiosensors and conventional analytical techniques employed for soil health monitoring

make them highly versatile in environmental fields (Jangra et al. 2024). Nanosensors are eco-friendly and reliable tools that enhance the sensitivity and functioning of biosensors. In nanobiosensors, various types of nanomaterials are used, such as nanoparticles, nanowires, nanotubes, and nanosheets, that act as detectors of signals, increasing the sensitivity of biosensors (Mahmoud and Fawzy 2021). Nanobiosensors are eco-friendly, reusable, biocompatible, and portable with ease of use at different sites (Fig. 10.2).

10.2.1 Components of Nanobiosensors

A typical nanobiosensor comprises:

(i) **Analyte:** A substance or chemical component (nutrient, pollutant, microbe, etc.) which is analyzed to determine its concentration within a sample (e.g., soil).
(ii) **Nano-bioreceptor:** It is a key component of nanobiosensors that combines nanomaterials (nanoparticles, carbon-based materials and quantum dots) with biological recognition components (e.g., cell, DNA, enzyme, antibodies, biomolecules and microorganisms) that recognizes and binds to a target molecule (e.g., analyte) to initiate a readable signal in transducer.
(iii) **Transducer:** It is a component (optical, electrochemical, thermal, electronic, gravimetric and piezoelectric) of a biosensor that plays a key role in conversion of biorecognition signal generated by interaction between an analyte and bio-

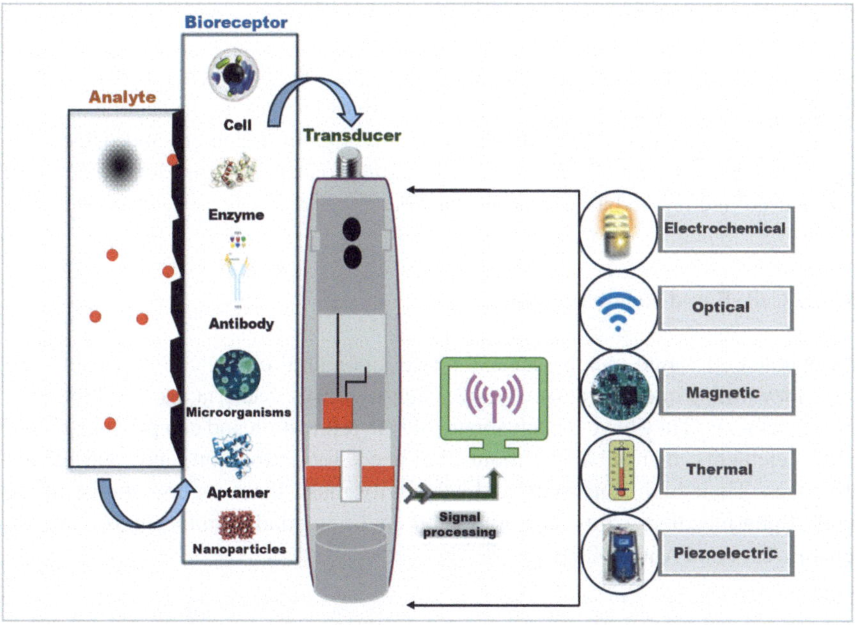

Fig. 10.3 Components of nanobiosensors and types of transducers based on the detection method

receptor into a quantifiable signal (e.g., electrical, optical, thermal or mechanical signal) that helps determine the presence or quantity of a chemical or biological component in sample (Fig. 10.3).

10.3 Classification of Nanobiosensors

10.3.1 Based on the Type of Bioreceptors

Biosensors are classified into catalytic (non-affinity) biosensors and non-catalytic (affinity) biosensors based on the type of specific bioreceptor used.

10.3.1.1 Catalytic Nanobiosensors

The type of biosensor in which bioreceptor and analyte undergo a catalytic reaction, and a new biochemical product is formed. It includes enzymes, organelles/cells, and microorganisms as bioreceptors (Naresh and Lee 2021).

Enzyme-Based Nanobiosensors

Enzymes are the biocatalysts that upregulate the rate of biochemical reactions, thereby enhancing the sensitivity and selectivity of biosensors. In nanobiosensors, an enzyme is attached to the transducer as a recognition element and detects the target analyte molecule (Öndeş et al. 2021). Electrochemical transducers are commonly used in enzyme-based nanobiosensors.

Whole Cell–Based Nanobiosensors

The whole cell–based nanobiosensors possess potential bio-detection capabilities, since they utilize microorganisms (algae, fungi, viruses, bacteria, etc.) as biorecognition elements. They have the advantage of self-replication and can produce recognition elements themselves without the need for extraction and purification processes. Their high sensitivity and selectivity make them a good choice in the field of pharmacology, drug screening, and environmental monitoring as nanobiosensors (Lim and Ahmed 2016).

10.3.1.2 Non-catalytic Nanobiosensors

This type of biosensor does not undergo any catalytic reaction and solely relies on a binding event between bioreceptor and analyte that generates a signal, and no new biochemical product is formed.

It includes nucleic acids (DNA/RNA), antibodies, aptamers, hormones, etc., as bioreceptors (Naresh and Lee 2021).

Antibody-Based Nanobiosensors

Antibodies are affinity-based biorecognition elements and are very effective nanobiosensors due to strong antigen-antibody interactions. In antibody-based nanobiosensors, antibodies are attached to the transducer as labeled or non-labeled immunosensors. In labeled immunosensors, a detectable label is attached to antibodies to measure the antigen-antibody complex, while non-labeled immunosensors are label-free sensors (Narang et al. 2024; Naresh and Lee 2021).

Aptamer-Based Nanobiosensors

Aptamer is a highly stable short single-stranded DNA or RNA molecule that acts as a high-affinity biorecognition element in sensing probes. It can be folded into 2-Dimensional and 3-Dimensional structures, enhancing surface density for binding of target molecules (Naresh and Lee 2021).

Table 10.1 Types of nanobiosensors based on transducers

Nanobiosensor	Transducer	Detection mechanism
Electrochemical	Amperometric	Measures the change in current produced by redox reaction between analyte and electrode
	Impedimetric	Measures impedance/electric resistance changes that occur due to biological interaction
	Potentiometric	Measures voltage changes due to change in electron or ion concentration in response to biological interaction
Optical	Fluorescence	Detects analyte based on change in fluorescence emission, providing high spatial resolution through imaging and spectroscopic techniques
	Surface Plasmon Resonance (SPR)	Measures fluctuations in refractive index on metallic thin film due to molecular binding of biomolecule to analyte
	Surface-Enhanced Raman Spectroscopy (SERS)	A spectroscopic technique in which electromagnetic interaction of light with metal NPs enhances Raman scattering for highly sensitive detection of biomolecules
Thermal	Thermal resistor	Measures changes in heat generated due to biochemical reaction between biomolecule and analyte
Piezoelectric/ Gravimetric	Piezoelectric crystals	Detects analyte by measuring mechanical changes in piezoelectric material that generates an electric charge in response to mass or pressure exerted on it
Magnetic	Magnetic nanoparticles	Detects target molecules based on changes in magnetic properties of nanoparticles that arise due to interaction between biomolecules and analyte

10.3.2 Based on the Type of Transducer

A transducer is a key component of a nanobiosensor that converts biological interaction into readable signals. Based on the transducer, nanobiosensor is classified into electrochemical, optical, magnetic, thermal, piezoelectric, and gravimetric nanobiosensors (Chakraborty et al. 2021; Narang et al. 2024) (Table 10.1).

10.4 Nanobiosensors for Assessment of Soil Health

Adverse changes in physical, chemical, and biological properties of soil caused by soil contamination, soil erosion, loss of soil organic matter, salinization, and climatic changes result in degradation of soil leading to about 20 million tons of grain loss per year (Rickson et al. 2015). About 80% of the agricultural land is affected by moderate to severe erosion worldwide, resulting in an annual economic loss of about $400 billion (Pimentel and Burgess 2013; Pimentel et al. 1995). Moreover, toxic substances transferred to the human body via agricultural inputs provoke various health defects such as metabolic and immunological disorders, infertility, respiratory infections, and disruption of numerous biological functions (Sharma et al. 2021). Eroded soils limit soil mass and depth, nutrient cycling, soil moisture, microbial respiration, and rooting depth, hence leading to poor soil fertility and reduced crop yield (Rickson et al. 2015). Soil health is not only affected by natural processes but also due to soil management practices adopted to improve soil fertility and crop yield that deteriorate soil quality. Consequently, assessment of soil health status and dynamic changes is essential to establish stress mitigation strategies and improve soil health and fertility. Currently, various chromatographic and spectroscopic techniques for monitoring soil health are available. However, time consumption, high cost, and the requirement of sophisticated equipment, laboratory experiments, and skilled personnel to operate these equipment restrict the deployment of these techniques (Kundu et al. 2019). There is a critical need to develop real-time sensors to collect information related to soil health that will enable the development of precision farming and smart agricultural systems. Incorporating nanotechnology into biosensors can increase their sensitivity and performance (Jangra et al. 2024). The tiny size and high surface area-to-volume ratio of nanoparticles help them deal with the target molecules and detect substances at the nanoscale. Nanobiosensors advance soil health assessment by considerably aiding in the real-time detection of soil conditions in terms of wetness level, pH, nutrients, chemical residues, pathogens, and toxic metals in soil, granting early exploration and timely interventions to achieve sustainable agriculture (Sharma et al. 2021).

10.5 Detection of Soil Parameters Using Nanobiosensors

10.5.1 Detection of Physical Parameters

Detecting soil quality, soil moisture, and temperature plays a vital role in agricultural practices to support sustainability. The optimum levels of irrigation and precipitation necessities for crop production are decided by soil temperature and moisture content. For the detection of soil water content, humidity, and temperature, diverse nanosensors are reported in literature (Kalita et al. 2016; Lin et al. 2012; Mahadeva et al. 2011), however, nanobiosensors for the evaluation of soil physical properties are relatively uncommon. Huang et al. (2016) developed a fluorescent bimetallic Au/Ag nanosensor for temperature sensing. A biomolecule, lipoic acid, was employed as a reducing and protecting agent for the green synthesis of highly red fluorescence-emitting bimetallic Au/Ag nanoclusters. The fluorescence response of the sensor to the environmental temperature ranged between 20 and 65 °C. Recently, carbon dots have been reported as effective temperature sensors, highlighting their potential as valuable components for use in nanothermometers (Hernández-Rodríguez et al. 2018). A micro electro-mechanical system

(MEMS)-based temperature/moisture sensor can measure both the soil moisture level and temperature using a wireless communication system (Jackson et al. 2008). Kalita et al. (2016) devised a highly sensitive soil moisture sensor following the MEMS fabrication process using graphene quantum dots (GQDs). Electrochemically synthesized GQDs were dropcasted on the top of gold-interdigitated electrodes (IDE), while microsensors were assembled on silicon wafer that contained IDE electrodes. Microsensors were wrapped using a sensiron filter cap attached to the top of IDE electrodes, which allows only soil moisture to penetrate the chamber and interact with microsensors. The sensor demonstrated a response time of 2–3 min. Similarly, Patil et al. (2014) developed a poly-alanine nanofiber-based piezoresistive microcantilever sensor for the simultaneous measurement of soil humidity and moisture level. A piezoresistive microcantilever was encapsulated with spin-coated SU-8 (structural layer) and carbon black–based nanocomposite. Its surface was then covered with a sensing layer of polyaniline nanofibers. Such highly sensitive, cost-effective, portable, and wireless networking devices can be utilized for efficient and real-time monitoring of the physical parameters of soil.

10.5.2 Detection of Chemical Parameters

Precision agriculture is a farming approach that employs tools and techniques to gather in situ information on soil health status to make data-driven decisions. Soil chemical parameters like pH, nutrients, hormones, pesticide residues, and heavy metals are crucial indicators of soil health.

10.5.2.1 Soil pH

Soil pH influences microbial activity and nutrient availability to plants (Yang et al. 2022). The optimum pH range for plants to uptake required nutrients is 5.5–6.5. Low soil pH in rhizospheric area increases the concentration of manganese and aluminum ions, which in turn reduces soil microbial activity, while high pH reduces nutrient translocation to plants (Siddiqui and Aslam 2023a, b). Detection of soil pH provides invaluable information about soil health for precision agriculture and fertilizer management (Yang et al. 2022). A simple, highly stable, and reusable soil pH sensor was designed by Siddiqui and Aslam (2023a, b) by incorporating a 3D porous graphene framework into quinizarin. This single electrode-based sensor could be employed for more than 500 cycles to detect the pH of different environments by simply washing the sensor with deionized water. In a parallel study by Siddiqui and Aslam (2023a, b), a 3D L-arginine functionalized graphene oxide sensor was designed through coating of reduced graphene oxide nanosheet on gold-interdigitated electrodes for soil pH measurement. Interestingly, high frequency improved the sensitivity of sensors to the hydrogen ions concentration in the sample, thereby enhancing the detection response to pH variations. Similarly, Yang et al. (2022) fabricated a dual copper-doped carbon dots-based nanobiosensor with functional groups of carboxyl and amine, which served as fluorescence nanoprobes for pH detection. In another research conducted by Ke et al. (2016), polymeric pH nanobiosensor using a synthetic peptide octa arginine was designed to enhance the measurement range of pH from 3.9 to 7.3 by simultaneously integrating two complementary pH-sensitive fluorophores in the same nanoparticle. An ultrasensitive DNA-based fluorescent nanodevice for pH detection with a sensing range of 5.3 to 7.5 has been reported, making it suitable for soil pH measurement (Halder and Krishnan 2015).

10.5.2.2 Soil Nutrients

Real-time detection of soil nutrient profile helps in the optimization of fertilization management strategies (Sharma et al. 2021; Yin et al. 2021). Enrichment of urea fertilization causes decreased soil pH and increases pest problems. Soil fertility data is specifically offered by the urea concentration and breakdown assessment. Öndeş et al. (2021) designed a potentiometric biosensor based on poly nanoparticles as carriers for entrapping the urease enzyme. The response time of this sensor was 30 seconds. Similarly, Kaushik et al. (2009) reported a urea sensor utilizing iron oxide-chitosan nanobiocomposite as a support, on which urease and glutamate dehydrogenase enzymes were co-immobilized to measure urea content with a detection limit as low as 0.5 mg/dL. The urease enzyme catalyzes the breakdown of urea into ammonia and carbon dioxide, thus making the analysis of ammonia content in soil a key step in determining fertilizer decomposition. Bertocchi et al. (1996) devised an amperometric enzyme probe that combines urease and glutamate dehydrogenase, immobilized on platinum electrodes, to measure ammonium and urea levels. When fertilizer decomposes, it primarily releases nitrate, which is important for plant growth. However, excessive nitrate will activate nitrification and denitrification, resulting in the discharge of greenhouse gases like nitrous oxide. In various cases, nitrate overload within the soil accelerates these processes in unpredictable ways, raising environmental issues. Therefore, real-time, on-site nitrate detection is essential for accurate fertilizer management. Kundu et al. (2022) fabricated a disposable nanobiosensor that detects and quantifies soil nitrate. This sensor utilized carbon nanotubes functionalized with hematite nanostructures, which were attached to screen-printed carbon electrodes and modified with nitrate reductase enzyme to create the NiR/CNT-α-Fe2O3/SCE enzymatic nanosensor. The non-targeted sites on the nanosensor were covered by treating the activated nanosensor with bovine serum albumin (BSA). The nanosensor showed high reproducibility and good specificity for nitrate with a detection limit of 0.09 mg/L. Similarly, Ahmad et al. (2017) created a highly sensitive nitrate nanobiosensor using zinc oxide nanorods assembled on the silver electrode. Nitrate reductase enzyme was immobilized on an electrode by the physical adsorption method. The nanobiosensor demonstrated a fast response time of 3 sec with minimum detection limit of 1.0 μM nitrate.

Micro- and macro-nutrients are the important components of soil that are required for plant growth and development. However, changes in concentration of these nutrients beyond their optimum range negatively impact soil health and plant growth (Shrivastav et al. 2020). Consistent soil testing for the detection of nutrient level is necessary to maintain their availability to plants. Zinc (Zn) is a transition metal ion that plays a significant biological role. However, excessive Zn ions in soil diminish soil microbial activity and induce phytotoxicity. Bothra et al. (2018) introduced a fluorescent nanobiosensor for turn-on sensing of zinc (II) in biological and environmental samples. Red fluorescent lysozyme cocooned gold nanoclusters (Lyso-AuNCs) were conjugated with vitamin B6 cofactor pyridoxal-5′-phosphate (PLP) for the development of PLP_Lyso-AuNCs sensor. The minimum detection limit of this nanosensor was 39.2 nM Zn^{2+} ions. Similarly, L-histidine-capped copper nanoparticles have been reported for the detection of iron ions (Fe^{3+}), with a detection limit of 82 nM (Lin et al. 2017). In a parallel study, bimetallic gold/silver (Au/Ag) nanosensor has been reported for iron (Fe^{3+}) sensing (Huang et al. 2016). They used a biomolecule lipoic acid as a reducing and protecting agent for the green synthesis of bimetallic (Au/Ag) nanoclusters, which emit a highly red fluorescence on the addition of Fe^{3+} ions. While for the detection of sulfide ions (S^{2-}), Chen et al. (2011) formulated fluorescent DNA-templated Au/Ag nanoclusters with a detection limit of 0.83 nM S^{2-}. Guo et al. (2017) developed a nanobiosensor for visual detection of silver (Ag) ions. Thiolated cytosine (C)-rich oligonucleotide probe immobilized on the AuNPs was designed to capture Ag ions through C-Ag-C coordination chemistry and produce a visible red band with the accumulation of Ag ions. The minimum detection limit of this visual sensor was 1.0 ppb Ag ions under normal conditions.

Similarly, FRET-based nanobiosensors for imaging intracellular Ca^{2+} were designed by Zamaleeva et al. (2015). For the determination of halide ions like iodide (I^-), Li et al. (2017) established a fluorometric and colorimetric device, within which DNA was restrained on the Au/Ag NPs as a biodetection component. The addition of I^- ions into DNA-Au/Ag nanocluster enforces it to emit fluorescence, permitting the detection of those ions with a detection limit of 0.3 μM/L I^- ions. In a similar study, a fluorescent conductor nanocluster was presented for synchronic detection of halide and bromide ions. To stabilize the conductor nanocluster, carboxymethyl dextran was introduced, and the detection limit of this device was 1.0×10^{-7} M (Fu et al. 2017). The capability of nanobiosensors to assess soil nutrients in real-time offers valued data regarding soil health and crop productivity.

10.5.2.3 Heavy Metals

Heavy metals (HMs) pollute agricultural soils, inflicting harmful impacts on crop yield and human health (Anas et al. 2024b). HMs are hazardous even in minimal concentrations and infiltrate agricultural soils through excessive fertilizers, pesticides, polluted irrigation water, sewage sludge, industrial and household discharges, plastic breakdown, and atmospheric deposition. Thus, real-time detection of HMs' levels in soil is necessary for their removal and to ensure soil health. Lead (Pb^{2+}) is a highly toxic metal that negatively affects crop growth through disruption of mineral assimilation, inhibition of seed germination, and reduction of photosynthetic activity (Anas et al. 2024a). Zhou et al. (2017) devised a sensitive, label-free electrochemical sensor for the specific detection of Pb^{2+}. Disodium–anthraquinone–2,6–disulfonate intercalated on Pb^{2+}-dependent GR-5DNAzyme served as an indicator, while nanoporous gold was employed for signal transduction and amplification. The Pb^{2+} detection limit of this sensor was 0.02 nM. In another study, a G-quadruplex DNA and gold NPs-based electrochemical nanobiosensor was reported for the detection of Pb^{2+} with a minimum limit of 0.0042 nM (Xu et al. 2018). Dali et al. (2018) developed a multiplex sensor for the simultaneous detection of Pb^{2+} and Cd^{2+} ions, employing *Trichoderma asperellum* fungi as the sensing element. The sensor was fabricated by incorporating a single-walled nanotube with fungal biomass extracted from metal-contaminated soils. The resulting suspension was subsequently applied to carbon electrodes, improving conductivity and facilitating the detection of trace metals in the analyte.

Analogously, Moyo et al. (2014) developed a nanobiosensor for detecting Pb^{2+} and Cu^{2+} ions. They immobilized peroxidase onto a maize tassel multiwalled nanotube composite for Pb^{2+} and Cu^{2+} sensing, achieving detection limits of 2.5 $\mu g\ L^{-1}$ for Pb^{2+} and 4.2 $\mu g\ L^{-1}$ for Cu^{2+}. Sun et al. (2023) introduced a DNA-electrochemical sensor for Cu^{2+} detection. They used an Fe-MIL-101 glassy carbon electrode (GCE) onto which AuNPs were deposited, forming an AuNPs/Fe-MIL-101/GCE composite. Subsequently, DNAzyme was applied to this composite as a recognition element for Cu^{2+} detection, with a detection limit of 10^{-12} M Cu^{2+}. The sensor exhibited high reliability, repeatability, and sensitivity, along with strong anti-interference properties against other anions. Dayra et al. (2024) recently devised an optical chemical sensor that not only detects but also removes ultra-trace levels of mercury ions. Phenanthraquinone monophenyl thiosemicarbazone (PPT) was inserted as a chelating agent into cellulose triacetate polymer to synthesize an optode membrane. Contaminated soil was filtered through a membrane, and mercury ions were selectively detected and adsorbed on the membrane. Likewise, an optical biosensor incorporating chitosan-capped gold nanoparticles on bovine serum albumin has been reported for the detection and quantification of mercuric ions, in the range 0.1–540 parts per billion (Sadani et al. 2019).

Arsenic (As), a naturally occurring and highly toxic metalloid present in soil, air, and water, can have many adverse effects on human health, including skin diseases, cancer, and heart disorders (Zhu et al. 2020). As toxicity in soil interferes with metabolic processes in plants, resulting in stunted growth and reduced crop production (Zhu et al. 2020). Ensafi et al. (2016) developed a fluorescent aptasensor for ultrasensitive detection of As(III) utilizing an aptamer to induce aggregation of positively charged cysteamine-stabilized cadmium tellurium zinc sulfide (CdTe/ZnS core/shell) quantum dots (QDs). In the presence of As(III), a complex forms with the aptamer, preventing QDs aggregation and releasing fluorescence signals. The fluorescence intensity increased with the increase in concentration of As(III). The aptamer used in this sensor is interchangeable and can be easily substituted with other aptamers to use it for the detection of other metals. Zhu et al. (2020) used a hairpin DNA and polydopamine nanosphere as a reversible and nonspecific inhibitor to design a nanobiosensor for highly sensitive detection of As(III) in environmental samples. Tan et al. (2016) presented a multidimensional sensor for the detection of nine metal ions (Ag^{+}, Hg^{2+}, Cr^{3+}, Sn^{4+}, Cd^{2+}, Cu^{2+}, Pb^{2+}, Zn^{2+}, and Mn^{2+}). They used DNA as a probe that was immobilized on AuNPs conjugates. Metal ions disrupt the excluded effect of negatively charged AuNPs and DNA, and the amount of DNA immobilized on AuNPs' surfaces increases, resulting in a fluorescence quenching effect. The detection limit of this highly sensitive sensor was 50 nM.

The release of cyanide, a highly toxic anion, from mining industries further pollutes the soil. For the detection of cyanide, Peng et al. (2018) designed a biosensor using a fluorescent DNA biosensor. DNA was mixed with Ag nanoclusters (NCs) to form a DNA-Ag-NCs complex. The composite was further used to add guanine-rich solution to amplify fluorescence and detect cyanide in the analyte. Like heavy metals, plastics also pollute the soil which has raised growing concern over the environmental issues. (Sadani et al. 2019). Bisphenol A (BPA) is a harmful substance that is widely used in the production of food containers and plastic bottles (Zou et al. 2019). It is released into the environment and infiltrates soil, posing great danger to human health and agriculture. Zou et al. (2019) reported the construction of a nanobiosensor for BPA detection by using a composite of graphene nanoplatelets and gold nanoparticles immobilized on a GCE electrode. This sensor has been proven to detect BPA with a minimum detection limit of 0.027 nM. With the addition of nanotechnology to biosensors, their detection sensitivity for soil metals is increased, which provides an environmentally friendly solution to soil pollution problems and more effective soil health monitoring (Table 10.2).

10.5.2.4 Pesticides

Pesticides play an important role in modern agriculture due to their efficacy in controlling pests and managing plant diseases. Nonetheless, these pest control practices may lead to soil degradation and increased contamination risk if practiced without caution. Farmers regularly use 2,4-D, DDT, atrazine, and cyclohexane alongside organophosphorus compounds, malathion and parathion chemicals to combat the growth of weeds and protect crops from rodents and insects (Arora 2018). Pesticides residues, left in agricultural field, can pose a serious threat to human health and agricultural sustainability (Singh et al. 2013). Nanobiosensors provide rapid and real-time detection of pesticides in soil (Table 10.3). For instance, atrazine is a pesticide which is highly toxic even at low concentration. Anti-atrazine antibody functionalized gold NPs (Singh et al. 2013) and zinc oxide nanofiber-based immunosensors (Supraja et al. 2020) have been reported to detect atrazine. Like atrazine, organophosphorus pesticides (isocarbophos, profenofos, omethoate, and phorate) are commonly used to kill pests and protect plants (Fu et al. 2019). While highly effective, they can contaminate soil, food, and drinking water, presenting significant risks to human health and the environment. Fu et al. (2019) developed an electrochemical aptasensor for the detection of organophosphorus pesticides. Aptasensor, graphene oxide, and copper NPs were electrodeposited on

Table 10.2 Summary of nanobiosensors for detection of nutrients, heavy metals, and soil pollutants with their composition and limit of detection

Nanobiosensor	Biomaterial	Nanomaterial	Analyte	Detection limit	Citation
Electrochemical	Urease enzyme	Poly (2-hydroxyethyl methacrylate-glycidyl methacrylate) NPs	Urea	0.77 μM	Öndeş et al. (2021)
Electrochemical	Urease and glutamate dehydrogenase enzymes	Superparamagnetic iron oxide NPs-chitosan-based nanobiocomposite film	Urea	0.5 mg dL^{-1}	Kaushik et al. (2009)
Electrochemical	Nitrate reductase enzyme	Nanocomposite of carbon nanotubes and hematite nanostructures	Nitrate	0.09 mg L^{-1}	Kundu et al. (2022)
Electrochemical	Nitrate reductase enzyme	Zinc oxide nanorods	Nitrate	1.0 μM	Ahmad et al. (2017)
Optical (FRET-based)	Cell-penetrating peptides	Commercially available QDs	Ca^{2+}	0.26 μM	Zamaleeva et al. (2015)
Optical (fluorescence)	Lipoic acid	Bimetallic gold/silver nanoclusters	Iron	–	Huang et al. (2016)
Optical (fluorescence)	L-histidine	Copper nanoparticles	Iron	82 nM	Lin et al. (2017)
Optical (fluorescence)	Vitamin B6 cofactor and lysozyme cocoon	Gold nanoclusters	Zinc (Zn^{2+})	39.2 nM	Bothra et al. (2018)
Optical (fluorescence)	Fluorescent DNA	Gold/silver nanoclusters	Sulfide ions	0.83 nM	Chen et al. (2011)
Optical (fluorescence)	Fluorescent DNA	Gold/silver nanoclusters	Iodide ions	0.3 μM L^{-1}	Li et al. (2017)
Optical (UV-induced fluorescence)	Carboxymethyl dextran	Silver nanoclusters	Iodide and bromide ions	1.0×10^{-7} M	Fu et al. (2017)
Optical	Oligonucleotide probe	Gold NPs	Silver (Ag^{+})	1.0 ppb	Guo et al. (2017)
Optical (fluorescence)	Lipoic acid	Gold NPs	Silver (Ag^{+})	9.5 nM	Wang et al. (2018)
Electrochemical	Cytosine-rich DNA	Mushroom-like polyaniline and gold NPs nanocomposite	Silver (Ag^{+})	10 pM	Yang et al. (2015)

Optical (fluorescence)	DNA	Copper/silver nanoclusters	Mercury (Hg^{2+})	2.4 nM	Mao and Wei (2019)
Optical	Bovine serum albumin	Chitosan capped gold NPs	Mercury (Hg^{2+})	0.1–540 ppb	Sadani et al. (2019)
Optical (fluorescence)	DNA	Silver nanoclusters	Mercury (Hg^{2+}) and copper (Cu^{2+})	5 nM and 10 nM	Li et al. (2014)
Optical (fluorescence)	DNA	Gold nanoclusters	Copper (Cu^{2+})	–	Qing et al. (2017)
Electrochemical	DNAzyme	Gold nanoclusters	Copper (Cu^{2+})	10^{-12} M	Sun et al. (2023)
Electrochemical	Horseradish peroxidase enzyme	Maize tassel multiwalled carbon nanotube	Lead (Pb^{2+}) and copper (Cu^{2+})	2.5 μg L^{-1} (Pb^{2+}) 4.2 μg L^{-1} (Cu^{2+})	Moyo et al. (2014)
Electrochemical	*Trichoderma asperellum* fungi biomass	Single-walled carbon nanotubes	Lead (Pb^{2+}) and cadmium (Cd^{2+})	10^{-8} M (Pb^{2+}) 10^{-7} M (Cd^{2+})	Dali et al. (2018)
Optical (fluorescence)	DNA	Silver nanoclusters	Lead (Pb^{2+})	1.0 pM	Wang et al. (2019)
Chronocoulometric aptasensor	DNAzyme	Gold NPs	Lead (Pb^{2+})	0.012 nM	Zhang et al. (2016)
Electrochemical	GR-5DNAzyme	Nanoporous gold	Lead (Pb^{2+})	0.02 nM	Zhou et al. (2017)
Electrochemical	G-quadruplex DNA	Silver NPs	Lead (Pb^{2+})	0.0042 nM	Xu et al. (2018)
Optical (fluorescence)	DNA	Gold NPs	Ag^{+}, Hg^{2+}, Cr^{3+}, Sn^{4+}, Cd^{2+}, Cu^{2+}, Pb^{2+}, Zn^{2+}, and Mn^{2+}	50 nM	Tan et al. (2016)
Optical (fluorescence)	DNA aptamers	Silica NPs	Arsenic (III)	0.9 ppb	Oroval et al. (2017)
Electro-chemiluminescent	Hairpin DNA	Polydopamine nanospheres	Arsenic (III)	1.2×10^{-3} ppb	Zhu et al. (2020)
Optical (fluorescence)	Aptamer	Cadmium tellurium/zinc sulfide (CdTe/ZnS) core/shell QDs	Arsenic (III)	1.3 pmol L^{-1}	Ensafi et al. (2016)
Optical (fluorescence)	ssDNA	Silver nanoclusters	Cyanide	25.6 nM	Peng et al. (2018)
Electrochemical	Graphene nanoplatelets	Gold NPs	Bisphenol A (BPA)	0.027 nM	Zou et al. (2019)

Table 10.3 Summary of nanobiosensors for detection of pesticides with their composition and limit of detection

Nanobiosensor	Bioelement	Nanomaterial	Analyte	Detection limit	Citation
Electrochemical	Mercaptomethamidophos substrate containing thiol group	Gold NPs	11 kinds of organophosphorus pesticides	0.019–0.077 ng mL^{-1}	Zhao et al. (2021)
Electrochemical	Aptasensor	Graphene oxide and copper NPs	Organophosphorus (profenofos phorate, isocarbophos, and omethoate)	0.003 nM (profenofos), 0.3 nM (phorate), 0.03 nM (isocarbophos) and 0.3 nM (omethoate)	Fu et al. (2019)
Electrochemical	DNA	Gold NPs	Organophosphorus (profenofos)	0.27 μM	Selvolini et al. (2018)
Electrochemical	DNA aptamer	Lipid bilayer nanopore	Organophosphorus (omethoate)	0.27 μM	Nobukawa et al. (2015)
Electrochemical	DNA	Gold and upconversion NPs	Insecticide (acetamiprid)	0.36 nM	Yang et al. (2019)
Bare-eye-based lateral flow immunoassay	Monoclonal antibodies	Silver NPs	Insecticide (imidacloprid chlorpyrifos-methyl isocarbophos)	50 μg L^{-1} (imidacloprid) 100 μg L^{-1} (chlorpyrifos-methyl) 100 μg L^{-1} (isocarbophos)	Wang et al. (2014)
Optical (fluorescence turn-on)	Aptamer	Zinc sulfide: manganese QDs	Insecticide (acetamiprid)	0.7 nM	Lin et al. (2016)
Optical (fluorescence)	Aptamer	Gold NPs and cadmium/tellurium QDs	Insecticide (acetamiprid)	7.29 nM	Guo et al. (2016)
Electrochemical	Aptamer	Nitrogen doped graphene nanocomposites decorated with silver NPs	Insecticide (acetamiprid)	3.3×10^{-14} M	Jiang et al. (2015)

Colorimetric	Aptamer (oligonucleotide ssDNA)	Silver NPs	Insecticide/acetamiprid (carbofuran)	89.8 nM	Jokar et al. (2016)
Electrochemical	Anti-atrazine antibody	Gold NPs	Herbicide (atrazine)	10.78×10^{-15} g mL^{-1}	Singh et al. (2013)
Electrochemical	Anti-atrazine antibody	Zinc oxide nanofiber and multiwalled carbon nanotubes	Herbicide (atrazine)	5.368×10^{-21} M	Supraja et al. (2020)
Electrochemical	Ascorbate phosphatase enzyme	Multiwalled carbon nanotube	Herbicide (2, 4-Dichlorophenoxy acetic acid)	1 ppb	Bollella et al. (2016)
Electrochemical (impedimetric spectroscopy)	Impedimetric aptasensor	Gold NPs	Fungicide (carbendazim)	0.5 pg mL^{-1}	Zhu et al. (2019)

screen-printed carbon electrodes, enhancing electrical conductivity. The detection limit of the sensor ranged from 0.003 nM to 0.3 nM. Zhu et al. (2019) employed the electrochemical impedance spectroscopy technique for sensing carbendazim fungicide. Carbon nanohorns were modified with aptasensor for their electrostatic attachment with AuNPs to synthesize a sensor with a detection limit of 3.7 mgL^{-1}. Glyphosate is a commonly used herbicide in agricultural fields to control weeds. However, overuse or mishandling of this herbicide can harm crops and soil beneficial microbes. 2, 4-Dichlorophenoxy acetic acid (2,4-D) is a widely used herbicide for controlling weeds in cereal crops. However, 2,4-D residues left in agricultural products and the environment can harm humans, crops, animals, and soil microbes (Bollella et al. 2016). Ascorbate phosphatase-based electrochemical nanobiosensor was developed in which the enzyme was spotted on screen-printed electrodes (graphene, graphite, and multiwalled carbon nanotube). The detection of 2,4-D concentration in soil was based on the principle of ascorbate phosphatase enzyme inhibition by 2,4-D. For quantitative detection and imaging of acetamiprid insecticide, ZnS: Mn quantum dots were conjugated with an acetamiprid-binding aptamer. Multiwalled carbon nanotube was used to turn off probe fluorescence, which was designed to turn on only after the introduction of acetamiprid due to specific binding between ZnS: Mn aptamer and acetamiprid (Lin et al. 2016). The development of these nanobiosensors contributes to advancements in pesticides detection to ensure soil health and environmental sustainability.

10.5.3 Nanobiosensors for Detection of Biological Parameters

10.5.3.1 Soil Microbes and Pathogen Detection

Early detection of plant pathogens, including bacteria, fungi, and viruses in soil, plays a key role in plant and soil health monitoring. Various antibody-based and DNA-based electrochemical and optical nanobiosensors have been reported that are highly sensitive and allow rapid detection of soil-borne pathogens as well as beneficial microbes (Khater et al. 2017) (Table 10.4). Kim et al. (2014) reported that aptamer-based nanobiosensors synthesized using quantum dots and gold electrode chips are very efficient for the detection of bacteria without requiring cell lysis. Aptamers are single-stranded oligonucleotides that can easily bind to different components, such as polysaccharides, proteins, or bacterial flagella, giving a high fluorescent signal. The high affinity and binding capacity of aptamers make them a good choice for the detection of complex targets (Kim et al. 2014). Unique components in bacterial cell walls, for example, lipopolysaccharides in Gram-negative bacteria and teichoic acid in Gram-positive bacteria, work as biomarkers that trigger changes in nanobiosensors, enabling signal transduction (Sharma et al. 2021). Recently, an innovative enzyme-based lipase/gold nanobiosensor was presented that provides rapid and sensitive detection of bacteria with human olfactory output (Duncan et al. 2017). Cebula et al. (2019) proposed an immunosensor for the detection of

Table 10.4 Summary of nanobiosensors for detection of soil microbes and pathogens with their composition and limit of detection

Nanobiosensor	Bioelement	Nanomaterial	Analyte	Detection limit	Citation
Fluorescence resonance energy transfer (FRET)	Antibody based on immunodominant membrane protein of phytoplasma	Thioglycolic acid–modified cadmium-telluride QDs	*Candidatus Phytoplasma aurantifolia*	5 ca. *P. aurantifolia* μL^{-1}	Rad et al. (2012)
Optical (fluorescent)	Pseudomonas aeruginosa monoclonal antibody immunoglobulin G (IgG)	Glucose stabilized silver NPs	*Pseudomonas aeruginosa*	1.5 CFU (colony forming units) mL^{-1}	Ellairaja et al. (2017)
Fluorescence resonance energy transfer (FRET)	Anti-*aspergillus* antibody IgG	FRET-based QDs	*Aspergillus amstelodami*	10^3 spores mL^{-1}	Kattke et al. (2011)
Optical (fluorescent)	Antibody developed against *Polymyxa* specific protein antigen	Cadmium-telluride QDs	*Polymyxa* species	96 samples (within less than 30 min)	Safarpour et al. (2012)
Colorimetric	Capture probe DNA	Gold NPs	*Escherichia coli K88*	1.35×10^2 CFU mL^{-1}	Wang et al. (2020)
Human olfactory output	Lipase enzyme	Gold NPs	*Escherichia coli*	10^2 CFU mL^{-1}	Duncan et al. (2017)
Polymerase chain reaction (PCR)	Thiolated single-stranded oligonucleotide	Gold NPs	*Ralstonia solanacearum*	15 ng	Khaledian et al. (2017)
Electrochemical	DNA	Zinc oxide NPs and a chitosan nanocomposite membrane	*Trichoderma harzianum*	1.0×10^{-19} Mol L^{-1}	Siddiquee et al. (2014)
Electrochemical	Oligonucleotide	Magnetic NPs	*Phytophthora palmivora*	0.30 ng DNA μL^{-1}	Franco et al. (2019)
Hybridization-based (lateral flow)	*P. infestans*-specific repetitive ssDNA sequence	Gold NPs	*Phytophthora infestans*	0.1 pg μL^{-1} genomic DNA	Zhan et al. (2018)

(continued)

Table 10.4 (continued)

Nanobiosensor	Bioelement	Nanomaterial	Analyte	Detection limit	Citation
Optical (localized surface plasmon resonance)	DNA sequences complementary to the analyte	Gold NPs	Multiple fungal pathogen	~0.8 particles μm^{-2}	Zopf et al. (2019)
Optical (immunomagnetic separation)	Anti-ToRSV antibody, anti-BPMV antibody, and anti-ArMV antibody	Upconversion NPs	*Tomato ringspot virus* (ToRSV), *bean pod mottle virus* (BPMV), *and Arabis mosaic virus* (ArMV)	10^{-4} mg/m L^{-1}	Zhang et al. (2013)

Pseudomonas Syringae subtype *Lachrymans* employing gold electrodes. Some soil-borne pathogens persist in the soil for extended periods, thus on-field detection of these pathogens is highly desirable for soil health. For instance, *Ralstonia solanacearum* is a persistent soil-borne bacterium that causes severe diseases in plants, including potato bacterial wilt. Khaledian et al. (2017) developed a biosensor with a nanoscale strategy that can locate *R. solanacearum* within soil samples. For identifying the genomic DNA of the designated bacteria, they used a probe consisting of gold nanoparticles attached to thiolated oligonucleotides. For the detection of fungi, certain substances released by them can provide valuable information about their population and contamination risks. Specifically, mycotoxins, which are the secondary metabolites of fungi, can serve as biomarkers of mycotoxin-producing fungi such as *Aspergillus*, *Alternaria*, *Fusarium*, and *Penicillium* in soil. Mak et al. (2010) reported that a magnetic nanotag-based enzyme-linked immunosensor demonstrates a high efficacy in the simultaneous detection of multiple mycotoxins. Likewise, Franco et al. (2019) devised an electrochemical nanobiosensor for the detection of genomic DNA of *Phytophthora palmivora* that causes black pod rot in *Theobroma cacao L.* (cacao). Oligonucleotide probe was labeled with electrically active magnetic nanoparticles, and the detection limit of the probe was 0.30 ng DNA μL^{-1}.

Unlike pathogenic microflora, beneficial microbes present in soil promote soil health and plant growth by enhancing nutrient uptake, improving soil fertility, and protecting plants from pathogens (Finkel et al. 2017). For instance, genus *Trichoderma* has been known as a good microorganism and is used as a biocontrol agent as well as a growth promoter to enhance plant protection and crop productivity. Siddiquee et al. (2014) constructed an electrochemical nanobiosensor for *Trichoderma harzianum* detection utilizing ZnO nanoparticles and chitosan nanocomposite. Like any other microbe, finding a soil virus that disrupts the soil ecosystem is important for soil health evaluation. A multiple antibody-based nanobiosensor has been reported for the detection of three distinct viruses (Zhang et al. 2013). Here, an antibody was covalently bound to the surface of magnetically responsive ($NH2\text{-}FeCl_3/FeSO_4$) nanoparticles. These magnetic NPs facilitate sample separation by magnetic forces while allowing simultaneous detection of multiple viruses. Likewise, for *P. infestans* detection, nanobiosensor mediated by AuNPs (Zhan et al. 2018), for *Candidatus Phytoplasma aurantifolia*, an immunobiosensor facilitated by CdTe-QDs (Rad et al. 2012), for *Polymyxa* species, CdTe-QDs-based fluorescent immunobiosensor (Safarpour et al. 2012), and for *Pseudomonas aeruginosa*, glucose stabilized silver NP-based fluorescent immunobiosensor (Ellairaja et al. 2017) have been reported (Table 10.4). The recent achievements in the field of nanobiosensors have greatly improved the specificity and sensitivity of pathogen detection, making it possible to monitor soil and plant conditions in real time.

10.5.3.2 Toxic Gas Sensing

The soil acts as a major source and sink of emissions, especially greenhouse gases (GHG) like nitrous oxide (N_2O), carbon dioxide (CO_2), methane (CH_4), and nitric oxide (NO). CH_4 is produced by methanogenesis under anaerobic conditions, while nitrification and denitrification processes drive N_2O and NO emissions from soil. These gases influnce soil health and climate change (Oertel et al. 2016). Thus, it is critical to better assess and quantify the GHG emissions from soil for monitoring soil degradation, soil microbial activity, carbon sequestration, and reducing their adverse environmental impacts. A wide range of nanosensors have been reported for gas sensing, but the implementation of nanotechnology-based biosensors in this field remains scarce. *Ochradenus baccatus* leaf extract-mediated zinc oxide nanoparticles have been employed for gas sensor fabrication. Two gold electrodes were deposited onto a glass substrate using DC sputtering, and then sensing material of ZnONPs was fabricated on the electrodes. The fabricated sensor was used for sensing humidity and gases such as CH_4, NO_2, H_2, and CO (Khan et al. 2024). Hwang et al. (2023) developed a multiplexed ssDNA-functionalized graphene sensor for the detection of NH_3, NO, NO_2, and H_2S gases. Likewise, Thamri et al. (2021) devised thiol-amine functionalized carbon nanotubes for the detection of aromatic and nonaromatic compounds as target gas molecules. Multiwalled carbon nanotubes loaded with platinum and palladium NPs were functionalized with a monolayer of thiol-amine for sensor fabrication. Similarly, cellulose acetate nanofibers have been reported for use as a highly sensitive ammonia sensor. Polyethylenimine (positively charged) and graphene oxide (negatively charged) were assembled following the electrostatic layer-by-layer method on the surfaces of cellulose acetate (negatively charged) nanofibers (Jia et al. 2016). All in all, nanosensors have revolutionized the soil monitoring by allowing more efficient, sensitive, and real-time assessment of soil conditions. As research on nanobiosensors continues to advance, they will become even more profound transformative tools, paving the way for ecological sustainability.

10.6 Concluding Remarks and Future Perspectives

Soil degradation not only reduces soil fertility but also compromises crop productivity, disrupts ecosystems, and threatens environmental sustainability. Soil contaminants, particularly heavy metals and pesticide residues, can enter food chains via plants and drinking water, posing serious threats to animals and human health. Thus, effective assessment of soil health can help enhance agricultural productivity and ecological sustainability. Nanotechnology has revolutionized soil health monitoring, offering enhanced sensitivity, selectivity, and specificity. Nanomaterial integration into biosensors enables real-time and rapid detection of soil physical, chemical, and biological parameters, providing valuable insights into soil health and fertility. However, conventional techniques employed for assessment of soil quality are laborious, expensive, and time-consuming, lacking on-site measurements. Nanobiosensors are eco-friendly, biocompatible, and portable with ease of use at different sites for early detection of changes in soil conditions, enabling informed decision-making and timely interventions to ensure agricultural sustainability. Despite significant advantages of nanobiosensors in soil health assessment, their scalability for large-scale applications and environmental impacts associated with the disposal of nanomaterials can limit their widespread adoption. Further research should focus on these challenges to make nanobiosensors a valuable tool for the assessment of soil health and environmental monitoring to ensure ecological sustainability.

References

Ahmad R, Bhat KS, Ahn M-S, Hahn Y-B (2017) Fabrication of a robust and highly sensitive nitrate biosensor based on directly grown zinc oxide nanorods on a silver electrode. New J Chem 41(19):10992–10997. https://doi.org/10.1039/C7NJ02526B

Anas M, Bashir MS, Saleem K, Noor A, Quraishi UM (2024a) Adaptive mechanisms of wheat cultivars to lead toxicity through enhanced oxidative defense, ionomic redistribution, and anatomical modifications. J Plant Physiol 303:154370. https://doi.org/10.1016/j.jplph.2024.154370

Anas M, Bibi S, Farwa U, Khan A, Quraishi UM (2024b) Manganese oxide nanoparticles elicit cadmium tolerance in wheat seedlings by ionomic and phenomic adjustment, regulation of AsA–GSH pathway, cellular thickness recovery, and antioxidant modulation. Environ Sci Nano 11(11):4588–4605. https://doi.org/10.1039/D4EN00490F

Arias ME, González-Pérez JA, González-Vila FJ, Ball AS (2005) Soil health: a new challenge for microbiologists and chemists. Int Microbiol 8(1):13–21

Arora K (2018) Advances in nano based biosensors for food and agriculture. In: Gothandam KM, Ranjan S, Dasgupta N, Ramalingam C, Lichtfouse E (eds) Nanotechnology, food security and water treatment. Springer, pp 1–52. https://doi.org/10.1007/978-3-319-70166-0_1

Bertocchi P, Compagnone D, Palleschi G (1996) Amperometric ammonium ion and urea determination with enzyme-based probes. Biosens Bioelectron 11(1):1–10. https://doi.org/10.1016/0956-5663(96)83708-0

Bollella P, Fusco G, Tortolini C, Sanzò G, Antiochia R, Favero G, Mazzei F (2016) Inhibition-based first-generation electrochemical biosensors: theoretical aspects and application to 2,4-dichlorophenoxy acetic acid detection. Anal Bioanal Chem 408(12):3203–3211. https://doi.org/10.1007/s00216-016-9389-z

Bothra S, Babu LT, Paira P, Ashok Kumar SK, Kumar R, Sahoo SK (2018) A biomimetic approach to conjugate vitamin B6 cofactor with the lysozyme cocooned fluorescent AuNCs and its application in turn-on sensing of zinc(II) in environmental and biological samples. Anal Bioanal Chem 410(1):201–210. https://doi.org/10.1007/s00216-017-0710-2

Cebula Z, Żołędowska S, Dziąbowska K, Skwarecka M, Malinowska N, Białobrzeska W, Czaczyk E, Siuzdak K, Sawczak M, Bogdanowicz R, Nidzworski D (2019) Detection of the plant pathogen pseudomonas Syringae pv. Lachrymans on antibody-modified gold electrodes by electrochemical impedance spectroscopy. Sensors 19(24)

Chakraborty U, Kaur G, Chaudhary GR (2021) Development of environmental nanosensors for detection monitoring and assessment. In: Kumar R, Kumar R, Kaur G (eds) New Frontiers of nanomaterials in environmental science. Springer, Singapore, pp 91–143. https://doi.org/10.1007/978-981-15-9239-3_5

Chen W-Y, Lan G-Y, Chang H-T (2011) Use of fluorescent DNA-templated gold/silver nanoclusters for the detection of sulfide ions. Anal Chem 83(24):9450–9455. https://doi.org/10.1021/ac202162u

Dali M, Zinoubi K, Chrouda A, Abderrahmane S, Cherrad S, Jaffrezic-Renault N (2018) A biosensor based on fungal soil biomass for electrochemical detection of lead (II) and cadmium (II) by differential pulse anodic stripping voltammetry. J Electroanal Chem 813:9–19. https://doi.org/10.1016/j.jelechem.2018.02.009

Dayra HA, Abdelaal MY, Khalifa ME, Abdallah AB (2024) Novel optical optode for selective detection and removal of ultra-trace level of mercury ions in different environmental real samples. Sci Rep 14(1):26400. https://doi.org/10.1038/s41598-024-76571-y

Duncan B, Le NDB, Alexander C, Gupta A, Yesilbag Tonga G, Yazdani M, Landis RF, Wang L-S, Yan B, Burmaoglu S, Li X, Rotello VM (2017) Sensing by smell: nanoparticle–enzyme sensors for rapid and sensitive detection of bacteria with olfactory output. ACS Nano 11(6):5339–5343. https://doi.org/10.1021/acsnano.7b00822

Ellairaja S, Krithiga N, Ponmariappan S, Vasantha VS (2017) Novel pyrimidine tagged silver nanoparticle based fluorescent immunoassay for the detection of Pseudomonas aeruginosa. J Agric Food Chem 65(8):1802–1812. https://doi.org/10.1021/acs.jafc.6b04790

Ensafi AA, Kazemifard N, Rezaei B (2016) A simple and sensitive fluorimetric aptasensor for the ultrasensitive detection of arsenic(III) based on cysteamine stabilized CdTe/ZnS quantum dots aggregation. Biosens Bioelectron 77:499–504. https://doi.org/10.1016/j.bios.2015.10.011

Finkel OM, Castrillo G, Herrera Paredes S, Salas González I, Dangl JL (2017) Understanding and exploiting plant beneficial microbes. Curr Opin Plant Biol 38:155–163. https://doi.org/10.1016/j.pbi.2017.04.018

Franco AJDM, Merca FE, Rodriguez MS, Balidion JF, Migo VP, Amalin DM, Alocilja EC, Fernando LM (2019) DNA-based electrochemical nanobiosensor for the detection of Phytophthora palmivora (Butler) Butler, causing black pod rot in cacao (Theobroma cacao L.) pods. Physiol Mol Plant Pathol 107:14–20. https://doi.org/10.1016/j.pmpp.2019.04.004

Fu L, Li C, Li Y, Chen S, Long Y, Zeng R (2017) Simultaneous determination of iodide and bromide using a novel LSPR fluorescent Ag nanocluster probe. Sensors Actuators B Chem 240:315–321. https://doi.org/10.1016/j.snb.2016.08.151

Fu J, An X, Yao Y, Guo Y, Sun X (2019) Electrochemical aptasensor based on one step co-electrodeposition of aptamer and GO-CuNPs nanocomposite for organophosphorus pesticide detection. Sensors Actuators B Chem 287:503–509. https://doi.org/10.1016/j.snb.2019.02.057

Guo J, Li Y, Wang L, Xu J, Huang Y, Luo Y, Shen F, Sun C, Meng R (2016) Aptamer-based fluorescent screening assay for acetamiprid via inner filter effect of gold nanoparticles on the fluorescence of CdTe quantum dots. Anal Bioanal Chem 408(2):557–566. https://doi.org/10.1007/s00216-015-9132-1

Guo Z, Zheng Y, Xu H, Zheng B, Qiu W, Liu G (2017) Lateral flow test for visual detection of silver(I) based on cytosine-ag(I)-cytosine interaction in C-rich oligonucleotides. Microchim Acta 184(11):4243–4250. https://doi.org/10.1007/s00604-017-2460-4

Halder S, Krishnan Y (2015) Design of ultrasensitive DNA-based fluorescent pH sensitive nanodevices. Nanoscale 7(22):10008–10012. https://doi.org/10.1039/C5NR01158B

Hernández-Rodríguez MA, Afonso MM, Palenzuela JA, Martín IR, Soler-Carracedo K (2018) Carbon dots as temperature nanosensors in the physiological range. J Lumin 196:313–315. https://doi.org/10.1016/j.jlumin.2017.12.062

Huang H, Li H, Feng J-J, Wang A-J (2016) One-step green synthesis of fluorescent bimetallic Au/Ag nanoclusters for temperature sensing and in vitro detection of Fe3+. Sensors Actuators B Chem 223:550–556. https://doi.org/10.1016/j.snb.2015.09.136

Huang X, Zhu Y, Kianfar E (2021) Nano biosensors: properties, applications and electrochemical techniques. J Mater Res Technol 12:1649–1672. https://doi.org/10.1016/j.jmrt.2021.03.048

Hwang YJ, Yu H, Lee G, Shackery I, Seong J, Jung Y, Sung S-H, Choi J, Jun SC (2023) Multiplexed DNA-functionalized graphene sensor with artificial intelligence-based discrimination performance for analyzing chemical vapor compositions. Microsyst Nanoeng 9(1):28. https://doi.org/10.1038/s41378-023-00499-y

Jackson T, Mansfield K, Saafi M, Colman T, Romine P (2008) Measuring soil temperature and moisture using wireless MEMS sensors. Measurement 41(4):381–390. https://doi.org/10.1016/j.measurement.2007.02.009

Jangra P, Kumar S, Rosin KG, Dwivedi N, Barnwal PP, Mishra AK, Kumari U (2024) Nanobiosensors for soil remediation. In: Yasheshwar AK, Mishra, Kumar M (eds) Recent advancements in sustainable agricultural practices: harnessing technology for water resources, irrigation and environmental management. Springer, Singapore, pp 353–374. https://doi.org/10.1007/978-981-97-2155-9_18

Jia Y, Yu H, Zhang Y, Dong F, Li Z (2016) Cellulose acetate nanofibers coated layer-by-layer with polyethylenimine and graphene oxide on a quartz crystal microbalance for use as a highly sensitive ammonia sensor. Colloids Surf B: Biointerfaces 148:263–269. https://doi.org/10.1016/j.colsurfb.2016.09.007

Jiang D, Du X, Liu Q, Zhou L, Dai L, Qian J, Wang K (2015) Silver nanoparticles anchored on nitrogen-doped graphene as a novel electrochemical biosensing platform with enhanced sensitivity for aptamer-based pesticide assay. Analyst 140(18):6404–6411. https://doi.org/10.1039/C5AN01084E

Jokar M, Safaralizadeh MH, Hadizadeh F, Rahmani F, Kalani MR (2016) Design and evaluation of an apta-nano-sensor to detect Acetamiprid in vitro and in silico. J Biomol Struct Dyn 34(11):2505–2517. https://doi.org/10.1080/07391102.2015.1123188

Kalita H, Palaparthy VS, Baghini MS, Aslam M (2016) Graphene quantum dot soil moisture sensor. Sensors Actuators B Chem 233:582–590. https://doi.org/10.1016/j.snb.2016.04.131

Kattke MD, Gao EJ, Sapsford KE, Stephenson LD, Kumar A (2011) FRET-based quantum dot immunoassay for rapid and sensitive detection of aspergillus amstelodami. Sensors 11(6):6396–6410

Kaushik A, Solanki PR, Ansari AA, Sumana G, Ahmad S, Malhotra BD (2009) Iron oxide-chitosan nanobiocomposite for urea sensor. Sensors Actuators B Chem 138(2):572–580. https://doi.org/10.1016/j.snb.2009.02.005

Ke P, Sun H, Liu M, Su Z, Hu K, Zhua H, Guo H, Sun H, Andresen TL, Olsen LF (2016) Polymeric pH nanosensor with extended measurement range bearing octaarginine as cell penetrating peptide. IET Nanobiotechnol 10(1):8–12. https://doi.org/10.1049/iet-nbt.2014.0070

Khaledian S, Nikkhah M, Shams-bakhsh M, Hoseinzadeh S (2017) A sensitive biosensor based on gold nanoparticles to detect Ralstonia solanacearum in soil. J Gen Plant Pathol 83(4):231–239. https://doi.org/10.1007/s10327-017-0721-z

Khan MW, Shaalan NM, Ahmed F, Sherwani S, Aljaafari A, Alsukaibi AKD, Alenezi KM, Al-Motair K (2024) Gas sensing performance of zinc oxide nanoparticles fabricated via Ochradenus baccatus leaf. Chemosensors 12(2)

Khater M, de la Escosura-Muñiz A, Merkoçi A (2017) Biosensors for plant pathogen detection. Biosens Bioelectron 93:72–86. https://doi.org/10.1016/j.bios.2016.09.091

Kim YS, Chung J, Song MY, Jurng J, Kim BC (2014) Aptamer cocktails: enhancement of sensing signals compared to single use of aptamers for detection of bacteria. Biosens Bioelectron 54:195–198. https://doi.org/10.1016/j.bios.2013.11.003

Kundu M, Krishnan P, Kotnala RK, Sumana G (2019) Recent developments in biosensors to combat agricultural challenges and their future prospects. Trends Food Sci Technol 88:157–178. https://doi.org/10.1016/j.tifs.2019.03.024

Kundu M, Krishnan P, Chobhe KA, Manjaiah KM, Pant RP, Chawla G (2022) Fabrication of electrochemical Nanosensor for detection of nitrate content in soil extract. J Soil Sci Plant Nutr 22(3):2777–2792. https://doi.org/10.1007/s42729-022-00845-5

Li S, Cao W, Kumar A, Jin S, Zhao Y, Zhang C, Zou G, Wang PC, Li F, Liang X-J (2014) Highly sensitive simultaneous detection of mercury and copper ions by ultrasmall fluorescent DNA–Ag nanoclusters. New J Chem 38(4):1546–1550. https://doi.org/10.1039/C3NJ01019H

Li Z, Liu R, Xing G, Wang T, Liu S (2017) A novel fluorometric and colorimetric sensor for iodide determination using DNA-templated gold/silver nanoclusters. Biosens Bioelectron 96:44–48. https://doi.org/10.1016/j.bios.2017.01.005

Lim SA, Ahmed MU (2016) Electrochemical immunosensors and their recent nanomaterial-based signal amplification strategies: a review. RSC Adv 6(30):24995–25014

Lin Q, Li Y, Yang M (2012) Investigations on the sensing mechanism of humidity sensors based on electrospun polymer nanofibers. Sensors Actuators B Chem 171–172:309–314. https://doi.org/10.1016/j.snb.2012.03.082

Lin B, Yu Y, Li R, Cao Y, Guo M (2016) Turn-on sensor for quantification and imaging of acetamiprid residues based on quantum dots functionalized with aptamer. Sensors Actuators B Chem 229:100–109. https://doi.org/10.1016/j.snb.2016.01.114

Lin SM, Geng S, Li N, Liu SG, Li NB, Luo HQ (2017) L-histidine-protected copper nanoparticles as a fluorescent probe for sensing ferric ions. Sensors Actuators B Chem 252:912–918. https://doi.org/10.1016/j.snb.2017.06.079

Ma B, Karimi MS, Mohammed KS, Shahzadi I, Dai J (2024) Nexus between climate change, agricultural output, fertilizer use, agriculture soil emissions: novel implications in the context of environmental management. J Clean Prod 450:141801. https://doi.org/10.1016/j.jclepro.2024.141801

Mahadeva SK, Yun S, Kim J (2011) Flexible humidity and temperature sensor based on cellulose–polypyrrole nanocomposite. Sensors Actuators A Phys 165(2):194–199. https://doi.org/10.1016/j.sna.2010.10.018

Mahmoud AED, Fawzy M (2021) Nanosensors and nanobiosensors for monitoring the environmental pollutants. In: Makhlouf ASH, Ali GAM (eds) Waste recycling technologies for nanomaterials manufacturing. Springer, pp 229–246. https://doi.org/10.1007/978-3-030-68031-2_9

Maikhuri RK, Rao KS (2012) Soil quality and soil health: a review. Int J Ecol Environ Sci 38(1):19–37

Mak AC, Osterfeld SJ, Yu H, Wang SX, Davis RW, Jejelowo OA, Pourmand N (2010) Sensitive giant magnetoresistive-based immunoassay for multiplex mycotoxin detection. Biosens Bioelectron 25(7):1635–1639. https://doi.org/10.1016/j.bios.2009.11.028

Mao A, Wei C (2019) Cytosine-rich ssDNA-templated fluorescent silver and copper/silver nanoclusters: optical properties and sensitive detection for mercury(II). Microchim Acta 186(8):541. https://doi.org/10.1007/s00604-019-3658-4

Moyo M, Okonkwo JO, Agyei NM (2014) An amperometric biosensor based on horseradish peroxidase immobilized onto maize tassel-multi-walled carbon nanotubes modified glassy carbon electrode for determination of heavy metal ions in aqueous solution. Enzym Microb Technol 56:28–34. https://doi.org/10.1016/j.enzmictec.2013.12.014

Narang C, Kant R, Mandal S (2024) Nanosensors for soil and nutrients analytics. In: Shahzad R, Fiaz S, Qayyum A, Ul Islam M, Lee I-J (eds) Revolutionizing agriculture: a comprehensive exploration of Agri-nanotechnology. Springer, Cham, pp 177–200. https://doi.org/10.1007/978-3-031-76000-6_8

Naresh V, Lee N (2021) A review on biosensors and recent development of nanostructured materials-enabled biosensors. Sensors 21(4)

Nobukawa A, Osaki T, Tonooka T, Morimoto Y, Takeuchi S (2015) Electrical detection of pesticide vapors by biological nanopores with DNA aptamers. In: 2015 28th IEEE international conference on micro electro mechanical systems (MEMS)

Oertel C, Matschullat J, Zurba K, Zimmermann F, Erasmi S (2016) Greenhouse gas emissions from soils—a review. Geochemistry 76(3):327–352. https://doi.org/10.1016/j.chemer.2016.04.002

Öndeş B, Akpınar F, Uygun M, Muti M, Aktaş Uygun D (2021) High stability potentiometric urea biosensor based on enzyme attached nanoparticles. Microchem J 160:105667. https://doi.org/10.1016/j.microc.2020.105667

Oroval M, Coll C, Bernardos A, Marcos MD, Martínez-Máñez R, Shchukin DG, Sancenón F (2017) Selective Fluorogenic sensing of As(III) using Aptamer-capped nanomaterials. ACS Appl Mater Interfaces 9(13):11332–11336. https://doi.org/10.1021/acsami.6b15164

Patil SJ, Adhikari A, Baghini MS, Rao VR (2014) An ultra-sensitive piezoresistive polymer nano-composite microcantilever platform for humidity and soil moisture detection. Sensors Actuators B Chem 203:165–173. https://doi.org/10.1016/j.snb.2014.06.110

Peng J, Ling J, Wen Q-L, Li Y, Cao Q-E, Huang Z-J, Ding Z-T (2018) The presence of a single-nucleotide mismatch in linker increases the fluorescence of guanine-enhanced DNA-templated Ag nanoclusters and their application for highly sensitive detection of cyanide. RSC Adv 8(72):41464–41471. https://doi.org/10.1039/C8RA07986B

Pimentel D, Burgess M (2013) Soil erosion threatens food production. Agriculture 3(3):443–463

Pimentel D, Harvey C, Resosudarmo P, Sinclair K, Kurz D, McNair M, Crist S, Shpritz L, Fitton L, Saffouri R, Blair R (1995) Environmental and economic costs of soil erosion and conservation benefits. Science 267(5201):1117–1123. https://doi.org/10.1126/science.267.5201.1117

Qing T, Bu H, He X, He D, Zhou B, Sun H, Jia R, Ma W, Wang K (2017) A selective nanosensor for ultrafast detection of Cu2+ ions based on C5 DNA-templated gold nanoclusters and Fenton-like reaction. Anal Methods 9(44):6222–6227. https://doi.org/10.1039/C7AY02247F

Rad F, Mohsenifar A, Tabatabaei M, Safarnejad MR, Shahryari F, Safarpour H, Foroutan A, Mardi M, Davoudi D, Fotokian M (2012) Detection of candidatus phytoplasma aurantifolia with a quantum dots fret-based biosensor. J Plant Pathol 94(3):525–534. http://www.jstor.org/stable/45156279

Rickson RJ, Deeks LK, Graves A, Harris JAH, Kibblewhite MG, Sakrabani R (2015) Input constraints to food production: the impact of soil degradation. Food Secur 7(2):351–364. https://doi.org/10.1007/s12571-015-0437-x

Sadani K, Nag P, Mukherji S (2019) LSPR based optical fiber sensor with chitosan capped gold nanoparticles on BSA for trace detection of Hg (II) in water, soil and food samples. Biosens Bioelectron 134:90–96. https://doi.org/10.1016/j.bios.2019.03.046

Safarpour H, Safarnejad MR, Tabatabaie M, Mohsenifar A (2012) Development of high-throughput quantum dot biosensor against Polymyxa species. Commun Agric Appl Biol Sci 77(3):7–13

Selvolini G, Băjan I, Hosu O, Cristea C, Săndulescu R, Marrazza G (2018) DNA-based sensor for the detection of an organophosphorus pesticide: profenofos. Sensors 18(7)

Sharma P, Pandey V, Sharma MMM, Patra A, Singh B, Mehta S, Husen A (2021) A review on biosensors and nanosensors application in agroecosystems. Nanoscale Res Lett 16(1):136. https://doi.org/10.1186/s11671-021-03593-0

Shrivastav P, Prasad M, Singh TB, Yadav A, Goyal D, Ali A, Dantu PK (2020) Role of nutrients in plant growth and development. In: Naeem M, Ansari AA, Gill SS (eds) Contaminants in agriculture: sources, impacts and management. Springer, pp 43–59. https://doi.org/10.1007/978-3-030-41552-5_2

Siddiquee S, Rovina K, Yusof NA, Rodrigues KF, Suryani S (2014) Nanoparticle-enhanced electrochemical biosensor with DNA immobilization and hybridization of Trichoderma harzianum gene. Sens Bio Sens Res 2:16–22. https://doi.org/10.1016/j.sbsr.2014.06.002

Siddiqui MS, Aslam M (2023a) Highly stable and reusable 3D graphene-quinizarin voltammetric pH sensor. J Electrochem Soc 170(4):047511

Siddiqui MS, Aslam M (2023b) Three-dimensional conductometric network based on reduced graphene oxide for soil pH sensors. ACS Appl Nano Mater 6(19):17376–17386. https://doi.org/10.1021/acsanm.3c02033

Singh A, Joshi Y, Singh M, Arora K (2013) Anti-atrazine functionalized gold nano structures for environmental monitoring. Biosens J 3(1):105

Sun S, Song P, Sun C, Wang J (2023) A novel electrochemical sensor based on MOFs composites, gold nanoparticles, and DNAzyme for Cu2+ sensitive detection. J Environ Chem Eng 11(1):109151. https://doi.org/10.1016/j.jece.2022.109151

Supraja P, Singh V, Vanjari SRK, Govind Singh S (2020) Electrospun CNT embedded ZnO nanofiber based biosensor for electrochemical detection of atrazine: a step closure to single molecule detection. Microsyst Nanoeng 6(1):3

Tan L, Chen Z, Zhao Y, Wei X, Li Y, Zhang C, Wei X, Hu X (2016) Dual channel sensor for detection and discrimination of heavy metal ions based on colorimetric and fluorescence response

of the AuNPs-DNA conjugates. Biosens Bioelectron 85:414–421. https://doi.org/10.1016/j.bios.2016.05.038
Thamri A, Baccar H, Casanova-Chafer J, Mejri MB, Llobet E, Abdelghani A (2021) Thiol-amine functionalized decorated carbon nanotubes for biomarker gases detection. Chemosensors 9(5)
Van Es H (2017) A new definition of soil. Csa News 62(10):20–21
Wang L, Cai J, Wang Y, Fang Q, Wang S, Cheng Q, Du D, Lin Y, Liu F (2014) A bare-eye-based lateral flow immunoassay based on the use of gold nanoparticles for simultaneous detection of three pesticides. Microchim Acta 181(13):1565–1572. https://doi.org/10.1007/s00604-014-1247-0
Wang G, Wang S, Yan C, Bai G, Liu Y (2018) DNA-functionalized gold nanoparticle-based fluorescence polarization for the sensitive detection of silver ions. Colloids Surf B: Biointerfaces 167:150–155. https://doi.org/10.1016/j.colsurfb.2018.04.004
Wang J, Zhang Z, Gao X, Lin X, Liu Y, Wang S (2019) A single fluorophore ratiometric nanosensor based on dual-emission DNA-templated silver nanoclusters for ultrasensitive and selective Pb2+ detection. Sensors Actuators B Chem 282:712–718. https://doi.org/10.1016/j.snb.2018.11.121
Wang Z, Lu Q, Xu T, Wang F, Huang F, Peng Y, Deng L (2020) G-quadruplex-based assay combined with aptamer and gold nanoparticles for Escherichia coli K88 determination. Microchim Acta 187(5):308. https://doi.org/10.1007/s00604-020-04291-x
Xu S, Chen X, Peng G, Jiang L, Huang H (2018) An electrochemical biosensor for the detection of Pb2+ based on G-quadruplex DNA and gold nanoparticles. Anal Bioanal Chem 410(23):5879–5887. https://doi.org/10.1007/s00216-018-1204-6
Yang Y, Zhang S, Kang M, He L, Zhao J, Zhang H, Zhang Z (2015) Selective detection of silver ions using mushroom-like polyaniline and gold nanoparticle nanocomposite-based electrochemical DNA sensor. Anal Biochem 490:7–13. https://doi.org/10.1016/j.ab.2015.08.010
Yang L, Sun H, Wang X, Yao W, Zhang W, Jiang L (2019) An aptamer based aggregation assay for the neonicotinoid insecticide acetamiprid using fluorescent upconversion nanoparticles and DNA functionalized gold nanoparticles. Microchim Acta 186(5):308. https://doi.org/10.1007/s00604-019-3422-9
Yang Y, Wang C, Shu Q, Xu N, Qi S, Zhuo S, Zhu C, Du J (2022) Facile one-step fabrication of Cu-doped carbon dots as a dual-selective biosensor for detection of pyrophosphate ions and measurement of pH. Spectrochim Acta A Mol Biomol Spectrosc 268:120681. https://doi.org/10.1016/j.saa.2021.120681
Yin H, Cao Y, Marelli B, Zeng X, Mason AJ, Cao C (2021) Soil sensors and plant wearables for smart and precision agriculture. Adv Mater 33(20):2007764. https://doi.org/10.1002/adma.202007764
Zamaleeva AI, Despras G, Luccardini C, Collot M, De Waard M, Oheim M, Mallet J-M, Feltz A (2015) FRET-based nanobiosensors for imaging intracellular Ca2+ and H+ microdomains. Sensors 15(9):24662–24680
Zhan F, Wang T, Iradukunda L, Zhan J (2018) A gold nanoparticle-based lateral flow biosensor for sensitive visual detection of the potato late blight pathogen, Phytophthora infestans. Anal Chim Acta 1036:153–161. https://doi.org/10.1016/j.aca.2018.06.083
Zhang M, Chen W, Chen X, Zhang Y, Lin X, Wu Z, Li M (2013) Multiplex immunoassays of plant viruses based on functionalized upconversion nanoparticles coupled with immunomagnetic separation. J Nanomater 2013(1):317437
Zhang C, Lai C, Zeng G, Huang D, Tang L, Yang C, Zhou Y, Qin L, Cheng M (2016) Nanoporous Au-based chronocoulometric aptasensor for amplified detection of Pb2+ using DNAzyme modified with Au nanoparticles. Biosens Bioelectron 81:61–67. https://doi.org/10.1016/j.bios.2016.02.053
Zhao G, Zhou B, Wang X, Shen J, Zhao B (2021) Detection of organophosphorus pesticides by nanogold/mercaptomethamidophos multi-residue electrochemical biosensor. Food Chem 354:129511. https://doi.org/10.1016/j.foodchem.2021.129511

Zhou Y, Zhang J, Tang L, Peng B, Zeng G, Luo L, Gao J, Pang Y, Deng Y, Zhang F (2017) A label-free GR–5DNAzyme sensor for lead ions detection based on nanoporous gold and anionic intercalator. Talanta 165:274–281. https://doi.org/10.1016/j.talanta.2016.12.069

Zhu C, Liu D, Chen Z, Li L, You T (2019) An ultra-sensitive aptasensor based on carbon nanohorns/gold nanoparticles composites for impedimetric detection of carbendazim at picogram levels. J Colloid Interface Sci 546:92–100. https://doi.org/10.1016/j.jcis.2019.03.035

Zhu X, Zhang S, Li W, Zhan Y, Yu L, Wu X, Li J, Xu H, Yang G (2020) Label-free and immobilization-free electrochemiluminescent sensing platform for highly sensitive detection of as(III) by combining target-induced strand displacement amplification with polydopamine nanospheres. Sensors Actuators B Chem 311:127818. https://doi.org/10.1016/j.snb.2020.127818

Zopf D, Pittner A, Dathe A, Grosse N, Csáki A, Arstila K, Toppari JJ, Schott W, Dontsov D, Uhlrich G, Fritzsche W, Stranik O (2019) Plasmonic nanosensor array for multiplexed DNA-based pathogen detection. ACS Sensors 4(2):335–343. https://doi.org/10.1021/acssensors.8b01073

Zou J, Zhao G-Q, Teng J, Liu Q, Jiang X-Y, Jiao F-P, Yu J-G (2019) Highly sensitive detection of bisphenol A in real water samples based on in-situ assembled graphene nanoplatelets and gold nanoparticles composite. Microchem J 145:693–702. https://doi.org/10.1016/j.microc.2018.11.040

Chapter 11
Nano-biosensors for Precision Fertilizer Management

Maida Mobeen, Muhammad Majeed, Aftab Umar, Muhammad Ramzan, Maryam Afzaal, Muhammad Waheed, Zonaira Saeed, Jamila Naz, Javed Iqbal, and Banzeer Ahsan Abbasi

Abstract Nano-biosensors are used for real-time monitoring and enhancing the efficiency of fertilizer application. Nano-biosensors combine nanotechnology with bio-sensing elements to detect soil nutrients such as nitrogen, phosphorus, and potassium. In the past, fertilization techniques had many drawbacks, such as nutrient runoff, acidic soil, and environmental pollution. However, nano-biosensors have introduced new techniques that allow us to manage nutrients on-site rather than conducting extensive work in the lab. There are different types of nano-biosensors: electrochemical (measure electrical signals from biomolecule interactions, common in glucose sensors), optical (use fluorescence, surface plasmon resonance (SPR), or quantum dots for detection), and piezoelectric sensors (detect changes in mass or mechanical properties, used in pathogen detection). Nanoparticles and quantum dots have made sensors more sensitive, stable, and useful in real-world settings. However, some challenges still exist, such as high production cost, short sensor life span, and difficulties in understanding the data they produce. These issues made it difficult to use nano-biosensors on a large scale. We can track nutrients in soil by using nano-biosensors. This leads to better crop growth, less waste, and lower harm to the environment by making farming more precise. These sensors support global efforts to grow food sustainably and protect the climate.

Keywords Nano-biosensors · Nanotechnology · Surface plasmon resonance · Nutrient management · Sustainability · Optimal growth · IoT

M. Mobeen · M. Majeed (✉)
Department of Botany, University of Gujrat, Gujrat, Punjab, Pakistan
e-mail: 23011706-011@uog.edu.pk; m.majeed@uog.edu.pk

A. Umar
Department of Botany, University of Agriculture, Faisalabad, Punjab, Pakistan

M. Ramzan
Department of Botany, Government Graduate College, Layyah, Punjab, Pakistan

J.-T. Chen (ed.), *Nanobiosensors for Crop Monitoring and Precision Agriculture*, Sustainability Sciences in Asia and Africa,
https://doi.org/10.1007/978-981-96-8335-2_11

11.1 Introduction

Agriculture is the foundation of global food supply but we do not have proper fertilizer management facilities, which is why it leads to soil depletion, financial losses to farmers, and environmental degradation. Traditionally, fertilizers were used on a broad spectrum, which results in water pollution, nutrient wastage, soil pollution and excessive nutrient runoff. Nowadays, due to the integration of nanotechnology in agriculture, these issues are resolved. Nano-biosensors are used for real-time monitoring of nutrients so the issues can be quickly assessed and resolved on-site and on time. All these practices and techniques will improve soil health and productivity. Traditional methods were useful and beneficial at that time, but they were harmful too. However, now due to the use of sensors, we can detect macro and micronutrients present in soil and manage it unlike traditional methods which are time consuming require certain sort of laboratory analysis. Now, we can use smart farming tools like Internet of Things (IoT), artificial intelligence (AI), and automatic irrigation systems that detect nutrients on-site and make sure that crops receive optimal nutrients on time. Our population is increasing day by day, and we require more food to meet the needs and demands of people in order to fulfill their requirements. We need to introduce smart farming on a large scale. Three types of nano-biosensors can be used in this regard: electrochemical, optical, and piezoelectric sensors. No doubt nanotechnology is helpful in agricultural production and a solution toward many problems, but there are certain barriers associated with this technology. Some of them are sensor stability, cost, and data interpretation, which are discussed in detail (Arif et al. 2021; Ali et al. 2023; Muhammad et al. 2024b; Rasool et al. 2025). By addressing every concern in detail, this chapter will help provide a guide on how nano-biosensors can revolutionize our agricultural system concerning sustainability and production.

M. Afzaal
Department of Biochemistry, Government College University, Faisalabad, Punjab, Pakistan

M. Waheed
Department of Botany, Government of Islamia Graduate College, Lahore, Punjab, Pakistan

Z. Saeed
Institute of Botany, University of the Punjab, Lahore, Punjab, Pakistan

J. Naz
Department of Forestry and Range Management, Bahauddin Zakariya University, Multan, Punjab, Pakistan

J. Iqbal
Department of Botany, Bacha Khan University, Charsadda, Khyber Pakhtunkhwa, Pakistan

B. A. Abbasi
Department of Botany, Rawalpindi Women University, Rawalpindi, Pakistan

11.1.1 Background

As we are going ahead, the population is increasing and their demands for food are also increasing. So this puts a lot of pressure on farmers to increase crop productivity while protecting the environment. Fertilizers help crops to grow and increase productivity, but they cause environmental damage: they harm soil, pollute water, and increase carbon emissions. Traditional farming methods apply fertilizers evenly and on a fixed schedule, without considering that crops need different nutrients at different stages of growth. This can lead to waste and environmental damage. As a result, nutrients leach into the water and cause eutrophication there, and disturb aquatic life. To overcome this, precision agriculture has become a modern approach that uses advanced technology for farming. Instead of using the traditional method, precision agriculture helps farmers to give an exact amount of nutrients when and where they are needed. One of the innovative ways for precision agriculture is to use nano-biosensors, which offer real-time monitoring and targeted nutrient delivery. The concept of nano-biosensors was given in the 1960s for the first time nano-biosensors were used in medical field. With the passage of time, nanotechnology has enabled the use of biosensors in agriculture for the enhancement of crop production. These sensors operate by using biological recognition elements, such as enzymes, antibodies, and DNA to detect soil nutrients and chemical composition. By integrating nanomaterials, such as graphene, gold nanoparticles, and specialized electrodes, nano-biosensors achieve higher sensitivity and faster response times (Arshad et al. 2022; Aziz et al. 2023a; Naveed et al. 2024c). The use of wireless sensor networks (WSNs) and IoT technology enables large-scale deployment, allowing farmers to monitor soil and crop conditions in real-time and optimize fertilizer application.

Despite their potential, nano-biosensors are still emerging in agriculture. Key challenges, including high costs, scalability, and accessibility, particularly in developing regions, must be overcome. Researchers are also working to enhance their durability and reliability in real-world farming conditions. Nevertheless, with continuous advancements and a growing focus on sustainable agriculture, nano-biosensors are expected to become a crucial tool for future farming practices.

11.1.2 Significance

The use of nano-biosensors in precision fertilizer management is a game-changer for modern agriculture, providing environmental, economic, and agronomic benefits. One of the biggest challenges in farming today is the overuse and improper application of fertilizers, which wastes valuable resources and harms the environment. By enabling real-time monitoring and precise nutrient delivery, nano-biosensors help reduce waste, improve crop yields, and minimize pollution, making agriculture more efficient and sustainable. Farmers can reduce fertilizer wastage by using this targeted approach and can improve crop productivity in efficient ways. From an environmental perspective, nano-sensors can mitigate major issues like water pollution, soil pollution, and carbon emission, which lead to climate change. Nano-biosensors can ensure food security by providing crop yield consistency with real-time detection of nutrient deficiencies, and farmers can correct that on time. Furthermore, nano-biosensors help reduce agriculture's carbon footprint and promote sustainable food production. This technology is particularly beneficial for smallholder farmers in resource-constrained regions, where improved fertilizer management can greatly increase food production and enhance income stability. Furthermore, the integration of nano-biosensors with IoT-based platforms strengthens precision farming by enabling remote monitoring, automated irrigation, and AI-powered decision-making (Aziz et al. 2023b; Bashir et al. 2023; Naveed et al. 2024b;). As agriculture continues to embrace smart farming and digital innovation, nano-biosensors are poised to play a crucial role in creating efficient, sustainable, and highly productive food systems.

11.1.3 Limitations

Despite their promising potential, nano-biosensors face several challenges that hinder their widespread adoption in fertilizer management. One major obstacle is their high production cost, as the advanced materials and manufacturing processes required make them expensive, particularly for small-scale farmers in developing regions. Additionally, durability remains a concern, as agricultural environments expose sensors to extreme temperatures, moisture, soil microbes, and chemical interactions, which can degrade their performance over time and necessitate frequent calibration and maintenance. Another key challenge is the complexity of data interpretation. While nano-biosensors generate real-time soil nutrient data, many farmers lack the technical expertise to analyze and apply this information effectively, highlighting the need for user-friendly and automated systems. Furthermore, environmental and regulatory concerns surrounding the long-term impact of nanoparticles on soil health, microbial ecosystems, and food safety must be thoroughly addressed before large-scale commercialization (Bhattacharya et al. 2023; Haq et al. 2022a; Barwant et al. 2024; Naveed et al. 2024a). Overcoming these limitations requires continuous research, interdisciplinary collaboration, and supportive policies to ensure that nano-biosensors can become a practical and sustainable tool for modern agriculture.

11.2 Nano-biosensors in Precision Agriculture

Precision agriculture changes the ways of agricultural farming by using advanced technology to use resources efficiently, increase crop yield, and protect the environment. One of the most important technologies in this field is nano-biosensors, which are tiny, highly sensitive devices that help monitor soil and plant health, especially for fertilizer management. Nano-biosensors utilize biological components (such as enzymes, DNA, or antibodies) combined with nanomaterials to instantly detect soil nutrients, environmental conditions, and plant health. This technology enables precise fertilizer application, ensuring nutrients are supplied exactly when and where they are needed, preventing both excess and deficiency that could harm crops and the environment (Haq et al. 2022b; Hussain et al. 2023; Majeed et al. 2024d). Unlike traditional soil testing, which involves sending samples to a lab-leading to delays and generalized recommendations, nano-biosensors provide real-time, location-specific insights. This allows farmers to make faster, data-driven decisions for more efficient and sustainable fertilizer use.

11.2.1 The Need for Precision Fertilizer Management

Fertilizers are essential for crop production, but their excessive use can cause various environmental issues and climate change. Traditional methods involve even distribution of fertilizers while considering site-specific requirements of the fertilizers. As a result, this leads to the over-fertilization of fertilizers at some sites and cause deficiency at other sites. Precision agriculture means the use of nano-biosensors, IoT-based monitoring, and AI-driven analytics to optimize nutrient application and crop productivity. The major reason for using precision agriculture is due to the loss of fertilizers as a result of runoff, volatilization, leaching, and a decline in nutrient use efficiency (NUE). Excessive use of nitrogen and phosphorus leads to eutrophication and algal bloom formation, which deplete the oxygen level, harming aquatic life. By using nanotechnology, farmers can apply fertilizers when and where they are needed and reduce damage to the environment. Overuse of chemical fertilizers has harmed soil health, depleting its organic matter and microbial diversity, leading to reduced productivity. Precision fertilizer management, which includes bio-based fertilizers, controlled-release nutrients, and site-specific soil amendments, offers a solution to this issue. This approach helps maintain soil fertility and promotes sustainable agriculture. It also leads to higher crop productivity, lower costs, improved environmental sustainability, and reduced reliance on synthetic chemicals (Hassan et al. 2022a; Majeed et al. 2024c). However, to implement these techniques widely, farmers need education, access to affordable technologies, and supportive policies, especially in developing regions, to ensure they can effectively adopt these advanced agricultural solutions.

11.2.2 Role of Nanotechnology in Agriculture

Nanotechnology, which involves working with tiny materials at the nanoscale, is revolutionizing agriculture and food production by improving efficiency and sustainability. This advanced field helps create better fertilizers, pesticides, biosensors, and food packaging. In precision agriculture, nanotechnology allows for smart delivery systems for fertilizers and pesticides, effective soil cleaning methods, and real-time monitoring tools like nano-biosensors to track crop health. These innovations lead to higher productivity, less waste, and more sustainable farming practices. Nanotechnology is transforming agriculture by developing nano-fertilizers and nano-pesticides, which help improve nutrient use and control pests more efficiently. Unlike conventional fertilizers that can wash away before plants absorb them, nano-fertilizers release nutrients in a controlled and targeted way, ensuring crops get what they need. This reduces nutrient runoff, improves soil health, and cuts down on fertilizer costs. Similarly, nano-pesticides offer precise control of pests and diseases, reducing harmful chemical residues. Another important use of nanotechnology in agriculture is nano-biosensors. These tiny, highly sensitive devices detect soil nutrients, harmful pathogens, toxins, and plant stress indicators in real time. Using nanomaterials like carbon nanotubes, quantum dots, and gold nanoparticles, these sensors provide detailed information about soil health. By integrating them with Internet of Things (IoT) and AI-based systems, farmers can get real-time insights to apply fertilizers and pesticides only where needed. However, there are challenges such as high costs, regulatory hurdles, and concerns about nanoparticle toxicity in the environment (Hassan et al. 2022b; Majeed et al. 2024b). Ongoing research is focused on developing biodegradable nanomaterials and cost-effective solutions to make nanotechnology a key driver of sustainable and smart agriculture in the future (Fig. 11.1).

11.2.3 Overview of Nano-biosensors and Their Working Mechanism

The nonmaterial-based transducers to detect specific analyses in soil, plants, and water systems, which are highly specialized with biological sensing such as a nano-biosensors. Biological interactions, which generate measurable signals that are converted into electrical, optical, and mechanical outputs, are recognized by the presence of target molecules such as nitrogen, phosphorus, potassium, and pathogens. These outputs are then processed and analyzed to provide real-time, exact agricultural data. The bio receptor is a biological molecule-such as an enzyme antibody DNA strand, a tamer that selectively binds with a target analyte. The transducer that converts this biological interaction into a measureable signal precedes a digital interface to provide a readable result. Optical biosensors use fluorescence surface of plasmon resonance, which is used for the detection of soil nutrient

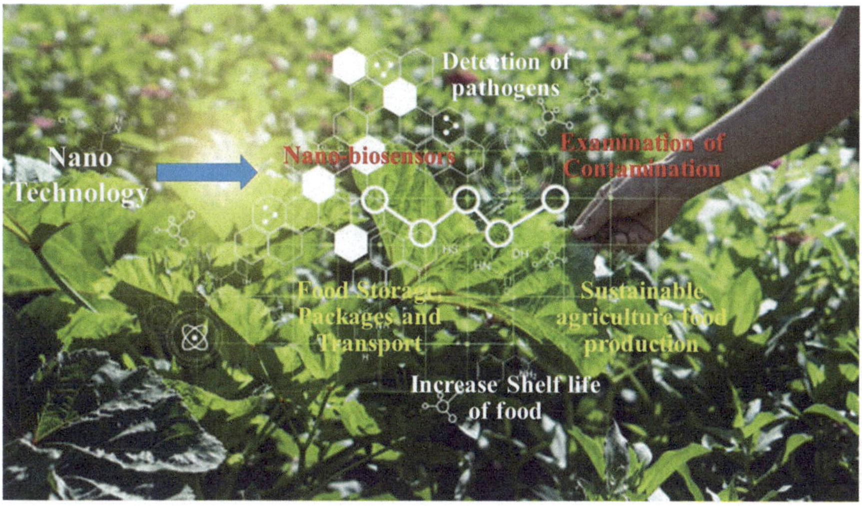

Fig. 11.1 Agricultural applications of nano-biosensors

analysis. Nano-biosensors play a crucial role in precision agriculture by providing real-time on-site soil diagnostics and optimizing fertilizer use. The valuable tool in sustainable and smart farming techniques also detect low concentration of nutrients with high specificity (Jamil et al. 2022; Khoja et al. 2022; Rasool et al. 2024; Majeed et al. 2024a). The number of challenges such as cost sensor stability and mass production must be addressed for the adoption in global agriculture systems.

11.3 Types and Mechanisms of Niño Biosensors in Fertilizer Management

Niño biosensors are highly sensitive analytical devices that combine biological recognition elements with nanotechnology-based transducers to detect and quantify specific soil nutrients. The real-time monitoring of soil condition enables targeted fertilizer application and reduces nutrient waste, which plays a vital role in the precision of fertilizer management. The Niño biosensor and receptor function as a working mechanism like enzyme antibodies that interact with a target nutrient. When the Niño biosensor is classified, they have different types like electrochemical, optical, and piezoelectric sensors that have a working mechanism, which is applied in smart agriculture. The nutrient concentration can be measured by detecting changes in potential and current resulting from chemical interactions in the soil. On the other hand, the Niño biosensor is used to rely on colorimetric surface plasmon resonance signals to detect soil components. Piezoelectric as a biosensor have mechanical properties for nutrient building and attention for potential use in field soil diagnostics. The most advanced in Niño biosensor is the integration with

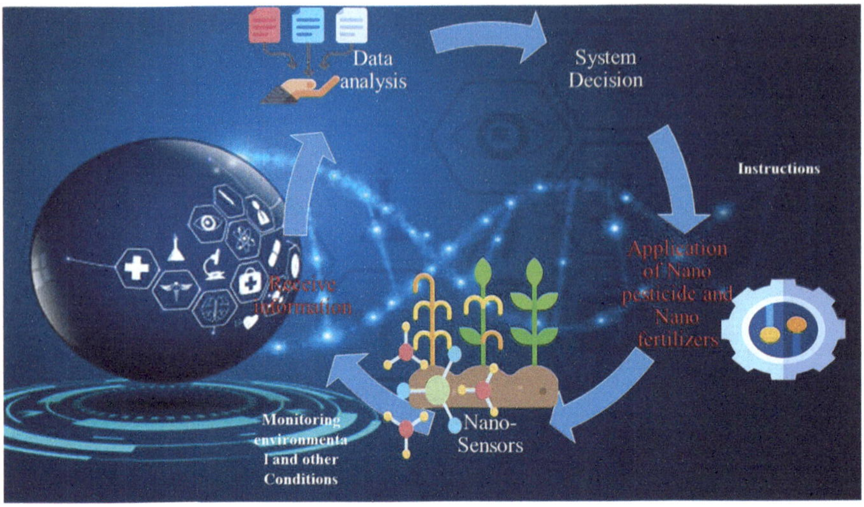

Fig. 11.2 The working process of nano-biosensors

Internet of Things and artificial intelligence. Niño biosensors are connected with a wireless data transmission system for continuous and remote access to soil nutrient information (Khan et al. 2022; Majeed et al. 2020; Shafqat et al. 2024; Rani et al. 2024). The use of fertilizer application and optimizing crop growth in a sustainable manner allows farmers to make data-driven decisions. The use of Niño biosensors has more benefits for fertilizer, which face challenges like stability, field adaptability, and large-scale development remain (Fig. 11.2).

11.3.1 Electrochemical Niño Biosensor for Soil Nutrient Detection

It is noted that the Niño biosensor gained significant attention in precision agriculture due to its ability to rapidly and accurately detect soil nutrients. The principle of converting biochemical interaction into an electrical signal, which determines the concentration of soil nutrients such as nitrogen, phosphorus, and potassium, can be possible by working on this principle. These transducers are employed for real-time soil analysis, where measurable changes in electrical conductivity and current are detected and processed. The key advantages of electrochemical biosensors are their high sensitivity and selectivity at low nutrient concentrations. The use of the potential metric is to measure change in potential due to nutrient binding and making them suitable for detecting soil pH and ion concentration. The change resulting from nutrient oxidation reduction makes them highly effective for nitrate and phosphate detection. The carbon annotates grapheme and metal nanoparticles, which further enhance the performance of electrochemical biosensors by increasing their electrode surface area conductivity and stability. Now a field sensor has been developed that enables farmers to conduct on-site soil testing without the need for expensive laboratory analysis. The electrochemical biosensor faces challenges like electrode fouling, sensor degradation over time, and environmental interference from soil moisture and pH fluctuations (Majeed et al. 2021a, b; Nasar et al. 2024b; Hussain et al. 2024). The use of Niño biosensor lies with wireless communication technologies and AI-driven predictive analytics, allowing for real-time automated nutrient management and improved agriculture productivity.

11.3.2 Optical Niño Biosensors and Their Applications

The most advanced sensing technologies are Niño biosensors in the precision fertilizer management, offering high sensitivity, fast response times, and noninvasive nutrient detection. They detect changes in the light properties such as fluorescence absorbance and surface plasmon resonance, which interact with specific soil nutrients. The nutrient concentration and soil health status may be analyzed to generate an optical signal. There are different optical biosensors that are widely used for detecting nitrate, phosphate, and potassium levels. The soil fertility can be analyzed by the intensity of fluorescence that correlates with the nutrient concentration. The quantum dots that change in response to specific soil nutrients allow for easy visual detection. Technologies like surface plasmon resonance–based sensor also detect real-time interaction at the nanoscale level. The imaging devices and smartphones enhanced their ability for on-field precision farming application. The high specificity and ability to provide continuous real-time monitoring of soil nutrient dynamics are major advantages of optical Niño biosensors (Majeed et al. 2021c; Majeed et al., 2021d; Khan et al. 2024; Nasar et al. 2024a). The development of low-cost user-friendly alternatives can easily be adopted by farmers, which are focused on miniaturizing the optical biosensors.

11.3.3 Piezoelectric and Other Advanced Biosensors

Piezoelectric is based on the principle of mechanical electrical energy conversion, which emerged as a novel of Niño biosensors for soil nutrient monitoring. These sensors rely on piezoelectric materials, such as quartz crystals or piezoelectric nanofilms, which generate an electrical signal in response to changes in mass or mechanical stress caused by nutrient binding. The mass-sensitive nature of piezoelectric biosensors makes them highly effective for detecting changes in soil chemistry, microbial activity, and fertilizer interactions. Apart from piezoelectric sensors, surface acoustic wave (SAW) biosensors and micro-cantilever biosensors are also being explored for soil nutrient analysis. SAW biosensors detect shifts in acoustic wave propagation caused by nutrient interactions, providing highly sensitive and label-free detection. Micro-cantilever biosensors, on the other hand, measure mechanical deflection when soil nutrients bind to their surface, offering ultra-sensitive and miniaturized detection capabilities. Despite their advanced sensing mechanisms, piezoelectric and other mechanical biosensors face challenges related to sensor calibration, environmental stability, and high fabrication costs (Majeed et al. 2022a; Waheed et al. 2023). Their integration with microfluidics and AI-based data processing is expected to enhance their performance and enable their widespread use in precision agriculture applications.

11.3.4 Integration of Niño Biosensors with Iota and AI

The transfer of fertilizer, which managed in modern agriculture, is the integration of Niño biosensor with the Internet of Things and artificial intelligence. The use of smart sensing allows for real-time automated monitoring of soil nutrient levels, which leads to precise and efficient fertilizer application. The farmer can access real-time data analytics and make data-driven decisions to optimize crop while minimizing environmental damage, which combines nanotechnology and wireless sensor network, cloud computing, and machine learning algorithms. The detection of soil nutrients at the molecular level and transmitting data wirelessly to centralized cloud-based platforms with Niño biosensors play a great role in smart agriculture. Communication models such as Bluetooth Zombie allow them to transmit real-time data to smartphones and tablets, which are used by the farmers and agronomists. The historical and real-time data predict nutrient deficiencies and plant health status and optimal fertilizer application rates. The correlation between soil nutrient properties, climate conditions, and crop requirements allows for personalized nutrient recommendations for a specific field. The use of an application through smart irrigation systems and autonomous farming equipment have key advantages to Niño biosensors that integrate with Internet of Things and artificial intelligence. A number of nutrients are absorbed by the crops while reducing chemical runoff into water bodies with the use of automated drones and robotic sprayers equipped with Internet of Things, which are connected as Niño biosensors where fertilizers are added. The blockchain technology can be integrated with Internet of Things, which enables Niño biosensors to create secure and transparent agriculture data networks. Limited technical expertise and high implementation costs are major barriers to the widespread adoption of smart farming systems. The power consumption and durability of Internet of Things in harsh environmental conditions remain a concern. The advancement in low-power sensor technologies and wireless communication protocols is continuously improving the feasibility of this technology for large-scale agricultural applications (Majeed et al. 2022b; Tassadduq et al. 2022; Ullah et al. 2022; Waheed et al. 2022). The government initiatives promoting smart agriculture and the integration of Niño biosensors with the Internet of Things and artificial intelligence play an important role in ensuring food security, sustainable resources management, and climate resilient farming.

11.4 Applications and Benefits of Nano-biosensors in Agriculture

Nino biosensors enable real-time, high-sensitivity monitoring of soil and plant health, thereby supporting the transformation of modern agricultural practices. It is noted that advanced devices play a role in enabling farmers to make data-driven decision about fertilizer application, irrigation, and crop protection. By integrating Nino biosensors with IoT technology, it becomes possible to monitor nutrient levels, soil conditions, and environmental parameters with exceptional precision, thereby supporting informed, data-driven decisions in agriculture. It is noted that soil testing method requires time-consuming lab analysis, which delays and generalizes fertilizer recommendations. Excessive fertilizer application, which often leads to soil degradation and water pollution, can be improved by nutrient use efficiency. The other application of Niño biosensors is in early disease detection and pest monitoring. The Niño biosensors that automate the irrigation system ensures that crops receive optimal water and nutrient based on real-time field conditions. The agricultural system and businesses can integrate smart Niño biosensors, which optimize fertilizer supply chains to monitor global soil health trends and support sustainable land management policies. The agriculture policies improve food security and address climate change challenges so that the government and research institutions can use this data (Majeed et al. 2021c; Sinha et al. 2023; Tariq et al. 2023;). It is noted that research in cost-effective Niño materials and artificial intelligence make Niño biosensors more accessible and impactful for farmers worldwide.

11.4.1 Real-Time Soil Nutrient Monitoring and Decision Support Systems

Nano-biosensor applications in precision agriculture gain essential power through their ability to monitor soil nutrients in real-time. The conventional method of testing soil includes laboratory work and sample collection, followed by analysis periods, which make it difficult for farmers to quickly determine their soil health status. Nano-biosensors remove delays because they conduct immediate on-site nutrient analyses to help farmers apply specific fertilizers, which leads to better crop yields. Nano-biosensors operate through the specific identification of nitrogen (N), phosphorus (P), and potassium (K) soil nutrients at their molecular scale. The sensors use bioreceptor elements like enzymes or antibodies that produce measurable electrical, optical, or piezoelectric signals by specific nutrient binding. Decision support systems (DSS) utilize AI-driven algorithms to process the data that comes from wireless transmission using the DSS platforms. Real-time soil health monitoring is available to farmers through mobile applications or web dashboards because nano-biosensors integrate with both IoT systems and cloud-based DSS platforms. The analysis of historical data together with the content of weather patterns, along with crop-relevant nutrient requirements, enables tailored fertilization recommendations. Proper resource management becomes more efficient through such systems because they distribute nutrients only during the required times. Large-scale sensor networks currently collect real-time soil nutrient data, which allows regional and national soil fertility mapping programs to operate effectively. Agencies that focus on agricultural development together with governments can utilize collected information to create lasting policies about fertilizers and identify future soil deterioration patterns through localized nutrient resource management measures (Majeed et al. 2023a; Muhammad et al. 2024a;). The many advantages of real-time nutrient monitoring require that researchers overcome challenges regarding sensor calibration together with power supply and affordability questions. Nano-biosensors strengthen their practicality and scalability for global agriculture through current technological advancements of low-cost nanomaterials and artificial intelligence analytics, and wireless communication.

11.4.2 Enhancing Fertility

The FUE improvement process depends heavily on nano-biosensors, which determine the accurate delivery of nutrients in specific soil areas during matching weather conditions. Auwdite application techniques create problems by using too much fertilizer, while loss of nutrients damages environmental quality, which costs farmers' money throughout the soil's deterioration process. The real-time soil nutrient availability analysis, which nano-biosensors offer, optimizes FUE through their application. The sensors operate at atomic levels to identify macro and micronutrient quantities, thus enabling precise fertilizer distribution. Nano-biosensor-based fertilizer application targets specific nutrient-deficient areas, whereas traditional methods spread fertilizers everywhere. Thus, this method protects against unnecessary substance application and soil-leached nutrients. Nano-biosensors work with smart nano-fertilizer delivery systems to transmit plant nutrients in carefully controlled amounts that respond to current environmental requirements. Nano-fertilizers have nanoparticles that use environmental trigger mechanisms to keep nutrients bioavailable during extended periods. The system reduces fertilizer misuse through enhanced plant nutrient capture while decreasing farmer expenses. The reduction of fertilizer waste produces environmental advantages because it decreases the quantity of nitrogen and phosphorus that runs off into water sources, thus preventing eutrophication and biodiversity depletion (Majeed et al. 2023b, c; Muhammad et al. 2023). Soil health preservation through sustainable nutrient management depends on the increased efficiency of fertilizer systems achieved by nano-biosensors.

11.4.3 Environmental Impact Reduction and Sustainable Farming

The high levels of chemical fertilizers alongside pesticides have produced major environmental problems, which feature soil destruction combined with water contamination, along with greenhouse gas release. Nano-biosensors serve as a sustainable management tool that controls nutrients precisely while reducing the amount of agrochemicals needed. Nano-biosensors provide the environment with a major advantage through their ability to minimize nutrient runoff losses (Pathania et al. 2023). Traditional agricultural fertilizer distribution allows nutrients to flow into aquatic environments, thus triggering eutrophication, which damages aquatic habitats. Through soil nutrient monitoring in real time, nano-biosensors enable needed fertilizer applications alongside minimized water and nutrient contamination from overuse. Nano-biosensors serve the purpose of checking soil health while simultaneously discovering contaminants, which include heavy metals together with pesticide residues and microbial pathogens. The immediate detection capabilities let intervention teams take prompt actions to avoid prolonged soil and water pollution. Smart irrigation systems that combine with nano-biosensors enable optimized water usage while achieving better water resource management without waste (Miguel-Rojas and Pérez-de-Luque 2023). Areas that struggle with water scarcity need precision irrigation systems used with site-specific fertilization methods to achieve maximum productivity without harming the environment.

11.4.4 Economic Benefits for Farmers and Large-Scale Agricultural Systems

The economic advantages of nano-biosensors help farmers decrease their fertilizer expenses and increase their yield production while protecting resources from unnecessary waste. The sensors guarantee that fertilizer applications will be used only at needed times to help farmers protect their expenses and achieve better nutrient use efficiency (Bakshi et al. 2024). Largely operational agricultural entities benefit from nano-biosensor AI-driven decision support systems because these systems enable effective resource management, which enhances business performance. Large farms can spread sensor networks throughout their wide fields to keep track of soil vitality as well as crop state and environmental elements in real-time. The sensor data allows farmers to exercise precision management over supplies and improve predictions of crop yields and supply chain management. The use of nano-biosensors qualifies farmers to receive benefits from government sustainability programs that support agrochemical reduction efforts. Precision farming obtains support from modern global agricultural policies, which aim to implement climate-smart agricultural strategies. Nano-biosensors lead to sustained economic benefits because they create higher productivity, together with lower loss rates and better farming systems efficiency (Yadav et al. 2023). Future sustainable agriculture will rely on nano-biosensors because their manufacturing cost reduction and broad acceptance as a foundational technology continues to advance.

11.5 Discussion

Nano-biosensors established themselves as a transformative technology in precision agriculture by deploying real-time evaluation along with more efficient nutrient utilization and data-aided decisions. These biosensors, using nanotechnology combined with biological detection elements, can change standard farming practices into intelligent and sustainable farming systems that enhance both production outcomes and protect environmental quality. Biological detection elements in traditional fertilizer application methods cause mineral imbalances because they depend on past soil tests and generic nutrient guidelines, while producing excessive fertilizer waste that harms environmental systems. The implementation of nano-biosensors removes traditional analysis inefficiencies because they deliver immediate site-specific nutrient tests, leading to controlled fertilizer implementation when needed. Multiple nano-biosensors demonstrate top-notch molecular-level sensitivity for both macro and micronutrients detection using electrochemical, optical, and piezoelectric functions, which assist farmers in individualizing fertilizer application through real-time soil measurements (Sharma et al. 2024). Nano-biosensors achieve improved effectiveness through their connectivity with IoT and AI-based decision support systems, which enables automated data acquisition together with sophisticated soil health parameter analysis, followed by predictive recommendations for optimal fertilizer use. The universal application of nano-biosensors encounters various difficulties since they have high manufacturing expenses and durability problems while requiring precise calibration and extensive training for farmers. Smaller-scale farmers in developing areas fail to achieve precision farming implementation through nano-biosensor techniques because they need vital infrastructure combined with financial capital and technological competence. Prior to wide manufacturing deployment, scientists need to establish a complete understanding of how nanomaterials impact microbial communities in soil and human health. Research into biodegradable nanomaterials that are eco-friendly will solve some existing problems while making nano-biosensors sustainably nontoxic. The successful adoption of nano-biosensor-based technologies in agriculture depends on governmental support. Recent progress in nanotechnology, especially in the areas of miniaturization and cost-effective sensor manufacturing, has paved the way for the broad application of nano-biosensors in modern agriculture. The rapid implementation of nano-biosensors for precision fertilizer management demands collaboration between policy interventions, farmer awareness programs, and industry collaboration efforts (Zain et al. 2023). The combination of research between farming experts and professionals who work in nanotechnology, artificial intelligence, and environmental sciences will help create advanced sensors with improved performance in different applications and better management functions for fields. Future research must prioritize the creation of autonomous data-driven wireless nano-biosensor networks that use self-powered intelligence to track environmental as well as plant and soil conditions in real-time. The critical role of smart technology-led agriculture becomes vital because the global population growth increases the importance of food security. Nano-biosensors establish themselves as a fundamental element of next-generation precision farming through their ability to support sustainable agriculture intensification, together with increased yield production and lower environmental impact (Samreen et al. 2024). The complete achievement of nano-biosensor capabilities depends on combined initiatives between research professionals, along with officials who make agricultural policies and stakeholders in the agricultural realm agricultural institutions to create framework protocols for their safe use and commercialization approval.

11.6 Future Recommendation

Widespread adoption and effective utilization of nano-biosensors for precision agriculture needs three fundamental elements: research-oriented development, strategic policy changes, and public-private industrial partnerships. The expense associated with producing nano-biosensors currently remains the main hurdle to their wide-scale implementation. Future studies need to develop low-price nanomaterials as well as biodegradable biosensors and miniaturization elements that will enable small-scale farmers to access affordable sensors. A program of mass nano-biosensor production must be developed through public-private partnerships that aims to bring innovations into agricultural markets at market-friendly costs. For effective implementation, nano-biosensors need to merge perfectly with AI-powered analytics systems, IoT farm monitoring instruments, and cloud processing tools, so that they can deliver instant information about soil quality and crop health (Bisht et al. 2024). Scientific research should prioritize the creation of independent wireless nano-biosensors that derive power from their environment while sending data to remote cloud-based analytics solutions.

Precision agriculture will benefit from nano-biosensor technology, which delivers immediate data-based agricultural information to optimize fertilizer utilization and farm control systems. Nanotechnology enables these advanced sensors to provide precise measurements of soil nutrients, plant health indicators, and environmental parameters because of their advanced detection capabilities. Nano-biosensors help agriculture deal with its key modern issues because they enhance resource efficiency and eliminate waste and environmental damage. The scientific community now acknowledges nano-biosensors as important tools to advance sustainable farming because they provide farmers with accurate current data instead of estimated results or time-tested experimental approaches for decision-making (Bharti et al. 2024). The ability of nano-biosensors serves to optimize the utilization of nutrients by agricultural systems. Current farming practices usually overuse fertilizers, causing both soil deterioration and water contamination, together with nutrient runoff. Soil nutrient levels get accurately measured through nano-biosensors, allowing farmers to distribute fertilizers specifically when required, along with proper ratios (Ullah et al. 2024). The specific application of nano-biosensors reduces fertilizer overuse and enhances farmer cost-effectiveness while minimizing agricultural operations' impact on the environment.

The analysis of sensor data using machine learning models for crop nutrient prediction requires further research investigation. Sensor manufacturers should focus on creating devices that improve both stability and precision through development, which addresses operational durability under environmental conditions. Research laboratories must create all-in-one detection sensors that analyze macro and micronutrients together with soil pH, moisture level, and microbial activity. Extended operational usability will be improved by nano-biosensors featuring self-calibrating capabilities and automatic cleaning functions. Scientists need to perform comprehensive studies regarding nano-biosensor impact on microorganisms in soil and food protection and health stability to fully realize their sustainable advantages (Rani et al. 2024). Research intelligence must focus on creating sustainable biodegradable nanomaterials through artificial fabrication combined with non-harmful production techniques and responsible waste disposal methods to address probable environmental concerns. A firm-wide deployment requires lasting regulatory approvals together with safety protocols approved by governmental bodies before starting widespread implementation.

Governments must establish public policies that promote precision agriculture by offering tax-reducing benefits, research grants, and subsidies to nano-biosensors. A synergy between agricultural research divisions and subject matter experts in nanotechnology and environmental science enables them to create more valuable technological solutions. The agricultural technology sector should provide additional investment to nano-biosensor technology for smart farming implementation. In order for precision agriculture initiatives to reach commercial success, venture capital firms plus government research agencies, and agribusiness companies should support startups along with research projects built on sensor technology. Real-time monitoring of soil health trends requires global agricultural databases to operate through extensive nano-biosensor networks spread worldwide. The network system provides enhanced value to global food security programs, along with climate change adaptation methods and agricultural disaster response systems. The support of government organizations together with universities and research institutions is necessary to maintain data cooperation programs for early prevention of disease and health outbreaks. Plant diseases and nutrient deficiencies become noticeable on a biochemical scale before visible symptoms become evident (Das et al. 2024). The technology enables nano-biosensors to recognize faint biochemical alterations, which lets farmers address the issue beforehand. By reducing pesticide use, these devices improve crop yield, support sustainability, and help minimize harvest losses. Widespread adoption of nano-biosensors in agriculture needs multiple challenges to be resolved, despite their many benefits. High cost stands as a main obstacle in this discussion. Nano-biosensor implementation demands advanced production techniques and particular substance compositions that create high monetary costs for basic agricultural operators. The reliability of nano-biosensors needs enhancement across agricultural environments with extreme weather to guarantee their long-term functionality Majeed et al. (2022c). The regulatory framework that shapes nanotechnology use in agriculture presents a major hindrance to its deployment. Detailed scientific evaluation is crucial to determine how molecular-level nanomaterial interactions affect human well-being, together with soil living creatures and environmental systems. Government policy makers need to create precise rules and security requirements to enable proper nano-biosensor applications in agricultural practices. The acceptance of new tools by farming communities stands as one of the main difficulties affecting the widespread adoption of nano-biosensors (Pavlović and Kupresanin 2024). Farmers in developing regions mostly lack the technical skills for effective nano-biosensor operation. The gap between farmer technical literacy needs to be addressed through training initiatives combined with government-backed programs and producer-stakeholder partnerships, which will provide farmers with comprehensive utilization abilities for this technology.

11.7 Conclusion

The achievement of total nano-biosensor-based precision farming adoption depends on worldwide standards about sensor performance, precision farming operational standards, and data privacy protocols. Professional training programs combined with widespread information dissemination about nano-biosensors represent the ideal approach for agricultural communities to utilize this technology. Future applications of nano-biosensors should concentrate on pesticide residue detection, medical disease diagnosis, real-time water quality assessment, and carbon storage monitoring, even though their main role remains as soil nutrient and fertilizer detectors. The advancements in nanotechnology, artificial intelligence (AI), and Internet of Things (IoT) systems will maximize the effectiveness of nano-biosensors in the upcoming years. AI systems analyze sensor measurements better than humans do to deliver immediate, actionable farm information. The integration of IoT systems allows farmers to monitor fields remotely through automated decisions and improved precision farming at large scales. Modern technology innovations enhance agricultural efficiency, profitability, and the sustainability of the environment.

References

Arif U, Bhatti KH, Ajaib M, Wagay NA, Majeed M, Zeb J, Hameed A, Kiani J (2021) Ethnobotanical indigenous knowledge of tehsil Charhoi, district Kotli, Azad Jammu and Kashmir, Pakistan. Ethnobot Res Appl 22:10.32859/ERA.22.50.1-24

Arshad F, Waheed M, Harun N, Fatima K, Khan BA, Fatima K, Abbas Z, Jabeen S, Majeed M (2022) Indigenous farmer's perception about fodder and foraging species of semi-arid lowlands of Pakistan: a case study of district Kasur, Pakistan. Taiwania 67(4):510–523. https://doi.org/10.6165/tai.2022.67.510

Aziz A, Anwar MM, Majeed M, Albanai JA, Almohamad H, Abdo HG (2023a) Quantifying the impacts of urbanization on urban green, evidences from Maga City, Lahore Pakistan. Discov Sustain 4(1):1–11. https://doi.org/10.1007/s43621-023-00169-z

Aziz A, Anwar MM, Majeed M, Fatima S, Mehdi SS, Mangrio WM, Elbouzidi A, Abdullah M, Shaukat S, Zahid N, Mahmoud EA, Casini R, Yessoufou K, Elansary HO (2023b) Quantifying landscape and social amenities as ecosystem services in rapidly changing Peri-urban landscape. Land 12(2). https://doi.org/10.3390/land12020477

Bakshi A, Dutta R, Kundu S, Sengupta S, Kanthal S, Sarkar T (2024) Prospect of nano-fertilizers in agriculture: an overview. Int J Agric Ext Soc Dev. https://doi.org/10.33545/26180723.2024.v7.i4sa.516

Barwant MM, Rasool A, Mannam VK, Hussain A, Abdullah M, Nisa Memon Z et al (2024) Nanotechnology applications in livestock farming and animal health. In: Harnessing nanoomics and nanozymes for sustainable agriculture. IGI Global, pp 413–442. https://doi.org/10.4018/979-8-3693-1890-4.ch020

Bashir SM, Altaf M, Hussain T, Umair M, Majeed M, Mangrio WM, Khan AM, Gulshan AB, Hamed MH, Ashraf S, Amjad MS, Bussmann RW, Abbasi AM, Casini R, Alataway A, Dewidar AZ, Al-Yafrsi M, Amin MH, Elansary HO (2023) Vernacular taxonomy, cultural and ethnopharmacological applications of avian and mammalian species in the vicinity of Ayubia National Park, Himalayan Region. Biology 12(4). https://doi.org/10.3390/biology12040609

Bharti AC, Jain U, Chauhan N (2024) From lab to field: nano-biosensors for real-time plant nutrient tracking. Plant Nano Biol 9:100079. https://doi.org/10.1016/j.plana.2024.100079

Bhattacharya R, Bhattacharya R, Majeed M, Bhandari S, Aziz R, Sinha D et al (2023) Environmental rehabilitation of industrial waste dumping site. In: Biohydrometallurgical processes. CRC Press, pp 44–89. https://doi.org/10.1201/9781003451457-4

Bisht M, Pooja LR, Prakash P, Patel PK (2024) A review on nanotechnology applications for agriculture and water quality management. Ecol Environ Conserv 30(Suppl):S299–S306. https://doi.org/10.53550/eec.2024.v30i07s.054

Das H, Singh SK, Priyadarshini E, Maurya CL, Tiwari A, Gulaiya S, Khokale SK, Panigrahi CK, Singh BV (2024) Nanotechnology facilitated real time soil monitoring for optimized crop production. Asian J Soil Sci Plant Nutr 10(2):582–607. https://doi.org/10.9734/ajsspn/2024/v10i2315

Haq SM, Tariq A, Li Q, Yaqoob U, Majeed M, Hassan M, Fatima S, Kumar M, Bussmann RW, Moazzam MFU, Aslam M (2022a) Influence of edaphic properties in determining forest community patterns of the Zabarwan Mountain range in the Kashmir Himalayas. Forests 13(8). https://doi.org/10.3390/f13081214

Haq SM, Yaqoob U, Majeed M, Amjad MS, Hassan M, Ahmad R et al (2022b) Quantitative ethnoveterinary study on plant resource utilization by indigenous communities in high-altitude regions. Front Vet Sci 9:944046

Hassan M, Haq SM, Ahmad R, Majeed M, Sahito HA, Shirani M, Mubeen I, Aziz MA, Pieroni A, Bussmann RW, Alataway A, Dewidar AZ, Al-Yafrsi M, Elansary HO, Yessoufou K (2022a) Traditional use of wild and domestic Fauna among different ethnic groups in the Western Himalayas—a cross cultural analysis. Animals 12(17). https://doi.org/10.3390/ani12172276

Hassan M, Haq SM, Majeed M, Umair M, Sahito HA, Shirani M, Waheed M, Aziz R, Ahmad R, Bussmann RW, Alataway A, Dewidar AZ, El-Abedin TKZ, Al-Yafrsi M, Elansary HO, Yessoufou K (2022b) Traditional food and medicine: ethno-traditional usage of fish Fauna across the valley of Kashmir: a Western Himalayan region. Diversity 14(6). https://doi.org/10.3390/d14060455

Hussain S, Raza A, Abdo HG, Mubeen M, Tariq A, Nasim W, Majeed M, Almohamad H, Al Dughairi AA (2023) Relation of land surface temperature with different vegetation indices using multi-temporal remote sensing data in Sahiwal region, Pakistan. Geosci Lett 10(1):33. https://doi.org/10.1186/s40562-023-00287-6

Hussain F, Salam IU, Memon ZN, Abdullah M, Abbas G, Akbar M et al (2024) Occurrence of fungal microbial contamination in drinking water of megacity of Karachi (Pakistan) and their physico-chemical control. Heliyon 10(7). https://doi.org/10.1016/j.heliyon.2024.e28926

Jamil MD, Waheed M, Akhtar S, Bangash N, Chaudhari SK, Majeed M, Hussain M, Ali K, Jones DA (2022) Invasive plants diversity, ecological status, and distribution pattern in relation to edaphic factors in different habitat types of district Mandi Bahauddin, Punjab, Pakistan. Sustainability (Switzerland) 14(20). https://doi.org/10.3390/su142013312

Khan AM, Li Q, Saqib Z, Khan N, Habib T, Khalid N, Majeed M, Tariq A (2022) MaxEnt modelling and impact of climate change on habitat suitability variations of economically important Chilgoza pine (Pinus gerardiana wall.) in South Asia. Forests 13(5). https://doi.org/10.3390/f13050715

Khan AM, Altaf M, Hussain T, Hamed MH, Safdar U, Ayub A et al (2024) Ethnopharmacological uses of fauna among the people of Central Punjab, Pakistan. Front Vet Sci 11:1351693. https://doi.org/10.3389/fvets.2024.1351693

Khoja AA, Haq SM, Majeed M, Hassan M, Waheed M, Yaqoob U, Bussmann RW, Alataway A, Dewidar AZ, Al-Yafrsi M, Elansary HO, Yessoufou K, Zaman W (2022) Diversity, ecological and traditional knowledge of Pteridophytes in the Western Himalayas. Diversity 14(8). https://doi.org/10.3390/d14080628

Majeed M, Bhatti KH, Amjad MS, Abbasi M, Id RWB, Nawaz F, Rashid A, Mehmood A, Id MM, Khan WM, Id SA (2020) Ethno-veterinary uses of Poaceae in Punjab, Pakistan. PLoS One:1–28. https://doi.org/10.1371/journal.pone.0241705

Majeed M, Bhatti KH, Amjad MS (2021a) Impact of climatic variations on the flowering phenology of plant species in jhelum district, Punjab, Pakistan. Appl Ecol Environ Res 19(5):3343–3376. https://doi.org/10.15666/aeer/1905_33433376

Majeed M, Bhatti KH, Pieroni A, Sõukand R, Bussmann RW, Khan AM, Chaudhari SK, Aziz MA, Amjad MS (2021b) Gathered wild food plants among diverse religious groups in Jhelum District, Punjab, Pakistan. Food Secur 10(3). https://doi.org/10.3390/foods10030594

Majeed M, Tariq A, Anwar MM, Khan AM, Arshad F, Mumtaz F, Farhan M, Zhang L, Zafar A, Aziz M, Abbasi S, Rahman G, Hussain S, Waheed M, Fatima K, Shaukat S (2021c) Monitoring of land use–land cover change and potential causal factors of climate change in Jhelum district, Punjab, Pakistan, through GIS and multi-temporal satellite data. Land 10(10). https://doi.org/10.3390/land10101026

Majeed M, Khan AM, Habib T, Anwar MM, Sahito HA, Khan N, Ali K (2022a) Vegetation analysis and environmental indicators of an arid tropical forest ecosystem of Pakistan. Ecol Indic 142. https://doi.org/10.1016/j.ecolind.2022.109291

Majeed M, Lu L, Haq SM, Waheed M, Sahito HA, Fatima S, Aziz R, Bussmann RW, Tariq A, Ullah I, Aslam M (2022b) Spatiotemporal distribution patterns of climbers along an abiotic gradient in Jhelum District, Punjab, Pakistan. Forests 13(8). https://doi.org/10.3390/f13081244

Majeed M, Tariq A, Haq SM, Waheed M, Anwar MM, Li Q, Aslam M, Abbasi S, Mousa BG, Jamil A (2022c) A detailed ecological exploration of the distribution patterns of wild Poaceae from the Jhelum District (Punjab), Pakistan. Sustainability (Switzerland) 14(7). https://doi.org/10.3390/su14073786

Majeed M, Lu L, Anwar MM, Tariq A, Qin S, El-Hefnawy ME, El-Sharnouby M, Li Q, Alasmari A (2023a) Prediction of flash flood susceptibility using integrating analytic hierarchy process (AHP) and frequency ratio (FR) algorithms. Front Environ Sci 10. https://doi.org/10.3389/fenvs.2022.1037547

Majeed M, Muhammad M, Hussain T, Ali M, Naz T, Nawaz S et al (2023b) Nanopriming-based management of biotic stresses for sustainable agriculture. In: Nanopriming approach to sustainable agriculture. IGI Global, pp 263–289. https://doi.org/10.4018/978-1-6684-7232-3.ch012

Majeed M, Muhammad M, Nawaz S, Naz T, Iqbal MM, Zahid N et al (2023c) Nanopriming for crop management for sustainable agriculture: an overview. In: Nanopriming approach to sustainable agriculture, pp 110–141. https://doi.org/10.4018/978-1-6684-7232-3.ch005

Majeed M, Gulshan AB, Muhammad M, Hussain M, Aziz R, Abdullah M et al (2024a) An investigation into the reactions of biochar in heavy metal polluted soils. In: Biochar-assisted remediation of contaminated soils under changing climate. Elsevier, pp 247–276. https://doi.org/10.1016/B978-0-443-21562-9.00010-4

Majeed M, Noor W, Rasool A, Saleem M, Bashir I, Fatima R et al (2024b) Nanotechnology for soil remediation and land restoration. In: Harnessing nanoomics and nanozymes for sustainable agriculture. IGI Global, pp 60–77. https://doi.org/10.4018/979-8-3693-1890-4.ch004

Majeed M, Gulshan AB, Muhammad M, Hussain M, Aziz R, Abdullah M, Tariq M (2024c) An investigation into the reactions of biochar in heavy metal polluted soils. In Biochar-assisted Remediation of Contaminated Soils Under Changing Climate (pp. 247–276). Elsevier. https://doi.org/10.1016/B978-0-443-21562-9.00010-4

Majeed M, Noor W, Rasool A, Saleem M, Bashir I, Fatima R, Rafique A (2024d) Nanotechnology for Soil Remediation and Land Restoration. In Harnessing NanoOmics and Nanozymes for Sustainable Agriculture (pp. 60–77). IGI Global. https://doi.org/10.4018/979-8-3693-1890-4.ch004

Miguel-Rojas C, Pérez-de-Luque A (2023) Nanobiosensors and nanoformulations in agriculture: new advances and challenges for sustainable agriculture. Emerg Top Life Sci 7:229–238. https://doi.org/10.1042/etls20230070

Muhammad M, Waheed A, Wahab A, Majeed M, Nazim M, Liu YH et al (2023) Soil salinity and drought tolerance: an evaluation of plant growth, productivity, microbial diversity, and amelioration strategies. Plant Stress 100319. https://doi.org/10.1016/j.stress.2023.100319

Muhammad M, Basit A, Arooj A, Dixit G, Majeed M, Sinha D et al (2024a) Roles of gibberellins in plant defense against biotic and abiotic stress. In: Plant growth regulators to manage biotic and abiotic stress in agroecosystems. CRC Press, pp 183–210

Muhammad M, Basit A, Majeed M, Shah AA, Ullah I, Mohamed HI et al (2024b) Bacterial pigments and their applications. In: Bacterial secondary metabolites. Elsevier, pp 277–298. https://doi.org/10.1016/B978-0-323-95251-4.00019-3

Nasar S, Muhammad M, Siddiqi EH, Majeed M, Ameen N, Ullah S et al (2024a) OMICS-based knowledge for achieving food and nutritional security. In: OMICs-based techniques for global food security. Wiley, pp 67–90. https://doi.org/10.1002/9781394209156.ch4

Nasar S, Rasool A, Ikram W, Shujat S, Majeed M, Muhammad M, Saleem S (2024b) Nano-osmics: a new approach to plant breeding. In: Sustainable agriculture: nanotechnology and biotechnology for crop production and protection, vol 129. https://doi.org/10.1515/9783111234694-008

Naveed M, Majeed M, Jabeen K, Hanif N, Naveed R, Saleem S, Khan N (2024a) Plant disease diagnosis with artificial intelligence (AI). In: Microbial data intelligence and computational techniques for sustainable computing. Springer, Singapore, pp 217–234. https://doi.org/10.1007/978-981-99-9621-6_15

Naveed M, Memon ZN, Abdullah M, Makhdoom SI, Azeem A, Mehmood S et al (2024b) Use cases and future aspects of intelligent techniques in microbial data analysis. In: Microbial data intelligence and computational techniques for sustainable computing. Springer, Singapore, pp 259–280. https://doi.org/10.1007/978-981-99-9621-6_17

Naveed M, Zahid N, Fatima I, Saleem A, Majeed M, Abid A et al (2024c) Sustainable AI-driven applications for plant care and treatment. In: Microbial data intelligence and computational techniques for sustainable computing. Springer, Singapore, pp 235–258. https://doi.org/10.1007/978-981-99-9621-6_16

Pathania D, Raizada P, Rustagi S, Singh P, Rani GM, Chaudhary V (2023) Panorama of biogenic nano-fertilizers: a road to sustainable agriculture. Environ Res 116456. https://doi.org/10.1016/j.envres.2023.116456

Pavlović Z, Kupresanin AM (2024) Enhancing NPK detection in agriculture through electrochemical sensors utilizing 2D nanomaterials. Meeting Abstracts MA2024-02(64):4273. https://doi.org/10.1149/ma2024-02644273mtgabs

Rani J, Rasool A, Sher H, Fatima R, Hussain J, Majeed M et al (2024) Precision farming with nanoscale sensors: monitoring soil, water, and crop health. In: Harnessing nanoomics and nanozymes for sustainable agriculture. IGI Global, pp 78–99. https://doi.org/10.4018/979-8-3693-1890-4.ch005

Rasool A, Tariq M, Asif M, Shah GM, Fatima R, Majeed M et al (2024) Nanotechnology for improved crop resilience in challenging environments. In: Harnessing nanoomics and nanozymes for sustainable agriculture. IGI Global, pp 149–175. https://doi.org/10.4018/979-8-3693-1890-4.ch008

Rasool A, Feroz E, Qayyum Khan Z, Qayyum Khan M, Majeed M, Kousar S (2025) Chapter 19: Sustainable agriculture: a nanoosmic-based management. In: Rajput V, Singh A, Ghazaryan K, Minkina T, Al-Tawaha A (eds) Sustainable agriculture: nanotechnology, biotechnology, management and food security. De Gruyter, Berlin/Boston, pp 411–430. https://doi.org/10.1515/9783111369754-019

Samreen T, Rasool S, Tahir A, Riaz U, Nazir MZ, Kanwal S, Muntaha S-T (2024) Nutrient management through nano-biofertilizers for smart agriculture, pp 228–243. https://doi.org/10.1201/9781003429289-11

Shafqat A, Rasool A, Fatima R, Majeed M, Rahman KU, Narejo AH et al (2024) Nanomaterials for sustainable pest and disease management in agriculture. In: Harnessing nanoomics and nanozymes for sustainable agriculture, pp 100–125. https://doi.org/10.4018/979-8-3693-1890-4.ch006

Sharma S, Yadav AK, Verma SB, Yadav M (2024) Revolutionizing agriculture with nanotechnology: advances, applications, and sustainability considerations. Int J Environ Clim Change 14(7):1–9. https://doi.org/10.9734/ijecc/2024/v14i74246

Sinha D, Maurya AK, Abdi G, Majeed M, Agarwal R, Mukherjee R, Ganguly S, Aziz R, Bhatia M, Majgaonkar A, Seal S, Das M, Banerjee S, Chowdhury S, Adeyemi SB, Chen JT (2023) Integrated genomic selection for accelerating breeding programs of climate-smart cereals. Genes 14(7) NLM (Medline). https://doi.org/10.3390/genes14071484

Tariq A, Mumtaz F, Majeed M, Zeng X (2023) Spatio-temporal assessment of land use land cover based on trajectories and cellular automata Markov modelling and its impact on land surface temperature of Lahore district Pakistan. Environ Monit Assess 195(1). https://doi.org/10.1007/s10661-022-10738-w

Tassadduq SS, Akhtar S, Waheed M, Bangash N, Nayab DE, Majeed M, Abbasi S, Muhammad M, Alataway A, Dewidar AZ, Elansary HO, Yessoufou K (2022) Ecological distribution patterns of wild grasses and abiotic factors. Sustainability (Switzerland) 14(18). https://doi.org/10.3390/su141811117

Ullah I, Aslam B, Shah SHIA, Tariq A, Qin S, Majeed M, Havenith HB (2022) An integrated approach of machine learning, remote sensing, and GIS data for the landslide susceptibility mapping. Land 11(8). https://doi.org/10.3390/land11081265

Ullah Q, Fatima F, Memon SUR, Khan MA, Ismail A, Rasool M, Ali HM, Kasim M, Ullah U (2024) Nanotechnological innovations in agrochemicals: enhancing efficacy and environmental stewardship in pesticide and herbicide applications. Middle East Res J Agric Food Sci 4(05):164–178. https://doi.org/10.36348/merjafs.2024.v04i05.002

Waheed M, Arshad F, Majeed M, Fatima S, Mukhtar N, Aziz R, Mangrio WM, Almohamad H, Al Dughairi AA, Al-Mutiry M, Abdo HG (2022) Community structure and distribution pattern of Woody vegetation in response to soil properties in semi-arid Lowland District Kasur Punjab, Pakistan. Land 11(12). https://doi.org/10.3390/land11122145

Waheed M, Arshad F, Majeed M, Haq SM, Aziz R, Bussmann RW, Ali K, Subhan F, Jones DA, Zaitouny A (2023) Potential distribution of a noxious weed (Solanum viarum Du-nal), current status, and future invasion risk based on MaxEnt modeling. Geol Ecol Landsc. https://doi.org/10.1080/24749508.2023.2179752

Yadav A, Yadav K, Ahmad R, Abd-Elsalam KA (2023) Emerging frontiers in nanotechnology for precision agriculture: advancements, hurdles and prospects. Agrochemicals 2(2):220–256. https://doi.org/10.3390/agrochemicals2020016

Zain M, Ma H, Chaudhary S, Nuruzaman M, Azeem I, Mehmood F, Rahman SU, Duan A, Sun C (2023) Nanotechnology in precision agriculture: advancing towards sustainable crop production. Plant Physiol Biochem 206:108244. https://doi.org/10.1016/j.plaphy.2023.108244

Chapter 12
Nano-biosensors for Precision Pesticide Management

Jasmine Kaur, Pawan Kumar, Ram Kumar, Amit Bhatia, Haramritpal Singh Sidhu, Anil Sindhu, Brijnandan Singh Dehiya, and Virender Yadav

Abstract This chapter explores the integration of nanotechnology and biosensors, specifically focusing on nano-biosensors as advanced tools for precision pesticide management in agriculture. Nano-biosensors, analytical devices leveraging nanomaterials, demonstrate remarkable sensitivity, specificity, and efficiency in detecting pesticides, pollutants, and soil conditions. The chapter introduces the fundamentals of nano-biosensors, detailing their components—bioreceptors, nanomaterials, and transducers—and the principles underlying their functionality. Emphasizing applications in pesticide residue detection, real-time monitoring, and sustainable farming, it highlights their role in optimizing pesticide usage, reducing environmental contamination, and enhancing food safety. Advanced nano-pesticides, such as nano-emulsions and polymer-based formulations, are discussed as eco-

J. Kaur · P. Kumar (✉) · R. Kumar · A. Bhatia
Department of Pharmaceutical Sciences and Technology, Maharaja Ranjit Singh Punjab Technical University, Bathinda, Punjab, India

R. Kumar
Department of Pharmaceutical Sciences and Technology, Maharaja Ranjit Singh Punjab Technical University, Bathinda, Punjab, India

Chitkara College of Pharmacy, Chitkara University, Rajpura, Punjab, India

H. S. Sidhu
Department of Mechanical Engineering, Giani Zail Singh Campus College of Engineering & Technology, Maharaja Ranjit Singh Punjab Technical University, Bathinda, Punjab, India

A. Sindhu
Department of Biotechnology, Deenbandhu Chhotu Ram University of Science and Technology, Murthal, Haryana, India

B. S. Dehiya
Department of Chemical Engineering, Deenbandhu Chhotu Ram University of Science and Technology, Murthal, Haryana, India

V. Yadav
Department of Chemical Engineering, Maharishi Kashyap Government Polytechnic, Jattal Panipat, Haryana, India

J.-T. Chen (ed.), *Nanobiosensors for Crop Monitoring and Precision Agriculture*, Sustainability Sciences in Asia and Africa,
https://doi.org/10.1007/978-981-96-8335-2_12

friendly alternatives. The chapter also addresses challenges like sensitivity, stability, environmental concerns, and ethical considerations, offering insights into regulatory frameworks and cost-effective designs. Case studies illustrate innovations, including graphene-based sensors and enzyme-functionalized systems, showcasing their potential for high-precision, low-cost solutions in modern agriculture. Future perspectives underline the integration of nano-biosensors with Internet of Things (IoT) and artificial intelligence (AI), paving the way for transformative advancements in precision farming and sustainable agriculture.

Keywords Nanotechnology · Nanomaterials · Biosensors · Pesticides · Agriculture

12.1 Introduction

Research and its results have become more well-defined with the development of a new scientific field, "nanotechnology." To a considerable degree, it is based on developing novel instrumental methods ensuring the possibility of working with nano-sized structures. Nanotechnology focuses on designing, producing, and utilizing structures, devices, and systems at the nanoscale. A nanometer is one-billionth of a meter, making this scale incredibly small—about 100,000 times smaller than the diameter of a human hair (Srilatha 2011). It is an interdisciplinary field that merges nanotechnology with biology to develop new devices at the nanometer scale (1–100 nm) for biological and medical applications. It focuses on manipulating biological molecules, structures, and devices at the nanoscale to create innovative solutions in diagnostics, therapeutics, and environmental management (Alshora et al. 2016). It involves two main concepts:

(a) Nanotechnology in Biology: It involves the use of nanoscale materials (e.g., nanoparticles (NPs), nanowires) to study and manipulate biological systems. It enables precise interaction with biomolecules such as DNA, proteins, etc.
(b) Biology for Nanotechnology: It uses biological molecules (e.g., enzymes, antibodies) as templates to fabricate nanoscale structures (Williams et al. 2021).

Nanobiotechnology has applications in various fields

(i) Medical and Healthcare: Diagnostics, Therapeutics, Regenerative Medicine, etc.
(ii) Environmental Monitoring: Detection of pollutants using nanodevices.
(iii) Food and Agriculture: Detecting contaminants and use of nanoscale pesticides.
(iv) Energy and Industry: Development of bio-nanomaterials for renewable energy technologies (Nasrollahzadeh et al. 2019).

Many developments, previously known as those related to biosensors, have now emerged as new fields called nano-biosensors, which deal with working with nano-biosensors (Reshetilov et al. 2011). The objective of this chapter is to study nano-biosensors and their application in the field of agriculture that are relevant to nano-biosensors for precision-making pesticides. The combination of nanostructured materials and biosensors is generally known as nano-biosensor technology. These sensors leverage the unique properties of nanomaterials for excellent sensitivity, specificity, and rapid response times (Kulkarni et al. 2022).

12.2 Fundamentals of Nano-biosensors

12.2.1 Definition and Principles

A biosensor is a device engineered to detect and quantify a specific substance by leveraging biological interactions and analyzing the resulting responses. Central to the operation of a biosensor is transduction mechanisms, which play a critical role in transforming the biochemical energy generated from bio-analyte interactions into an electrical signal for assessment and measurement (Malik et al. 2013). A nano-biosensor is an advanced sensing device that combines nanotechnology and biological components to detect and analyze specific biological or chemical substances at the nanoscale. These sensors utilize nanomaterials to enhance sensitivity, specificity, and detection limits, enabling the precise measurement of target analytes through the interaction of biological recognition elements (such as enzymes, antibodies, or nucleic acids) with nanostructures (Zhu et al. 2021). The nanoscale involves the precise modification and fabrication of materials, structures, and devices at the nanometer level, typically ranging from 1 to 100 nm. This scale enables the design and synthesis of nanoscale structures with high precision, which can be integrated into larger micro- and macroscopic systems. Such advancements play a crucial role in various fields, including nanotechnology, materials science, and biomedical applications, allowing for the development of highly functional and efficient devices with tailored properties (Bellucci 2008).

Nano-biosensors operate based on the interaction between a sensing element, known as a bioreceptor, and the target analyte, resulting in a measurable response. The bioreceptor, which can be an enzyme, antibody, nucleic acid, or other biomolecule, selectively binds to the analyte, generating a signal that is subsequently processed by a transducer component. This transducer converts the biological or biochemical interaction into a quantifiable signal through optical, electronic, thermal, or magnetic detection methods, enabling precise analyte measurement and monitoring (Zhu et al. 2021). They play an essential role in the different fields of biotechnology, nanotechnology, biology, chemistry, physical sciences, and molecular engineering. It is possible to detect and manipulate various molecules and atoms with nano-biosensors, which have great potential in domestic and industrial applications (Bhattacharya et al. 2022). At their core, nano-biosensors rely on biological recognition elements, such as enzymes, antibodies, or nucleic acids, to interact selectively with the analyte of interest. These interactions are captured and amplified by nanoscale transduction mechanisms, which convert the biological response into a measurable signal, often electrical, optical, or mechanical in nature. The use of nanomaterials, such as nanoparticles, nanowires, or carbon nanotubes, significantly enhances the performance of these sensors by increasing the surface area, improving signal-to-noise ratios, and enabling detection at extremely low concentrations. This unique combination of biological specificity and nanotechnology-driven sensitivity makes nano-biosensors a powerful tool for applications in healthcare, environmental monitoring, and biotechnology (Ramesh et al. 2023; Valenzuela-Amaro et al. 2023).

12.2.2 *Types of Nano-biosensors*

12.2.2.1 Electrochemical Nano-biosensors

Electrochemical nano-biosensors are a class of biosensors that utilize electrochemical transduction mechanisms to detect and analyze specific biological or chemical substances at the nanoscale (Fig. 12.1). They combine the specificity of biological recognition elements, such as enzymes, antibodies, or DNA, with the sensitivity of electrochemical techniques and the enhanced properties of nanomaterials. In these sensors, biological interactions, such as binding or enzymatic reactions, generate a measurable electrochemical signal, such as a change in current (amperometric), voltage (potentiometric), or impedance (Baranwal et al. 2022).

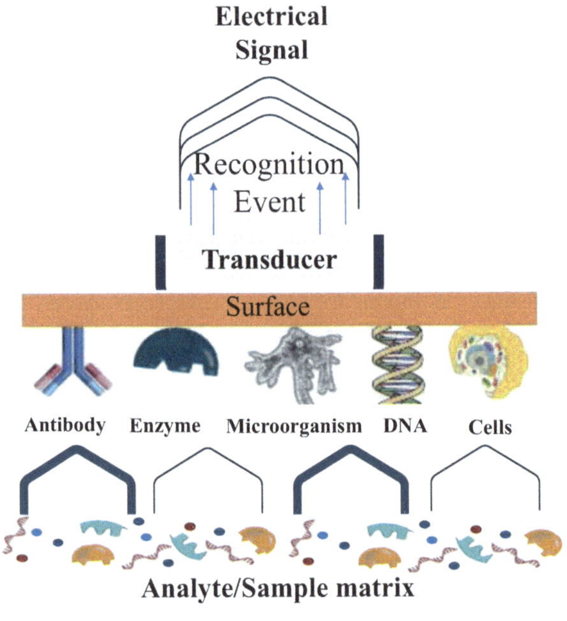

Fig. 12.1 Electrochemical sensing of biological species (Singh et al. 2016)

Various metal electrodes, including platinum, mercury, gold, and carbon, are now widely utilized as working electrodes in electrochemical sensing applications. To enhance analyte and biomarker recognition, bioreceptor molecules such as enzymes, nucleic acids, antibodies, dyes, and metal ions are immobilized onto these electrodes. This immobilization improves specificity and sensitivity, enabling more efficient and precise detection in biosensing technologies (Bhattacharya et al. 2022). When a potential is applied to the working electrode in relation to a reference electrode, such as Ag/AgCl or a saturated calomel electrode, various electrochemical reactions can be induced. Applying a negative potential allows electrons to flow from the working electrode into the solution, leading to the reduction of the analyte present. Conversely, when a positive potential is applied, electrons move from the solution to the working electrode, facilitating the oxidation of the analyte. This interplay of electron transfer reactions forms the basis of different voltametric techniques, which can be employed to analyze and quantify the electroactive species present in a sample. By utilizing modes such as cyclic voltammetry, differential pulse voltammetry, or square wave voltammetry, researchers can gather detailed information about the kinetics and thermodynamics of the electrochemical processes occurring at the electrode interface (Singh et al. 2016).

12.2.2.2 Optical Nano-biosensors

Optical nano-biosensors are advanced devices that utilize light-based techniques to measure various properties of biological samples at the nanoscale. These sensors assess key parameters such as amplitude, energy, polarization, decay time, and decay phase of different optical phenomena, including absorption, phosphorescence, fluorescence, Raman scattering, refraction, interference, and dispersion. By analyzing these characteristics, optical nano-biosensors can provide valuable insights into the molecular composition and interactions within biological systems, enabling applications in diagnostics, environmental monitoring, and biomedical research (Nosheen et al. 2019; Munawar et al. 2019).

12.2.2.3 Piezoelectric Nano-biosensors

Piezoelectric nano-biosensors are advanced analytical tools. They use the piezoelectric effect to detect biological and chemical analytes with high sensitivity and specificity. They are widely utilized in medical diagnostics, environmental monitoring, and agricultural applications due to their ability to transduce mechanical or chemical stimuli into measurable electrical signals (Uchino 2017). These sensors typically consist of piezoelectric materials, such as ZnO or quartz, integrated with nanomaterials to increase their surface area, mechanical properties, and sensitivity (Uchino 2017). Piezoelectric nano-biosensors operate on the piezoelectric effect, which generates electrical charge in response to applied mechanical stress. When integrated into nanoscale devices, this principle transduces physical, chemical, or biological changes into electrical signals (Zhao et al. 2023).

12.3 Parts and Mechanisms of Nano-biosensor Action

Parts of the nano-biosensor are (Fig. 12.2):

(a) *Bioreceptor*: A specific molecule that interacts selectively with the target analyte (the substance to be measured). Examples: Antibodies, enzymes, DNA/RNA aptamers, or even whole cells. Its role is to provide the biosensor with specificity, as it ensures that the sensor reacts only to the desired analyte (Kumar et al. 2022).
(b) *Nanomaterial*: Nanomaterials are nano-sized materials often used to support bioreceptors and enhance signal detection. Examples: Nanoparticles (e.g., gold, silver), nanowires, carbon nanotubes, graphene, or nano-shells. Nanomaterials help immobilize bioreceptors effectively and amplify the signal due to their high surface area, unique electronic properties, and capability to interact with transducers (Kumar et al. 2022).
(c) *Transducer*: A component that converts the biological interaction (such as the binding of the analyte to the bioreceptor) into a measurable signal. Some common types of transducers are given below:

- *Electrochemical*: Measures changes in current, voltage, or impedance.
- *Optical*: Detects changes in light properties (e.g., intensity, wavelength, or polarization).
- *Mass-based*: Monitors changes in mass using piezoelectric crystals or cantilevers.

These components work together to enable sensitive, specific, and efficient detection in applications like medical diagnostics, environmental monitoring, and

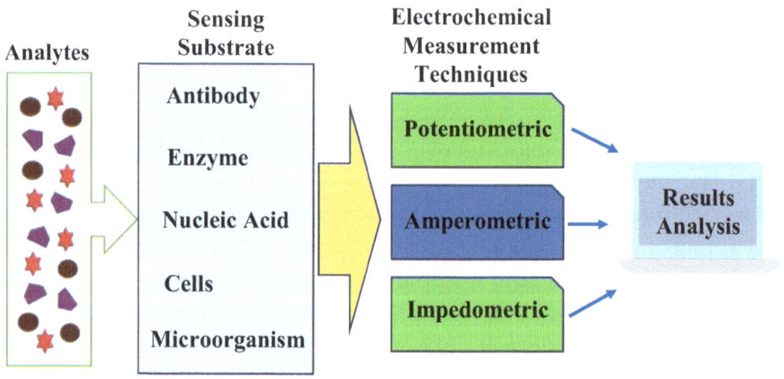

Fig. 12.2 Parts of a biosensor (Chamorro-Garcia and Merkoçi 2016)

food safety (Kumar et al. 2022). A nano-biosensor works on the principle of utilizing a biological recognition element, typically an antibody, enzyme, or nucleic acid, to specifically bind to a target analyte, which then induces a measurable change in a nanostructured transduce and converts the biological interaction into a detectable signal like an electrical current, optical properties, or mass, allowing for the quantification of the target analyte (Bellucci 2008).

Mechanism

1. *Analyte Binding*: The target analyte in a sample interacts with the biorecognition element immobilized on the nanostructured transducer and causes a specific binding reaction.
2. *Signal Transduction*: This binding event induces a change in the physical properties of the nanomaterial, like a change in electrical conductivity, optical properties (e.g., fluorescence), or mass.
3. *Signal Detection*: The change in the transducer's properties is detected by a suitable readout system, which translates the signal into a measurable value representing the analyte concentration (Hamzah and Nadzirah 2023).

12.4 Applications of Nano-biosensors in Pesticide Management

12.4.1 Monitoring Pesticide Residues in Crops and Soil

Detecting pesticides in soil at trace concentrations poses a significant challenge due to their low levels and complex matrices. In recent years, nanobiotechnology has emerged as a transformative technology, offering innovative and sustainable solutions to address this challenge. Its applications in agriculture enable precise, reliable, and efficient pesticide monitoring, thereby promoting environmental sustainability and enhancing agricultural productivity (Panoth et al. 2021). Integration of nanobiotechnology enables ultra-sensitive detection of toxins, contributing to sustainable farming practices by reducing environmental contamination and improving food safety (Wahab et al. 2024).

Pesticide residue analysis is a critical tool for monitoring soil quality, mitigating health risks, and safeguarding ecosystems from the detrimental effects of excessive pesticide use. Nano-sensor technology offers a cutting-edge alternative to traditional chromatographic methods for pesticide detection. Nano-biosensors, in particular, demonstrate exceptional sensitivity, enabling the detection of ultra-low pesticide concentrations with enhanced stability and reliability, making them ideal for applications in smart agriculture. The ongoing advancements in nano-sensors and nano-biosensors are pivotal in transforming agricultural practices, promoting precision farming, and fostering sustainable agricultural development (Panoth et al. 2021).

Nano-biosensors for evaluating pesticide residues in soil operate on the fundamental principle of interaction between the sensing layer and the analyte. This interaction is broadly classified into two categories: *bio-affinity interactions* and *biocatalytic interactions*.

- *Bio-affinity interactions* involve the direct binding of the analyte to the recognition element without any catalytic transformation. Examples include antigen-antibody or receptor-ligand binding.
- *Biocatalytic interactions*, on the other hand, utilize biocatalysts such as enzymes or tissues to catalyze a specific reaction with the analyte, producing a detectable signal.

The recognition elements in nano-biosensors can be integrated with various types of transducers to convert these interactions into measurable signals. Common transducers used for pesticide residue detection include:

1. *Optical Transducers*: Measure changes in light properties (e.g., fluorescence, absorbance) due to analyte interaction.
2. *Electrochemical Transducers*: Detect changes in current, voltage, or impedance resulting from the biochemical reaction.
3. *Thermometric Transducers*: Monitor heat generated or absorbed during the interaction.
4. *Biomechanical Transducers*: Detect changes in mechanical properties, such as mass or elasticity, induced by the analyte interaction.

These advanced transduction mechanisms enable nano-biosensors to achieve high sensitivity, specificity, and efficiency, making them invaluable tools for monitoring pesticide residues in soil and supporting sustainable agricultural practices (Muenchen et al. 2016).

12.4.2 Real-Time Pesticide Dispersion Control

Nano-biosensors provide a groundbreaking approach to pesticide management by enabling precise, real-time monitoring and dispersion control. This technology supports sustainable agricultural practices, reducing environmental impact while maximizing crop yield (Wahab et al. 2024). Nano-biosensors can help with pesticide management in agriculture by providing real-time monitoring and dispersion control.

12.4.2.1 Detect Pollutants

Nano-biosensors can detect pesticides, heavy metals, endocrine disruptors, and other pollutants at nanomolar to picomolar levels. The mechanism of detection includes steps like:

(a) Recognition of Pollutants: The biological component of the nano-biosensor binds specifically to target pollutants, such as heavy metals, pesticides, hydrocarbons, or pathogens in the soil.
(b) Signal Transduction: The interaction between the biological component and the pollutant generates a measurable signal.
(c) Data Analysis: The signal is processed to quantify the concentration and type of pollutant in the soil (Huang et al. 2023).

12.4.2.2 Monitor Crop Conditions

Nano-biosensors can monitor soil moisture, nutrients, pH, and temperature in real time. Key Applications of Nano-biosensors in Monitoring Crop Conditions include (John et al. 2022):

(a) Soil Quality Monitoring: Detect levels of essential nutrients (e.g., nitrogen, phosphorus, potassium) in soil, measure pH and salinity to optimize crop growth conditions, and monitor microbial activity and soil health for sustainable farming practices.
(b) Detection of Plant Stress: Identify early signs of biotic stress (e.g., pest attacks, fungal infections) and measure abiotic stress factors like drought, salinity, and temperature extremes. Also helps to monitor plant hormones such as abscisic acid to detect stress responses (Amjad and Serajuddin 2022).
(c) Water Quality and Irrigation: Ensure irrigation water is free from contaminants like heavy metals and pathogens, and optimize water usage by detecting soil moisture levels and evapotranspiration rates (Gupte and Pradeep 2022).
(d) Pesticide Residue Analysis: Detect pesticide residues on crops in real time, ensuring food safety and compliance with regulatory standards.
(e) Nutrient Delivery Systems: Monitor nutrient uptake by crops and release fertilizers in a controlled manner (Narang et al. 2024).

12.4.2.3 Detect Pathogens

Nano-biosensors can detect crop diseases caused by pests and pathogens. Nano-biosensors help in disease detection of fungal infections (detects spores of fungi such as *Fusarium* or *Aspergillus*), bacterial pathogens (identifies bacterial species like *Xanthomonas* or *Pseudomonas* that cause crop blights), viral diseases (detects viral particles or RNA of plant viruses like *Tobacco Mosaic Virus*), and pest monitoring by recognizing pest-specific proteins, pheromones, or metabolites, aiding in pest management strategies (Mandal et al. 2020).

(a) *Improve Resource Usage*: Nano-biosensors can help optimize resource usage and reduce resource wastage. Nano-biosensors play a vital role in detecting and recovering valuable metals from wastewater, offering an environmentally friendly and efficient solution for resource management (Javaid et al. 2021).

(b) *Support Sustainable Agriculture*: Nano-biosensors can help reduce the negative environmental impact of agricultural activities. Nano-biosensors detect soil moisture levels and help optimize irrigation, preventing overuse and conserving water resources. Sensors monitor soil nutrient levels, ensuring precise application of fertilizers, reducing runoff, and minimizing nutrient wastage (Bhagat et al. 2023).

(c) *Support Precision Farming*: Nano-biosensors can help with precision farming practices (Kumar et al. 2024). Precision farming, or precision agriculture, uses advanced technologies to monitor, manage, and optimize crop production systems. Nano-biosensors are integral to this approach, providing real-time data and insights at the molecular level to enable site-specific farming practices. By integrating nanotechnology with biosensing mechanisms, nano-biosensors revolutionize farming efficiency, sustainability, and productivity (Dar et al. 2020).

12.4.3 Nano-biosensors for Field Deployments

Food production has made the agricultural ecosystems face a lot of challenges, like the persistence of residues of different pesticides, accumulation of heavy metals, and contamination with toxic elemental particles, which have negatively influenced the agricultural environment. The entry of such toxic elements into the human body through agricultural products engenders human health by having effects such as nerve and bone marrow disorders, metabolic disorders, infertility, disruption of biological functions at the cellular level, and respiratory and immunological diseases (Sharma et al. 2021). Nano-biosensor is an emerging tool for field deployments, offering advanced monitoring capabilities for environmental and agricultural applications. These devices are compact and sensitive and enable real-time detection of chemical and biological agents in diverse field sets. By integrating nanobiotechnology, we address the need for portable and user-friendly systems for on-site diagnostics (Lopez and Aguilar 2023).

12.4.4 Enhancing Eco-friendly Pesticide Practices (Nano-pesticides)

A diverse array of eco-friendly pesticides—encompassing biological, botanical, and mineral types—is examined, highlighting their distinct modes of action and evaluating their effectiveness compared to synthetic counterparts (Verma et al. 2023). Nanotechnology presents distinctive properties that provide an alternative approach, incorporating diverse nano-formulations and encapsulation techniques. Nano-pesticides enable sustained release, improved permeability, and enhanced stability, significantly boosting pest control efficiency over short periods. Nanoparticles

(NPs) function as "magic bullets," capable of carrying various pesticides and delivering them precisely to target sites. Smart delivery systems for nano-pesticides play a crucial role in regulating pesticide dosage, ensuring optimal efficacy (Kapinder et al. 2021). There are different types of nano-pesticides:

12.4.4.1 Nano-emulsion

A nano-emulsion is a type of nano-pesticide where the active pesticide ingredient is dispersed as tiny droplets within a continuous aqueous phase, creating a very fine emulsion with particle sizes in the nanometer range, allowing for better penetration and efficacy compared to traditional pesticide formulations; essentially, it is a nano-sized version of an emulsion, often categorized as "oil-in-water" (O/W) where oil droplets are suspended in water, with the help of surfactants to stabilize the mixture. The benefits of using nano-emulsions are (Kah and Hofmann 2014):

- Reduced Pesticide Usage: Lower application rates of the active ingredient are often needed due to improved efficacy.
- Enhanced Penetration: Smaller droplets penetrate plant tissues and reach pest-hiding spots.
- Targeted Delivery: This can be designed to target certain pests specifically.

12.4.4.2 Polymer-Based Nano-pesticides

Polymer-based nano-pesticides are a promising method for delivering pesticides, and they can improve the sustainability of agricultural operations. They have many benefits, including:

- Environmental Impact: When applied in lower doses, they cause less environmental contamination.
- Efficacy: They have better efficacy because they release their active ingredients in a controlled manner.
- Biocompatibility: They are biocompatible, biodegradable, and nontoxic.
- Soil Health: They can increase soil permeability and aeration, increasing soil microorganisms' activity.
- Crop Yields: They can improve crop yields.
- Tolerance to Stress: They can help plants tolerate abiotic stress conditions (Sun et al. 2020).

Polymer-based nano-pesticides can be made from various natural and synthetic polymers, like chitosan, sodium alginate, albumin, starch, etc. These nano-pesticides can be prepared in different tri-dimensional structures, such as nano-capsules, nanospheres, micelles, nanogels, etc. (Kah and Hofmann 2014).

Nano-spheres

Nano-spheres are spherical particles smaller than 1 micrometer in diameter and made of a polymeric matrix. They can slowly release active substances like vitamins, fragrances, drugs, genes, and enzymes (Kah and Hofmann 2014). Hollow porous silica nano-spheres (HSNs) have been synthesized as a potential carrier for pesticide delivery. HSNs were synthesized with a single large hole or pore opening on shells (15.95 nm) (Nuruzzaman et al. 2019). Lignin nano-spheres (LNPs) are an emerging high-value nano-pesticide. The modification of new functions to LNPs is of great significance. This has been prepared by using lignin–Xylan (Xylan is a polysaccharide found in plant cell walls) hybrid nano-spheres (LXNPs) to achieve the application as pesticide carriers for enzyme-mediated controlled release. Hybrid LXNPs with various weight ratios (lignin to xylan, 3:1, 1:1, 1:3) have been obtained using deep solvent-extracted condensed lignin and water-insoluble xylan fragments, which exhibited a nanosphere size of about 166–210 nm. LXNPs with lignin-to-xylan ratios of 3:1 and 1:1 showed better core-shell structures with enriched hydroxyl groups on the surface. It has been proposed that lignin could anchor xylan fragments through van der Waals forces and hydrophobic interactions between lignin phenylpropanes and xylan molecules, thus facilitating the self-assembly process for forming this specific spherical structure. The resulting hydrophobic LXNPs core enabled the 57.9–67.0% efficiency in pesticide delivery (Jiang et al. 2020).

Nanogels

Nanogels have emerged as a promising technology for the controlled release of pesticides, offering precise delivery, prolonged activity, and reduced environmental impact. These nanoscale hydrogel particles can encapsulate active pesticide ingredients and release them in response to specific environmental triggers or conditions (Soni et al. 2016). Nanogels are three-dimensional, cross-linked hydrogel particles that are used in drug delivery, imaging, and other biomedical applications (Kah and Hofmann 2014). Nanogels have many unique properties that make them useful, including:

- Biocompatibility: Nanogels are mostly hydrophilic and highly biocompatible.
- Drug-Loading Capacity: Nanogels have a high capacity for loading guest molecules.
- Stimuli-Responsive Behavior: Nanogels can be designed to respond to environmental factors like pH, temperature, and ionic strength.
- Adaptability: Nanogels can be made from a variety of natural or synthetic polymers or a combination of the two (Badwaik et al. 2022).

A common, well-known example is poly(*N*-isopropylacrylamide) (PNIPAM)-based nanogels with high deformability ability, which were successfully constructed for smart pesticide delivery and effective pest control (Shen et al. 2022).

Electro-spun Nanofibers

Electrospinning is a highly precise method for producing nanofibers, offering control over their morphology by adjusting parameters, making it a better choice for efficiently producing and managing nanofiber properties. Nanofibers have been increasingly recognized for the possibilities within drug delivery systems that they possess and offer distinctive features such as high surface area (10 to 1000m^2/g), efficient drug loading, and the ability to provide sustained drug release. Fabricating nanofibers from biopolymers is particularly favored due to many advantages over nano-spheres and nanogels, including biocompatibility, capacity to replicate extracellular matrix, biodegradability, non-toxicity, and renewability (Shastri et al. 2024). There are many advantages of electro-spun nanofibers in pesticide release:

- Controlled Release: Ensures sustained delivery of pesticides, minimizing the need for frequent applications.
- Environmental Safety: Reduces pesticide runoff and leaching into soil and water bodies.
- Enhanced Stability: Protects active ingredients from degradation due to light or temperature fluctuations.
- High Surface Area: Facilitates efficient pesticide loading and release (Colín-Orozco et al. 2024).

12.5 Challenges and Limitations

12.5.1 Technical Challenges in Sensitivity and Stability

Nano-biosensors show exceptionally high sensitivity for detecting molecular changes. Despite their potential, technical challenges related to sensitivity and stability hinder their widespread application. These issues arise from material properties, biocompatibility concerns, fabrication inconsistencies, and operational limitations (Javaid et al. 2021). Sensitivity-related challenges include material properties (quantum dots or carbon nanotubes), signal amplification and noise, target selectivity (Complex biological matrices [e.g., serum or whole blood] can mask target molecules), and detection limits (Kulkarni et al. 2022).

Stability challenges include material degradation due to harsh environments, biocompatibility, biofouling (Prolonged contact with biological fluids often leads to biofouling), and mechanical and chemical wear of nanomaterials (Kulkarni et al. 2022).

12.5.2 Environmental and Ethical Considerations

Fluctuations in temperature, humidity, or ionic strength can destabilize sensor response. Long-term use in variable environments may lead to drift in signal output (Kulkarni et al. 2022). Certain nanomaterials in biosensors are not biodegradable, contributing to nano-waste pollution (Mohamed and Awad 2022). Accumulation in the environment may harm microorganisms, plants, and animals. While nano-biosensors have significant potential for improving pesticide management, ethical considerations must guide their use (Srivastava 2023). Adopting nano-pesticides introduces several ethical considerations that balance their technological benefits with potential risks to ecosystems, human health, and societal equity. Below are key ethical issues identified in recent research by Ijaz et al. (2024):

- Environmental Impacts and Long-Term Effects: The potential accumulation of nano-pesticides in the soil and water raises concerns about unknown ecological consequences and long-term toxicity (Mishra et al. 2024).
- Human Health Risks: The small size of nanoparticles can enable them to penetrate biological barriers, potentially causing harm to farmers and consumers. Insufficient testing of these products before market introduction poses health and safety risks (Ijaz et al. 2024).
- Equity and Accessibility: High costs and advanced technologies associated with nano-pesticides may restrict access to small-scale and subsistence farmers, exacerbating economic inequalities. (Shah et al. 2024) emphasized the need for ethical technologies that prioritize inclusivity and affordability in agriculture.
- Regulatory and Safety Challenges: The lack of robust regulatory frameworks to monitor and evaluate nano-pesticides creates uncertainty in their ethical deployment. It is important to assess nano-pesticide toxicity, scalability, and environmental safety (Sun et al. 2024).

Recommendations for Ethical Implementation: (Chávez-Hernández et al. 2024)

- Regulatory Oversight: Establish comprehensive guidelines to ensure nano-pesticide safety and effectiveness.
- Transparency: Maintain open communication about nano-pesticide risks and benefits with stakeholders.
- Research Investments: Conduct extensive studies on long-term ecological and health impacts.
- Global Standards: Harmonize international standards to prevent inequities and ensure responsible use.

12.5.3 Cost-Effectiveness of Nano-biosensor

Nano-biosensors represent an innovative analytical tool class that leverages nanotechnology for precise, efficient, and cost-effective diagnostics. Their cost implications depend on material choices, fabrication techniques, and the intended application. While nanomaterials like graphene, quantum dots, or carbon nanodots offer excellent sensitivity and specificity, their scalable and economical production significantly reduces costs. Emerging trends focus on developing low-cost, mass-producible nano-biosensors, particularly for healthcare and environmental applications (Ghosh et al. 2024). Metal-based nanoparticles (NPs), including those composed of gold, silver, palladium, manganese, and zinc, with dimensions smaller than 100 nm, are now synthesized using plant-derived extracts through various methodologies. The physicochemical, catalytic, and toxicological properties of these nanoparticles are highly dependent on their size and morphology. Biological synthesis, commonly referred to as green synthesis, is an eco-friendly and efficient approach for NP production that circumvents the need for extreme pressure, high temperatures, and hazardous chemicals. This sustainable method can be facilitated using a diverse range of biological entities, including bacteria, fungi, yeast, viruses, microalgae, and plant-derived biomass or extracts (Kumar et al. 2022).

12.6 Recent Advances and Case Studies

Chemical pesticide residues pose significant risks to human health and the environment. Reliable and efficient detection methods have been developed to address these concerns. A recent innovation involves the use of biosensors based on carbon quantum dots (CQDs), synthesized from waste tea, for detecting pesticides commonly used in tea cultivation. These CQDs are produced via a straightforward one-pot hydrothermal process and undergo comprehensive characterization using advanced techniques such as transmission electron microscopy (TEM), ultraviolet-visible (UV) spectroscopy, Raman spectroscopy, X-ray diffraction (XRD), atomic force microscopy (AFM), and X-ray photoelectron spectroscopy (XPS) (Sinha and Ray 2024).

Shelar et al. (2024) developed surface plasmon resonance (SPR) sensors, demonstrating their utility as effective tools for detecting challenging compounds, agro-toxins, and pesticides across various applications. Nano-sensors, in particular, offer a promising approach for identifying agro-toxin and pesticide residues due to their user-friendly design, label-free operation, compact size, high specificity and sensitivity, and real-time measurement capabilities.

Recent studies highlight that prolonged and excessive use of chlorpyrifos (CPS) has resulted in significant pollution, contaminating crops, vegetables, fruits, and water sources. While conventional methods were previously used to detect CPS in food and water samples, their practical application has been hindered by limitations in size, portability, and cost. Electrochemical sensors have now emerged as a preferred alternative for CPS detection, offering high sensitivity, on-site usability, user-friendliness, reliability, and rapid detection. Among these, graphene-based electrochemical sensors have garnered considerable attention due to their exceptional physicochemical and electrochemical properties. Their high surface area and superior conductivity enable remarkable sensitivity, selectivity, and quick response times (Krishna Perumal et al. 2024).

Deep et al. explored the creation of a printed carbon electrode that was modified with styrene sulfonic acid and doped with polyaniline to develop an immunosensor for detecting atrazine pesticides. The sensor operates based on the specific interaction between atrazine and the anti-atrazine antibody that is immobilized on its surface. This method exhibited remarkable specificity and sensitivity, achieving a detection limit as low as 0.01 ng/mL for atrazine within a concentration range of 0.01 to 50 ng/mL (Deep et al. 2014).

Parathion is a widely used global pesticide that serves as an acetylcholinesterase inhibitor, leading to varied poisoning symptoms such as nausea, vomiting, diarrhea, muscle cramping or twitching, and shortness of breath upon exposure. Mehta et al. designed graphene-based impedimetric immunosensors for the highly sensitive and specific detection of parathion. These sensors utilize graphene's exceptional conductivity and surface characteristics to deliver accurate and dependable measurements, making them a powerful tool for monitoring parathion residues. The fabrication process involved modifying screen-printed carbon electrodes (SPE) with graphene sheets, which were then functionalized with 2-aminobenzyl amine (2-ABA) through an electrochemical reaction. Following this, the amine-functionalized graphene electrodes were bio-interfaced with anti-parathion antibodies, enabling targeted detection of parathion. In impedimetric mode, the biosensor exhibited a wide linear detection range of 0.1–1000 ng/L, underscoring its sensitivity and effectiveness for monitoring parathion residues (Mehta et al. 2016).

Immuno-biosensors utilizing antibodies, such as monoclonal, polyclonal, and recombinant antibodies, and antigens as recognition elements are widely employed for atrazine detection. These biosensors are favored over enzymatic biosensors because of their superior specificity, selectivity, and stability. However, their adoption is limited by the high cost and low availability of antibodies, which pose significant challenges for large-scale or cost-effective applications (Steffens et al. 2022).

A cutting-edge cantilever nano-biosensor was functionally developed with 30 μL of anti-atrazine and antibody solution of glutaraldehyde for atrazine detection by Steffens et al. (2022). The interaction between the herbicide atrazine and its specific antibody (anti-atrazine) was studied using a cantilever-based nano-biosensor. The findings demonstrated that this biosensor could accurately detect atrazine even at low concentrations. The measured interaction revealed an adhesion strength of **172 ± 26 nN**, significantly higher than that of other nonspecific herbicides, indicating the biosensor's remarkable specificity and sensitivity for atrazine detection.

Enzyme-based nano-biosensors have been developed to detect organophosphorus (OP) pesticides. They use enzyme-based optical detection. Enzymes acetylcholinesterase and organophosphate hydrolase are used as OP pesticides and show high selectivity for these enzymes (Kaur and Singh 2020).

Liposome-based nano-biosensors were engineered for the detection of two commonly used organophosphorus pesticides, dichlorvos and paraoxon. The evaluation focused on assessing total toxicity in drinking water samples. The detection mechanism employed a fluorescence-based transducer, which demonstrated remarkable sensitivity, low detection limits, and an extended dynamic detection range. Acetylcholinesterase, the enzymatic component, was encapsulated within the internal nano-environment of the liposomes, a strategy that significantly enhanced enzyme stabilization (Muenchen et al. 2016).

Inhibition-based biosensors have been developed for the rapid, straightforward, and cost-effective detection of atrazine. The detection mechanism relies on the inhibition of the enzyme tyrosinase (Tyr), derived from mushrooms, which is immobilized on screen-printed electrodes (SPEs). To enhance biosensor performance, various carbon-based materials, including graphite (G), graphene (GP), and multi-walled carbon nanotubes (MWCNTs), were investigated as electrode modifiers. Tyrosinase immobilization on the electrode surface was achieved using either physical entrapment with polyvinyl alcohol or Nafion membranes or chemical immobilization with bovine serum albumin and glutaraldehyde. In the presence of catechol as a substrate, atrazine was successfully detected through its inhibitory effects on tyrosinase activity (Tortolini et al. 2016).

Kumar, Jha, and D'Souza developed an optical microbial biosensor for the detection of the pesticide methyl parathion. This biosensor employed bacterial cells of *Flavobacterium* sp., which were immobilized by entrapment within a glass-fiber filter and integrated into an optical fiber system. *Flavobacterium* sp. naturally expresses the enzyme organophosphorus hydrolase, which catalyzes the hydrolysis of methyl parathion into p-nitrophenol, a detectable product. The immobilized microbial biocomponent was designed to be disposable, cost-effective, and exhibited high reproducibility with consistent operational performance. This biosensing approach enabled a simple, single-step, and direct detection of methyl parathion, even at very low concentrations, making it a practical and efficient tool for pesticide monitoring (Kumar et al. 2006).

Moreover, A fluorescence-based enzymatic biosensor was recently developed for detecting organophosphate (OP) pesticides. This biosensor utilized a dual-enzyme complex of acetylcholine and choline oxidase, with cadmium telluride quantum dots (CdTe-QDs) serving as the recognition element. The CdTe-QDs facilitated sensitive and accurate detection, demonstrating the biosensor's effectiveness for identifying OP pesticide residues in real-world samples. This innovative design highlights its potential for practical applications in food and environmental monitoring (Meng et al. 2013).

12.7 Future Perspectives

The field of nano-biosensors is rapidly evolving, driven by advancements in nanotechnology, material science, and biochemistry. These sensors are set to transform diagnostics, environmental monitoring, and industrial applications through their unique ability to detect and quantify minute biological and chemical changes with high precision and sensitivity. The future of nano-biosensors lies in their innovation, integration, and adoption across diverse fields (Hassan 2022). There are various key fields in the future scope of nano-biosensors. These include the various points.

- *Advancement in Material Science*: This includes next-generation nanomaterials. Continued innovation in materials like graphene, carbon nanotubes, quantum dots, and 2D materials will enhance sensitivity, biocompatibility, robustness, and sustainable nanomaterials (development of eco-friendly and biowaste-derived materials to promote greener technologies and reduce costs) (Schiavo et al. 2024).
- *Miniature and Portability*: Nano-biosensors have become more portable, facilitating on-site diagnostics in remote and resource-constrained settings due to their compact designs. Integration of nano-biosensors into wearable technologies such as smartwatches, fitness bands, and patches for continuous monitoring of health parameters like glucose, lactate, and hydration levels is possible due to its compact design (Palacios and Bracamonte 2022).
- *Integration of Nano-biosensors with Advanced Technology*: Nano-biosensors will increasingly connect to Internet of Things (IoT) networks, enabling real-time data transfer and monitoring in applications like telemedicine, environmental sensing, and smart agriculture (Chandrashekhar et al. 2022). Artificial intelligence (AI) algorithms will enhance the interpretation of sensor data, allowing for accurate and predictive analytics in healthcare and industrial monitoring (Adir et al. 2020).

12.8 Conclusion

Nano-biosensors are at the forefront of a technological revolution, merging biology, nanotechnology, and data science. Their trajectory points toward smarter, faster, and more affordable solutions that will enhance our ability to diagnose, monitor, and respond to biological and environmental changes. The continuous evolution of this field holds the promise of a healthier, safer, and more sustainable future (Bhattacharya et al. 2022). Nano-biosensors represent a transformative technological leap, offering unparalleled sensitivity, specificity, and rapid response times for detecting and monitoring biological and chemical changes. Their integration into healthcare, agriculture, environmental monitoring, and industrial applications has already begun to address critical challenges, such as early disease diagnosis, sustainable farming practices, and pollution control (Darwish et al. 2024). Nano-biosensors demonstrate exceptionally low detection limits, often below the Maximum Residue Levels

(MRLs) set by regulatory authorities, enabling the identification of trace pesticide residues. Employing bio-inspired or natural materials reduces environmental impact, aligning with sustainability goals. Advanced nano-biosensors incorporate real-time monitoring and reporting of pesticides, which is crucial in precision agriculture. This integration supports proactive management strategies for pest control too.

Nano-biosensors show a wide range of applications and are an advance field which is been is a hot topic in recent research but much in this field is yet to be explored.

References

Adir O et al (2020) Integrating artificial intelligence and nanotechnology for precision cancer medicine. Adv Mater 32(13):1901989

Alshora DH, Ibrahim MA, Alanazi FK (2016) Chapter 6: Nanotechnology from particle size reduction to enhancing aqueous solubility. In: Grumezescu (ed) A.M.B.T.-S.C. of N. William Andrew Publishing, pp 163–191. https://doi.org/10.1016/B978-0-323-42861-3.00006-6

Amjad S, Serajuddin M (2022) Role of nanosensors and bionanosensors in crop abiotic stress. In: Augmenting crop productivity in stress environment. Springer, pp 1–12

Badwaik HR et al (2022) A review on challenges and issues with carboxymethylation of natural gums: the widely used excipients for conventional and novel dosage forms. Int J Biol Macromol 209:2197–2212. https://doi.org/10.1016/j.ijbiomac.2022.04.201

Baranwal J et al (2022) Electrochemical sensors and their applications: a review. Chemosensors. https://doi.org/10.3390/chemosensors10090363

Bellucci S (2008) Nanoscale science and technology. J Phys Condens Matter. John Wiley and Sons, Ltd. https://doi.org/10.1088/0953-8984/20/47/470301

Bhagat R, Ingle AP, Chen H (2023) Nanosensors and nanobiosensors for sustainable agriculture. In: Nanotechnology in agriculture and agroecosystems. Elsevier, pp 93–112

Bhattacharya M et al (2022) Nanobiosensors: principles, techniques, and innovation in nanobiosensors. In: Biosensors in food safety and quality. CRC Press, pp 85–110

Chamorro-Garcia A, Merkoçi A (2016) Nanobiosensors in diagnostics. Nano 3:1849543516663574. https://doi.org/10.1177/1849543516663574

Chandrashekhar KG et al (2022) Emerging nanomaterials for advanced technologies. In: Introduction to functional nanomaterials. CRC Press, pp 157–165

Chávez-Hernández JA et al. (2024) Safe nanomaterials: from their use, application and disposal to regulations. Nanoscale Advances [Preprint]

Colín-Orozco J, Colín-Orozco E, Valdivia-Barrientos R (2024) Production of nanofibers by electrospinning as carriers of agrochemical. Fibers. https://doi.org/10.3390/fib12080064

Dar FA, Qazi G, Pirzadah TB (2020) Nano-biosensors: NextGen diagnostic tools in agriculture. In: Nanobiotechnology in agriculture: an approach towards sustainability. Springer, Cham, pp 129–144

Darwish MA et al (2024) Advancements in nanomaterials for nanosensors: a comprehensive review. Nanosc Adv. [Preprint]

Deep A et al (2014) Styrene sulphonic acid doped polyaniline based immunosensor for highly sensitive impedimetric sensing of atrazine. Electrochim Acta 146:301–306

Ghosh A et al (2024) Ultrasensitive and selective detection of chromium(VI) using biowaste derived fluorescent Nanoprobes of hydrothermally synthesized carbon-dots. ChemistrySelect 9(46):e202404198. https://doi.org/10.1002/slct.202404198

Gupte T, Pradeep T (2022) Nanosensors for water quality monitoring. In: Separation science and technology. Elsevier, pp 37–53

Hamzah AA, Nadzirah S (2023) In: R.B.T.-E. of S. and B. (First E. Narayan (ed) Biosensor development. Elsevier, Oxford, pp 209–217. https://doi.org/10.1016/B978-0-12-822548-6.00112-6

Hassan RYA (2022) Advances in electrochemical nano-biosensors for biomedical and environmental applications: from current work to future perspectives. Sensors. https://doi.org/10.3390/s22197539

Huang C-W et al (2023) A review of biosensor for environmental monitoring: principle, application, and corresponding achievement of sustainable development goals. Bioengineered 14(1):58–80

Ijaz S et al (2024) Regulatory and ethical concerns of nanotechnology in agriculture. In: Revolutionizing agriculture: a comprehensive exploration of Agri-nanotechnology. Springer, pp 395–427

Javaid M et al (2021) Exploring the potential of nanosensors: a brief overview. Sensors Int 2:100130. https://doi.org/10.1016/j.sintl.2021.100130

Jiang Y et al (2020) Fabrication and characterization of lignin–xylan hybrid nanospheres as pesticide carriers with enzyme-mediated release property. Soft Matter 16(39):9083–9093

John SA et al (2022) Nanosensors for plant health monitoring. In: Nanosensors for smart agriculture. Elsevier, pp 449–461

Kah M, Hofmann T (2014) Nanopesticide research: current trends and future priorities. Environ Int 63:224–235

Kapinder, Dangi K, Verma AK (2021) Efficient & eco-friendly smart nano-pesticides: emerging prospects for agriculture. Mater Today Proc 45:3819–3824. https://doi.org/10.1016/j.matpr.2021.03.211

Kaur J, Singh PK (2020) Enzyme-based optical biosensors for organophosphate class of pesticide detection. Phys Chem Chem Phys 22(27):15105–15119

Krishna Perumal P et al (2024) Graphene-based functional electrochemical sensors for the detection of chlorpyrifos in water and food samples: A review. J Food Sci Technol 61(4):631–641

Kulkarni MB, Ayachit NH, Aminabhavi TM (2022) Recent advancements in nanobiosensors: current trends, challenges, applications, and future scope. Biosensors. https://doi.org/10.3390/bios12100892

Kumar J, Jha SK, D'souza SF (2006) Optical microbial biosensor for detection of methyl parathion pesticide using Flavobacterium sp. whole cells adsorbed on glass fiber filters as disposable biocomponent. Biosens Bioelectron 21(11):2100–2105

Kumar N, Goel G, Goel S (2022) Nanomaterials based biosensing: methods and principle of detection, pp 1–27. https://doi.org/10.1007/978-981-16-3645-5_1

Kumar P et al (2024) Trends of Nanobiosensors in modern agriculture systems. Appl Biochem Biotechnol:1–24

Lopez A, Aguilar J (2023) A data analysis smart system for the optimal deployment of nanosensors in the context of an eHealth application. Algorithms 16(2):81

Malik P et al (2013) Nanobiosensors: concepts and variations. In: Hwang KC, Pissuwan D (eds) ISRN nanomaterials, p 327435. https://doi.org/10.1155/2013/327435

Mandal N, Adhikary S, Rakshit R (2020) Nanobiosensors: recent developments in soil health assessment. In: Soil analysis: recent trends and applications. Springer, Singapore, pp 285–304

Mehta J et al (2016) Graphene modified screen printed immunosensor for highly sensitive detection of parathion. Biosens Bioelectron 83:339–346

Meng X et al (2013) A simple and sensitive fluorescence biosensor for detection of organophosphorus pesticides using H2O2-sensitive quantum dots/bi-enzyme. Biosens Bioelectron 47:402–407

Mishra R et al (2024) Innovations and future trends in storage pest management. J Exp Agric Int 46(5):155–165

Mohamed EF, Awad G (2022) Chapter 6: Development of nano-sensor and biosensor as an air pollution detection technique for the foreseeable future. In: Turan NB, Engin GO, Bilgili

MSBT-CAC (eds) Environmental nanotechnology: implications and applications. Elsevier, pp 163–188. https://doi.org/10.1016/bs.coac.2021.11.003

Muenchen DK et al (2016) Pesticide detection in soil using biosensors and nanobiosensors. Biointerface Res Appl Chem 6(6)

Munawar A et al (2019) Nanosensors for diagnosis with optical, electric and mechanical transducers. RSC Adv 9(12):6793–6803. https://doi.org/10.1039/C8RA10144B

Narang C, Kant R, Mandal S (2024) Nanosensors for soil and nutrients analytics. In: Revolutionizing agriculture: a comprehensive exploration of Agri-nanotechnology. Springer, pp 177–200

Nasrollahzadeh M et al (2019) Chapter 4: Applications of nanotechnology in daily life. In: Nasrollahzadeh M et al (eds) An introduction to green nanotechnology. Elsevier, pp 113–143. https://doi.org/10.1016/B978-0-12-813586-0.00004-3

Nosheen E et al (2019) In: Ozkan SA, A.B.T.-N.D. in N. for P.A. Shah (eds) Chapter 4: Optical nanosensors for pharmaceutical detection. Academic, pp 119–140. https://doi.org/10.1016/B978-0-12-816144-9.00004-3

Nuruzzaman M et al (2019) Hollow porous silica nanosphere with single large pore opening for pesticide loading and delivery. ACS Appl Nano Mater 3(1):105–113

Palacios LRG, Bracamonte AG (2022) Development of nano-and microdevices for the next generation of biotechnology, wearables and miniaturized instrumentation. RSC Adv 12(20):12806–12822

Panoth D et al (2021) Chapter 11. In: Nanosensors for pesticide detection in soil. https://doi.org/10.1016/B978-0-12-824554-5.00002-1

Ramesh M et al (2023) Nanotechnology-enabled biosensors: a review of fundamentals, design principles, materials, and applications. Biosensors. https://doi.org/10.3390/bios13010040

Reshetilov A et al (2011) Nanobiosensors and their applications. In: Metal nanoparticles in microbiology. Springer, pp 269–283

Schiavo L et al (2024) An overview of the advanced nanomaterials science. Inorg Chim Acta 559:121802

Shah F et al (2024) Innovative nanotechnological solutions for sustainable agriculture amidst rising population demands. Food Hum 3:100425. https://doi.org/10.1016/j.foohum.2024.100425

Sharma P et al (2021) A review on biosensors and nanosensors application in agroecosystems. Nanoscale Res Lett 16(1):136. https://doi.org/10.1186/s11671-021-03593-0

Shastri SS, Varma P, Kandasubramanian B (2024) Enhancing drug delivery with electrospun biopolymer nanofibers. Biomed Mater Devices. [Preprint]. https://doi.org/10.1007/s44174-024-00218-9

Shelar A et al (2024) Recent advances on highly sensitive plasmonic nanomaterial enabled sensors for the detection of agrotoxins: current progress and future perspective. Comput Electron Agric 227:109545. https://doi.org/10.1016/j.compag.2024.109545

Shen Y et al (2022) Temperature-dependent nanogel for pesticide smart delivery with improved foliar dispersion and bioactivity for efficient control of multiple pests. ACS Nano 16(12):20622–20632

Singh P et al (2016) Biomedical perspective of electrochemical nanobiosensor. Nano Micro Lett 8:193–203

Sinha N, Ray S (2024) Application of carbon quantum dots derived from waste tea for the detection of pesticides in tea: a novel biosensor approach. ACS Omega. [Preprint]

Soni KS, Desale SS, Bronich TK (2016) Nanogels: an overview of properties, biomedical applications and obstacles to clinical translation. J Control Release 240:109–126

Srilatha B (2011) Nanotechnology in agriculture. J Nanomed Nanotechnol 2(7):5

Srivastava S (2023) Ethical issues regarding the use of nanobiotechnology-based products. In: Nanobiotechnology for the livestock industry. Elsevier, pp 435–473

Steffens C et al (2022) Advances of nanobiosensors and its application in atrazine detection in water: a review. Sens Actuators Rep 4:100096. https://doi.org/10.1016/j.snr.2022.100096

Sun C et al (2020) Polymer-based nanoinsecticides: current developments, environmental risks and future challenges-a review. Biotechnol Agron Soc Environ:24

Sun Y et al (2024) Nano-enabled strategies for greenhouse gases emission mitigation: a comprehensive review. Nano Today 57:102378. https://doi.org/10.1016/j.nantod.2024.102378

Tortolini C et al (2016) Inhibition-based biosensor for atrazine detection. Sensors Actuators B Chem 224:552–558

Uchino K (2017) Advanced piezoelectric materials: science and technology. Woodhead Publishing

Valenzuela-Amaro HM et al (2023) Emerging applications of nanobiosensors in pathogen detection in water and food. Biosensors. https://doi.org/10.3390/bios13100922

Verma NS et al (2023) A review on eco-friendly pesticides and their rising importance in sustainable plant protection practices. Int J Plant Soil Sci 35(22):200–214

Wahab A et al (2024) Agriculture and environmental management through nanotechnology: eco-friendly nanomaterial synthesis for soil-plant systems, food safety, and sustainability. Sci Total Environ 926:171862. https://doi.org/10.1016/j.scitotenv.2024.171862

Williams RM et al (2021) Harnessing nanotechnology to expand the toolbox of chemical biology. Nat Chem Biol 17(2):129–137

Zhao Z et al (2023) Chemical sensor based on piezoelectric/triboelectric nanogenerators: a review of the modular design strategy. Chemosensors. https://doi.org/10.3390/chemosensors11050304

Zhu R et al (2021) In vivo nano-biosensing element of red blood cell-mediated delivery. Biosens Bioelectron. https://doi.org/10.1016/j.bios.2020.112845

Chapter 13
Nanobiosensors for Real-Time and High-Resolution Crop Monitoring

Sanam Ashraf, Javed Iqbal, Banzeer Ahsan Abbasi, Zubaria Haakim, Wadie Chalgham, and Tariq Mahmood

Abstract In recent years, nanotechnology has been considered to be the most noteworthy technological invention and is being utilized in different sectors, mostly in the crop sector, due to its forthcoming uses. Crop sectors consider as an emerging nation's economic growth, and crop technology breakthroughs are necessary to address worldwide problems like the rise in population, starvation, and atmospheric warming. The use of nanotechnology practices is one of the potential strategies for sustainable crop production. One essential use of nanotechnology is the construction of smart biosensors, which results in the formation of petite components that are far more potent and well-organized, known as nanobiosensors. These nanoscale dimensions combine nanotechnology's and biosensors' competences, which ensure more effective and precise monitoring of various properties that are essential for crop management. Metal nanoparticles (NPs), metal oxide nanoparticles, carbon nanotubes (CNTs), quantum dots (QDs), graphene-based materials, and polymer-based nanomaterials, all have nanoscale properties that enhance sensitivity through signal amplification and incorporate a number of groundbreaking transduction principles into nanosensors, including upgraded electrochemical, mass detection, and optical nanosensors' features. The incorporation of nanobiosensors with advanced technology such as Internet and data transfer with the help of wireless nanosensors

S. Ashraf · Z. Haakim · T. Mahmood
Department of Plant Sciences, Faculty of Biological Sciences, Quaid-i-Azam University, Islamabad, Pakistan
e-mail: zubaria@bs.qau.edu.pk; tmahmood@qau.edu.pk

J. Iqbal
Department of Botany, Bacha Khan University, Charsadda, Khyber Pakhtunkhwa, Pakistan

B. A. Abbasi (✉)
Department of Botany, Rawalpindi Women University, Rawalpindi, Pakistan

W. Chalgham
Department of Mechanical and Aerospace Engineering, University of California, Los Angeles, CA, USA
e-mail: Wadie.chalgham@ucla.edu

J.-T. Chen (ed.), *Nanobiosensors for Crop Monitoring and Precision Agriculture*, Sustainability Sciences in Asia and Africa,
https://doi.org/10.1007/978-981-96-8335-2_13

has transformed sensing with multimodal and real-time sensing abilities in the crop sectors. Nanobiosensors have made a name for themselves as agricultural innovators, facilitating revolutionary developments in agro-technologies that increase crop durability and productivity. By improving drainage systems, raising fertilizer efficiency, and improving food manufacturing, packaging, and storage processes, their use has revolutionized conventional procedures. These innovations have provided a priceless increase in agricultural resilience and output, satisfying the growing needs of a growing global population. Farmers today use cutting-edge digital nanotechnology to maintain farming practices that can quickly increase crop growth and productivity. The significance of nanobiosensors in enhancing crop productivity is emphasized in this chapter.

Keywords Agricultural resilience · Metal nanoparticles · Nanosensors · Nanotechnology

13.1 Introduction

With the world's population expected to approach nine billion by 2050, consumer appetite for agricultural products must rise dramatically. Climate fluctuation, dwindling arable land, and environmental deterioration, all highlight the critical importance of sustainable agricultural techniques. Precision agriculture has emerged as a disruptive method, leveraging data and technological innovations to improve crop health and optimize the usage of resources (Berners-Lee et al. 2018). Nanotechnology has established itself as an innovator in agriculture, enabling groundbreaking advances in agro-technologies that improve production and durability. Its application has transformed traditional practices by optimizing drainage systems, increasing fertilizer efficiency, and boosting food manufacturing, wrapping, and storage techniques. These technologies have offered an invaluable boost to agricultural productivity and resilience, meeting the rising requirements of an increasing number of people worldwide, as shown in Fig. 13.1 (Neethirajan and Jayas 2011). Nanotechnology has a wide range of agricultural uses, including nanofertilizers, nanoherbicides, nanopesticides, nanosensors, and nano-aptamers.

These techniques not only address difficulties, such as insufficient nutrients, water shortage, and eroded soil, but they also address critical environmental issues and the unpredictability of weather and excessive rainfall, all of which impact seedling development and the wellness of plants (Johnson et al. 2021). Farmers who incorporate nanotechnology utilize more advanced and precise methods for controlling these challenges, resulting in increased crop output and the preservation of resources. Advancements in nanotechnology have been important in the improvement of the sensitivity and functioning of nanosensors and nanobiosensors, which permit real-time monitoring of farm data. These sensors are important, as they can identify very small changes in the state of the soil, water purity, and ambient variables and give farmers this useful information. With their compact size, huge

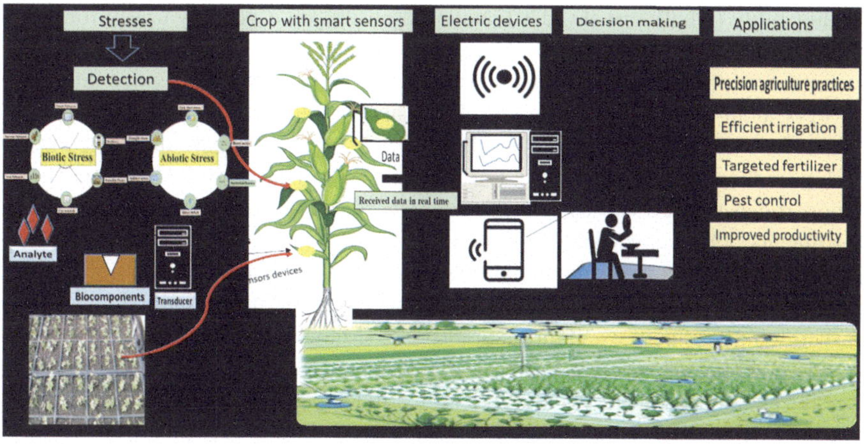

Fig. 13.1 Schematic demonstration of the nanobiosensors utilized in precision agriculture applications

surface area, and unrivaled, sophisticated material qualities, they simply work in an extraordinary way on a surprising level for farmers tackling farming difficulties (Momin et al. 2013). Nanotechnology bridges the gap between tradition and science by offering new, modern solutions for the improvement in the use of resources and the protection of the environment. The use of these tools for agricultural purposes has gained substantial speed, and they have become important in dealing with the complex problems relating to advanced farming (Mishra et al. 2019). Changes in science and technology due to the introduction of nanosensors have dramatically advanced multiple fields, particularly agriculture. However, as they are already used in products such as cellphones and electrical devices, these sensors have an adaptability to a tiny size, large surface area, and bulk object detection qualities (Guisbiers et al. 2012). Nanosensors have their own features; particularly, the nanosensors fabricated with highly complex nanomaterials like graphene, metal oxides, and carbon nanotubes show excellent performances of sensitivity, accuracy, reaction speed, and so on. They have a wide potential of use in agricultural, medical, and manufacturing processes (Joseph and Morrison 2006). With nanotechnology, the advances that have taken place regarding the improvement in agriculture have solved longstanding problems. Crops' physical, chemical, and morphological properties are obtained by various nanomaterials, including nanocomposites, nanotubes, nanowires, and nanosheets. Together, these components help in tracking nutrient intake, biologically active compounds, and microorganisms, such as pesticides, pests, and bacteria, as part of an all-encompassing crop maintenance system (Bhandari et al. 2023). The use of nanobiosensors is a critical issue in the context of precision agriculture, as they are intended to determine the parameters of soil, pH, cost, moisture content, pathogens, pesticides, chemical insecticides, herbicides, and chemical fertilizers. Nanobiosensors offer farmers the chance to make familiar selections, enhance routing and utilization of assets, and enhance yields with elevated environmental sustainability (Ghaffar et al. 2020). Nanobiosensors are especially useful for dealing with the agricultural issues of the environment. For example, these kinds of sensors incorporate the use of nanoparticles (NPs) in order to recognize and alert growers about the levels of hazardous chemical compounds, such as nitrogen oxides and sulfur oxides, which can negatively impact crops and decrease oxygen concentrations in the air (Zhu et al. 2019). For instance, carbon nanotubes (CNTs) and gold nanoparticles increase the sensitivity of organic sensors so as to detect plant stresses or ailments before any visible symptom appears (Fu et al. 2024). One of the frequently utilized metal oxide nanoparticles for soil water content material measurement and nutrient analysis was made up of zinc oxide and titanium dioxide because of their long-term nature and huge surface area (Maheshwaran et al. 2024).

The potential of nanobiosensors to collect impulses and transfer them through transducers allows for the accurate identification of agricultural contaminants and stress elements. For example, they can comprehend variations in plant metabolites, presenting initial notification of eventual problems and permitting preventive actions to be completed (Johnson et al. 2021). Beyond evaluation, nanobiosensors drastically decrease resource waste by enabling more precise management of fertilisers, insecticides, and moisture. They make contributions to the fight against environmental problems, along with soil deterioration, dryness, and unfavorable conditions, all of which have traditionally hampered agricultural production. Farmers can use those present-day units to implement sustainable farming strategies that lessen pollutants and keep valuable sources (Mondal et al. 2022). Local farmers may additionally see expanded sales from stepped-forward crop management and tracking. Higher yields result in great quantities of crop readily accessible for trade, which increases profit. This chapter investigates the remodeling function of nanobiosensors inside the farming industry, demonstrating their capability to revolutionize farming strategies. These sensors, which integrate nanotechnology and biology, offer accurate, actual-time monitoring of essential parameters required for best crop management. The use of biosensors augmented with nanomaterials presents extremely good precision for figuring out atmospheric and agricultural fitness parameters. This mixture positions nanobiosensors as a critical gear for reinforcing cutting-edge agriculture, tackling global concerns, and selling sustainable structures of food manufacturing (Wang et al. 2022).

13.2 Biosensor

A biosensor is a compact analytical tool that assimilates a biological or biomimetic sensing detail with a transducer device. It operates on the principle of selective binding between the goal analyte and a complementary biorecognition element constant onto a suitable aid medium. This relationship leads to adjustments in one or more physicochemical properties, such as pH forms, electron transfer, mass changes, temperature adjustments, absorption or discharge of gases or particular ions, all of which can be detected and measured via the transducer. Essentially, a biosensor

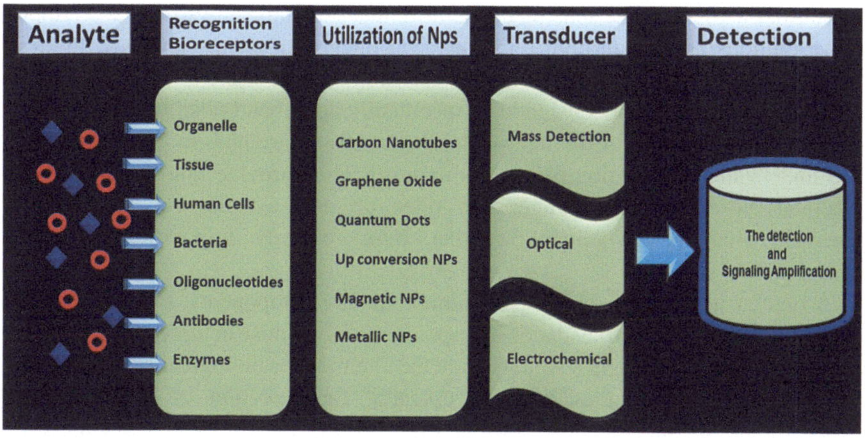

Fig. 13.2 Diagrammatic illustration of the primary elements of biosensors and their working standards

functions by combining a biological element, like an enzyme, antibody, or nucleic acid, with a transducer to classify substances. The biological interplay is transformed into an optical, thermal, or electrical signal through the transducer. This signal is then turned into a measurable electric property, along with voltage or current (Kulshreshtha et al. 2017). The output from the transducer can be digitized and displayed in real time on a screen by the use of a microcontroller and a signal-conditioning circuit (Polat et al. 2022). For effective operation, it is far important to immobilize the organic molecule near the transducer, which may be achieved via either chemical attachment or physical entrapment, as Fig. 13.2 demonstrates how a biosensor functions (Odobašić et al. 2019).

Biosensors provide essential facts about plant health by way of detecting physiological indicators like pH levels, the presence of pathogens, volatile organic compounds, and temperature changes. Key features for biosensors include robustness, linearity, value-effectiveness, specificity, repeatability, sensitivity, stability, and good selectivity (Krishnan et al. 2024). There is a huge variety of applications for biosensors in agriculture and food processing, all with unique necessities concerning analyte attention, precision of output, sample extent, assay period, reusability, and cleansing protocols. The choice of biosensor may also be influenced by the potential market size, as some types are easier to massproduce than others (Velasco-Garcia and Mottram 2003).

13.3 Nanobiosensors

In the twenty-first century, scientific development has led to the improvement of notably specialized miniature sensors called nanobiosensors, which leverage nanotechnology standards (Rai et al. 2012). Researchers have succeeded in developing biosensors with awesome sensing capabilities that offer amazing spatial and temporal resolutions, alongside high reliability, via integrating nanosciences with electronics, computing, and biology. Nanobiosensors are nanoscale sensors presenting immobilized bioreceptors designed to specifically apprehend positive analyte molecules. Generally built at the nanoscale, these sensors are able to amass, process, and understand information on an atomic level. They provide innovative gear for practical bio-analytical packages and open up new avenues for fundamental research, attaining responsibilities that were formerly impossible (Velasco-Garcia 2009).

A regular nanobiosensor includes numerous key components: the sample analyte, the bioreceptor, a transducer incorporated with nanostructures, and detectors (arranged from left to right). Analytes are the chemical or biological entities identified using the bioreceptor and are essential for detection or quantification. Bioreceptors can include numerous biologically derived molecules, inclusive of structural or functional proteins, oligonucleotides like aptamers, cells, specific tissues, subcellular organelles, or even complete microbes and their additives. The organic response is conveyed to the detector through a transducer, which can be included with or functionalized by the usage of nanostructures, together with metallic nanoparticles, magnetic nanoparticles, upconversion nanoparticles, quantum dots (QDs), and carbon-based substances like graphene oxide (GO) and carbon nanotubes (Srivastava et al. 2018).

A nanobiosensor basically includes three components: a biological probe that consists of primarily affinity-based substances like the ones concerned with the interactions among antibodies and antigens, enzymes and substrates, nucleic acids, and cells; a transducer that converts biological indicators into virtual formats; and a data-recording unit that captures and transmits the information (Verma and Rani 2021). This integration of biological probes with various nanoparticles—metal, magnetic, quantum dots, graphene oxide, and carbon nanotubes—can correctly identify analytes. Transducers that generate electrochemical indicators make use of methods like amperometric, voltametric, potentiometric, and optical alerts, which include colorimetric techniques, floor plasmon resonance, metal fluorescence, and optical fibers (Shawon et al. 2020). By developing present-day diagnostic tools and strategies, nanobiosensors are remodeling the sector of agriculture. These sensors offer particular, efficient, and price-powerful solutions to address numerous challenges in food production, farming practices, and environmental problems. Their programs in agriculture encompass the fast tracking of temperature, traceability, and humidity, along with the detection of heavy steel ions, contaminants, microbial loads, and sicknesses (Kumar et al. 2024).

13.4 Different Types of Nanobiosensors in Agriculture

Nanobiosensors are categorized primarily based on their detection mechanisms and applications, all of them providing particular advantages for tracking precise crop health parameters (Sharma et al. 2021). In Table 13.1, we discuss numerous forms of nanobiosensors relevant to agriculture, their operational concepts, and field packages.

Table 13.1 Several types of nanobiosensors relevant to agriculture, their operating principles, and field applications

Nanobiosensor	Nanomaterial	Sensor type	Applications	Advantage ((LOD) limit of detection, cost/time)
Acetylcholinesterase (AChE) biosensor	Calcium phosphate	Electrochemical	Pathogenic detection	0.30 ng/ml
Pesticide detection nanobiosensor	Graphene-based, with molecular imprinted polymers	Electrochemical	Pesticide screening for chlorothalonil and chlorpyrifos methyl	0.13 mg/L with very cost-effective technology, \$2 per 100 mm^2
QD nanosensor	Quantum dots	Fluorescence	Detection of pathogens	5 ca/μL
QD nanosensor	Quantum dots	Fluorescence	*Phytoplasma aurantifolia*	–
Quantum dot nanosensor	Gold particles	Immuno-based sensor	Detection of mycotoxins ZEA (zearalenone), DON (deoxynivalenol), and FB1/FB2 (fumonisin B1/B2) in corn, oats, and barley	Maize DON (deoxynivalenol) 80 μg/kg, wheat DON (deoxynivalenol) 400 μg/kg, 5 min
Surface plasmon resonance (SPR)	Multi-walled carbon nanotubes	SPR	Detection of cymbidium mosaic virus	
Artificial nose	Carbon nanomaterials	Volatile organic compound profile	Pathogen detection based on released organic chemicals	30 min with 80%–90% accuracy
Acetylcholinesterase	Cholinergic enzyme	Amperometric	Detection of chlorpyrifos	0.01 μg/L, 12 min
Acetylcholinesterase on white paper with indophenol acetate	Enzyme	Colored reaction	Detection of chlorpyrifos	3.3 μg/L, 10 min

13.5 Role of Nanobiosensors in Agriculture

Nanobiosensors play a vital role in agriculture by representing some of the latest technological advancements essential for sustainable agriculture. Within the agri-food industry, these sensors offer a highly sensitive and cost-effective detection method. They bring significant advantages, such as enhanced sensitivity and specificity, making them valuable across various fields, including agriculture, biodefense, food quality assurance, environmental management, and, importantly, healthcare applications, as demonstrated in Fig. 13.3. Our primary focus here is the role of nanobiosensors in agriculture and related products. Below is a list of potential applications for these sensors.

13.6 Nanobiosensors in Managing Biotic Stress

Plant pathogens are a significant factor contributing to reduced crop yields, which can lead to food shortages equally for individuals and livestock. It is estimated that foodstuff manufacture needs to rise from 2012 to 2050 by 35–50% to support the growing global population. Approximately 16% of worldwide yield damages are

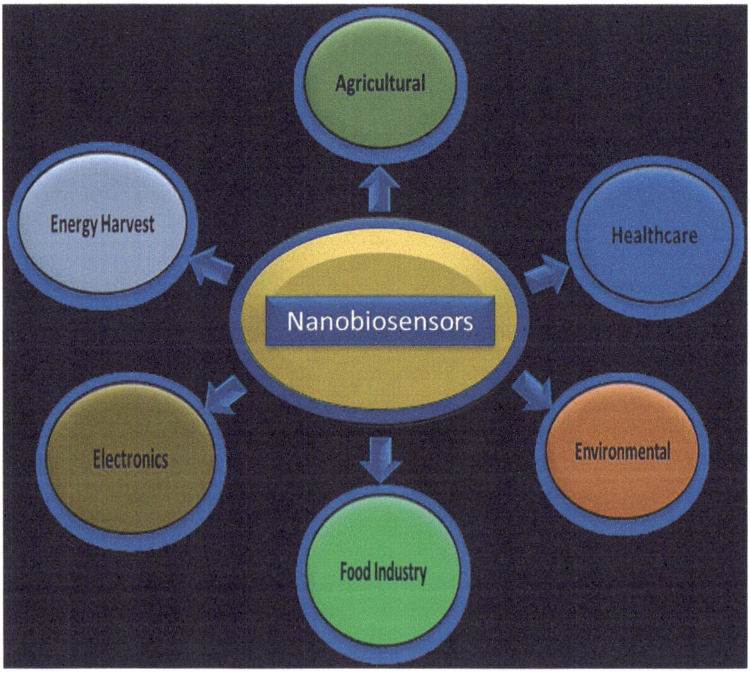

Fig. 13.3 Numerous applications of nanobiosensors in various fields

attributed to plant pathogens (FAO 2019). Various pesticides, insecticides, and fungicides can be applied to the soil or to the foliage to control plant diseases (PDs). However, immoderate use of these harmful chemicals poses extreme ecological concerns and might cause the development of resistance to pathogens over time. Consequently, the introduction of green and the introduction of smart sensing tools has become a critical cognizance inside the area of plant disease management. Nanomaterial-based biosensors are being used more and more in contemporary agriculture to identify diseases quickly. In addition to electrochemical sensing, immunosensing, and aptasensing, several biosensing techniques have been developed. For instance, an impedimetric immunosensor, a label-free electrochemical biosensor, was developed to identify *Pseudomonas syringae pv. Lachrymans* (Psl) with a detection limit of 337 CFU/mL (Cebula et al. 2019). Furthermore, a colorimetric detection method was developed to identify the bacteria that cause potato wilt disease, *Ralstonia solanacearum* (Khaledian et al. 2017). Researchers additionally created a lateral glide assay by the employment of gold nanoparticles for the detection of fungus (*Phytophthora infestans*) and the Potato leafroll virus (Zhan et al. 2018).

13.7 Nanobiosensors for Managing Abiotic Stress

Plants face diverse forms of ecological abiotic stress, along with drought, salinity, extreme temperatures, and exposure to heavy metals. These unfavorable situations can cause large physiological and biochemical modifications in plant life, consisting of alterations in membrane composition, adjustments in phytohormone levels, decrease in photosynthetic efficiency, accumulation of sugars and other compounds, stomatal closure, formation of small molecules and free radicals, and modifications within the modifications in the activity of antioxidant enzymes. Consequently, those elements can make a contribution to a decline in agricultural yield. To mitigate crop losses as a result of abiotic pressure, it is crucial to perceive and respond to stress at the earliest possible stage. Researchers have developed strategies to reveal plant responses to abiotic stress by detecting phytohormones.

To detect salicylic acid (SA) and indole-3-acetic acid (IAA) in soybean seedlings exposed to salt stress, for example, an electrochemical biosensor was developed, with detection limits of 3.30 μM and 1.99 μM, respectively (Ansari et al. 2020). At detection thresholds of 18.5 nmol/L for cadmium and 25.6 nmol/L for lead, an electrochemical immunosensor was also developed to identify vitronectin (VN)-like proteins, which are biomarkers found on plant cell surfaces under heavy steel pressure (Wang et al. 2019). Other electrochemical biosensors have also been advanced to identify phytohormones under various stress situations (He et al. 2020).

13.8 Soil Health and Disease Evaluation

Maintaining soil health and quality is essential for achieving sustainable agriculture, environmental balance, and global food security. Although traditional methods of evaluating soil attributes are often labor-intensive and time-consuming, they often lack the function to conduct real-time tracking. Due to that, a nanobiosensor-based creation affords a promise towards speedy, in situ, and continual monitoring of the soil health signs (Parameswari et al. 2024). These nanosensors can be used to sense the soil illness caused by dangerous microorganisms like viruses, bacteria, and fungi via the quantitative analysis of oxygen intake upon inhaling the useful versus harmful bacteria in the soil. Nanobiosensors have increasingly become progressively useful for utility (Mandal et al. 2020) due to their specificity, accuracy, stability, and short responsiveness that result from unique physicochemical properties and catalytic activity.

There are plenty of such sensors that would work to monitor soil pH, moisture tiers, and loads of pathogens and detect insecticides, herbicides, pesticides, fertilizers, and so on. Nanobiosensors can also be used properly and in a controlled manner that will bolster sustainable agricultural practices and increase the yields of crops while also assisting with pollutant control and material control (Ghaffar et al. 2020). For example, if nitrates and ammonia levels are increased in the soil, then this indicates microbial imbalances and the presence of soil feeding insects. Soil is home to a diverse range of microorganisms that have a significant impact on plant growth and soil health. Nitrogen exists in diverse forms, including ammonia, nitrates, and nitrites. Researchers have designed versatile, totally lipid-based nanobiosensors capable of detecting nitrites in soil samples, with a detection limit as low as 2.1 μg/L (Mondal et al. 2022). Zeolites can preserve vitamins within the root zone, making them available to flora when required, thereby enhancing the soil's nutrient retention talents and yields. When merged with nanobiosensors, zeolites can revolutionize farming by sensing nutrient or plant shortages and facilitating the controlled release of water and vitamins stored in the zeolite (Lauterwasser 2005).

13.9 Nanobiosensors for Toxic Chemical Detection

Managing pests and viruses in agriculture is a large challenge, with early recognition being critical to prevent important crop losses. To improve crop yields, diverse chemical compounds, which include pesticides, fungicides, and insecticides, are frequently applied. However, those substances can remain in the surroundings for extended intervals, and at high concentrations, they may have intense toxic consequences on nontarget organisms (Talari et al. 2021). Nanobiosensors have emerged as effective equipment for sensing pathogens and pests, presenting high levels of specificity and sensitivity. For example, primarily liposome-based biosensors can screen organophosphorus pesticides like dichlorvos and paraoxon even at very low concentrations (Vamvakaki and Chaniotakis 2007). A pyrolytic graphite electrode

(PGE) made of a variety of materials, including tap water, orange juice, and apples, was used by Porto et al. (2022) to create an incredibly sensitive electrochemical sensor that uses silver nanoparticles and carbon nanotubes to detect a few insecticides, specifically malathion (MLT), diazinon (DZN), and chlorpyrifos (CLPF). For DZN, MLT, and CLPF, the limits of detection were 0.35 μmol/L, 0.89 μmol/L, and 0.53 μmol/L, respectively.

Biosensors created using photosystem II (PSII), which can bind to different herbicide groups isolated from photosynthetic organisms, show potential for monitoring harmful chemicals. This could make it easier to create inexpensive, easy-to-use tools that can identify particular herbicides as well as a variety of organic contaminants present in sewage, groundwater, landfill leachate, industrial and urban waste, and irrigation water (Giardi and Piletska 2006). Additionally, organophosphate pesticides have been detected using enzyme-based biosensors, including AChE (acetylcholinesterase), OPH (organophosphorus hydrolase), and OPAA (organophosphorus acid anhydrolase) (Songa and Okonkwo 2016). These sensors can promptly identify pest infestations, enabling timely pest control interventions and reducing pesticide usage. When integrated into Internet of Things (IoT) networks, these technologies allow farmers to receive real-time updates about pest and pathogen activity, empowering them to act quickly to protect crops and enhance yield outcomes (Dar et al. 2020). Due to the usage of agrochemicals, such as pesticides, fungicides, herbicides, and insecticides, which can release metal ions into the field and later leak into water bodies, agriculture is the sector that is most impacted. Metal ions pose a serious risk to human health since they are carcinogenic, have a prolonged half-life, do not biodegrade, and accumulate over time (Ahmed et al. 2020). Nanobiosensors have emerged as an alternative to conventional methods of analysis to monitor heavy metals, which are being highlighted by detecting substances at the micro- and nanoscale through the use of sensor layers (Rigo et al. 2020). According to the World Health Organization (WHO), heavy metal ions, based on their toxicity, are considered to be "environmental health hazard" substances (Aragay et al. 2011). Noble nanoparticles such as AuNPs (Gold Nanoparticles) and carbon nanotubes contributed to the development of a smartphone-based detection platform for mercury(II) at concentrations as low as 2.53 nM and a quick detection tool for mercury, cadmium, palladium, arsenic, etc. (Jin et al. 2018). Advances in nanobiosensors for heavy metal detection directly imply a better control of industrial processes, the earlier and accurate detection of contamination problems, the real-time monitoring of treatment procedures, and remediation. Cantilever nanobiosensors are very versatile devices in terms of their detection and monitoring; require a small amount of sample; are easy-to-use, ultrasensitive, and miniaturized; and present a fast response (Muenchen et al. 2016). The choice of biological material is one of the major challenges for nanobiosensor construction, where the sensitive layer allows a high selectivity for particular analyte species (Nugaeva et al. 2005). The enzyme urease is one promising biological sensing layer, which is specific and able to recognize many heavy metals (mercury, cadmium, cobalt, lead, etc.) by their inhibition resulting from the reaction with sulfhydryl groups of the active site (Berezhetskyy et al. 2008).

13.10 Nanobiosensors for Crop Health Monitoring

Crop productivity has a significant impact on world food security. Production losses brought on by climate change can happen directly or indirectly. Plant diseases (PDs) cause organisms to affect the economy by reducing their capacity to generate food, fiber, and biofuels. Clean seeds, crop rotation, using resistant varieties, seed, foliar fungicidal treatment, and other cultivation and chemical practices are some of the methods by which PD is controlled. (Mancini and Romanazzi 2014). The ultimate goal of precision agriculture is to enhance both the yield and quality of crops, and nanobiosensors contribute significantly to achieving this goal. By providing accurate and timely information about various aspects of crop health, these sensors enable farmers to optimize growing conditions and make informed decisions about crop management (Kumar et al. 2024).

Nanofertilizers could be used to reduce nitrogen loss due to leaching, emissions, and long-term assimilation by soil microorganisms. This suggests that new nutrient delivery systems that exploit the nanoscale porous domains on plant surfaces can be developed. But the nanofertilizers should show a sustained release of nutrients on demand while preventing them from prematurely converting into chemical/gaseous forms that cannot be absorbed by plants. To achieve this, the biosensor could be attached to the nanofertilizer that allows selective nitrogen release linked to time and environmental and soil nutrient conditions. Slow controlled release of fertilizers may also improve soil by decreasing toxic effects associated with fertilizer overapplication (DeRosa et al. 2010). A plant-wearable biosensor based on laser-induced graphene technology was created by Zhao et al. (2020) for the electrochemical detection of pesticides such as methyl parathion. AuNPs (Gold Nanoparticles) can be used to improve this detection process, and the smartphone can wirelessly receive the detection output. Mathivanan (2021) applied DNA-based sensors to screen plant responses to various biotic and abiotic stresses in wheat plants, enhancing each disease and yield. By integrating nanobiosensors with virtual platforms, farmers can also track crop development in real time, allowing for adjustments in irrigation, pest management, and fertilization schedules to optimize both yield and crop fine. This integration not only enhances productivity however additionally helps sustainable practices through reducing the overuse of inputs such as water, fertilizers, and insecticides.

13.11 Principles of Nanobiosensor Function

The capability of nanobiosensors to monitor plants is based on key concepts that ensure the precise and timely detection of various factors essential to agricultural well-being. Essential components include fact amassment and evaluation, signal enhancement, sensitivity, and specificity, in addition to the mechanisms of signal transmission.

13.11.1 Signal Transduction Mechanisms

Nanobiosensors utilize a mechanism known as signal transduction to discover specific biological interactions and convert them into measurable records (Bhalla and Singh (2016)). Biosensors may be categorized into numerous kinds based totally on their transducers and recognition constituents. According to the transducer type, a biosensor may be electrochemical (such as photometric, impedimetric, or potentiometric), calorimetric, piezoelectric (including acoustic waves and quartz crystal microbalance), or optical (involving reflection, refraction, absorption, transmission, fluorescence, surface plasmon resonance, and waveguides) (Malhotra and Ali 2018). For instance, ion exchange induces variations in current or voltage when a target analyte interacts with the nanobiosensor's surface, a process referred to as electrochemical transduction. In contrast, optical transducers rely on alterations in fluorescence, absorbance, or luminescence to produce detectable optical signals upon the adhesion of a biological molecule to a nanomaterial's surface. Meanwhile, piezoelectric sensors detect mass changes associated with analyte binding, providing high sensitivity for identifying infections and other health indicators in crops through variations in resonant frequency (Kilic et al. 2023; Rai et al. 2012).

13.11.2 Detection Sensitivity and Specificity

For nanobiosensors to reliably perceive tiny concentrations of target analytes in complex agricultural settings, sensitivity and specificity are vital. While specificity lowers the possibility of false positives from non-target molecules, sensitivity guarantees that the sensor can identify even the smallest amounts of a selected nutrient or disease (Malik et al. 2023; Kulshreshtha et al. 2017). Contaminant sensors are sensors primarily based on nanotechnology. By growing surface vicinity and binding efficiency, new traits in nanomaterials like graphene, carbon nanotubes, and gold nanoparticles have significantly improved the sensitivity and specificity of agricultural biosensors. By assisting nanobiosensors in differentiating between chemical substances that are intently associated, those materials allow for the correct tracking of pathogen presence, moisture content, and dietary stages (Bobrinetskiy et al. (2021)).

13.11.3 Methods for Enhancing Signal Detection

Signal amplification is preferred to be used at low concentrations, as it raises the detectable signal for target analytes, which will increase more precise measurements (Perdomo et al. 2021). Techniques that include enzyme-linked amplification as well as nanoparticle-enhanced amplification have been applied to enhance the efficiency of nanobiosensors in agronomy (Arora 2018). Quantum dots can increase optical alerts via strong fluorescence, or gold nanoparticles can advance electrochemical signals through the binding of analytes to enhance electrochemical signals (Pingarrón et al. 2008; Peterson and Krauss 2006). In general, enzyme-linked signal amplification, is commonly utilizing in pathogen detection where the reactions are catalyzed through generating a greater signal to enhance the sensor detection limits and reliability for early disease analysis in vegetation (Ali et al. 2021).

13.11.4 Obtaining and Examining Data

With that, nanobiosensors need to collect as well as analyze field measurement information in order to efficaciously screen crops in real time. It's crucial to integrate these sensors with digital solutions including wireless transmission of the data along with the cloud based analytics platforms for real-time data gathering (Mishra et al. 2024). On the other hand, IoT-equipped nanosensors may directly send data to mobile devices that will allow farmers to respond quickly to the fitness of their crops. Moreover, device learning and underscoring enhanced data evaluation strategies smoothen the grasp of complicated datasets, improving choice-making accuracy, and help in forecasting future crop health trends, taking into consideration historic data (Bharti et al. 2024).

13.11.5 Advances in Nanomaterials for Biosensing Applications

Nanomaterials are frequently accountable for the enhanced sensitivity, specificity, and application of biosensors in agriculture. By providing innovative solutions for high-resolution, real-time crop monitoring, the development of novel nanomaterials, such as steel oxide nanoparticles, carbon nanotubes, quantum dots, primarily graphene-based substances, and primarily polymer-based nanomaterials, has notably stepped forward biosensing applications.

13.11.6 Carbon Nanotubes

The nanomaterials largely elevated the sensitivity, specificity, and effectiveness of biosensors in agriculture. The improvement of recent nanomaterials, such as iron oxide nanoparticles, carbon nanotubes, quantum dots, primarily graphene-based materials, and primarily polymer-based nanomaterials, has significantly advanced biosensing programs by way of imparting groundbreaking results for high-resolution, real-time crop tracking. Carbon-primarily based nanomaterials, such as carbon nanotubes, or CNTs, have attracted a lot of interest in plant technological study due to their superior potentials and likely applications. Carbon nanotubes (CNTs) are cylindrical structures made of rolled-up graphene sheets. They are ideal for a variety of plant science applications due to their exceptional mechanical, electrical, and thermal properties (Schrlau 2011).

Carbon nanotubes are nanomaterials that store energy, help in purification, and package particles made of carbon molecules. They have different wall construction sizes. Their usefulness is explained by their higher elasticity and van der Waals force strength. At the agronomic level, they facilitate the flow of nutrients from the soil and the rate of growth. The size of the carbon tube has an impact on the culture's profit as well. According to Shi et al. (2014), sensors are useful in farming since they are composed of materials with increased sensitivity. It has been investigated whether CNTs can enhance plant growth, nutrient uptake, and stress tolerance. Revisions state that employing CNTs can improve root evolution, increase seed germination, and increase total plant biomass (Lawal 2016). CNTs are suitable electronic materials for electrochemical sensing due to their remarkable electrical, optical, mechanical, thermal, and electrochemical properties, as well as their remarkably low-noise electronic features (Park et al. 2013). Supercapacitors, clinical chemistry, medicine, fuel cells, batteries, photocatalysis, photovoltaic (solar cells), chemical and biosensors, gas sensors (including greenhouse gas sensors), photonics, light-emitting diodes (LEDs), optical sensors, energy systems, optoelectronics, biofuels, thin-film transistors, and field effect transistors are just a few of the exciting applications that numerous groups worldwide are working on with CNTs and their hybrids as shown in Fig. 13.4 (Lawal 2016).

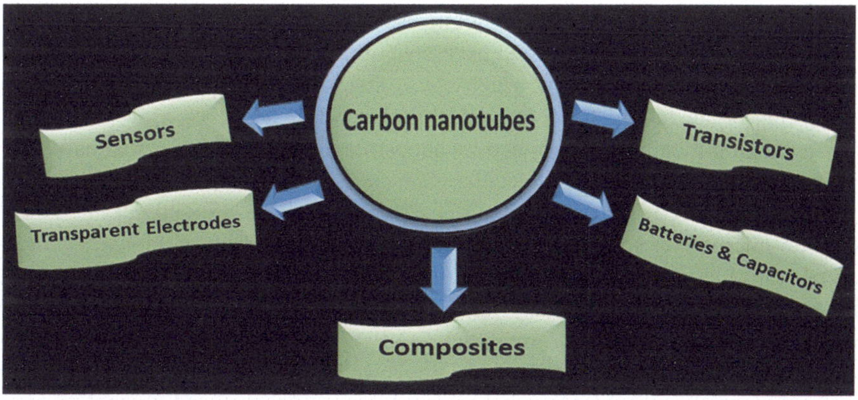

Fig. 13.4 Various applications of carbon nanotubes

13.11.7 Quantum Dots

Often referred to as "artificial atoms," quantum dots are nanostructured materials that have attracted a lot of attention because of their distinctive optical and electrical properties. Quantum dots, a unique and optically stable class of fluorophores, offer a wide range of uses in agriculture and medicine and have the potential to drastically alter the industrial landscape. In live organisms, including growing plants, quantum dots have been used to create extremely dependable nanosensors for the real-time monitoring of metabolic signaling, cellular processes, and stress biomarkers (Brkić 2016; Nsibande and Forbes 2016). Recent development in QD-primarily based fluorescent nanosensors for the detection and monitoring of heavy metal ions, food herbs, pesticide residues, veterinary drug residues, other chemical components, mycotoxins, foodborne pathogens, humidity, temperature, and unstable components at trace levels. Strong anti-interference capability, speedy and easy operation, exact linear reaction, and an extensive detection variety are some of the blessings of QD sensors (Jia et al. 2023). Using their ability to adjust their length and surface properties to maximize drug shipping to specific goal locations, QDs can serve as carriers for pharmaceuticals or therapeutic constituents, growing therapeutic efficacy and lowering side effects. By the usage of QDs for in vivo imaging, scientists and clinical specialists may additionally monitor biological processes within living organisms. This is essential for researching the direction of the ailment and assessing the effectiveness of treatment (Koul et al. 2024).

13.11.8 Metal Oxide Nanoparticles

Lately, nanostructured metal oxides (NMOs) have attracted interest as materials that provide a surface that effectively immobilizes biomolecules with the required orientation, advanced conformation, and excessive organic advanced conformation, and high biological activity, all of which cause stepped-forward sensing properties. Interesting organizations for connecting biorecognition additives with transducers for signal amplification are supplied by nanostructured metal oxides that have different optical, electrical, and molecular characteristics molecular characteristics in addition to required functions and surface charge qualities. (Solanki et al. 2011). The sub-micrometer size of an NMO enables the improvement of numerous novel signal transduction technologies in biosensors that may be applied to rapid and smooth in vivo evaluation (Rahman et al. 2010). Owing to their antifungal, antibacterial, and antiviral qualities, nanoparticles (NPs) created from a number of metals, which include Zn, Cu, Fe, Ni, Mg, and Ag, as well as their oxides or combinations with organic polymers, which include zein, chitosan, alginate, or different biopolymers, have become surprisingly popular. They have additionally been discovered to be very helpful in preventing pest damage to flora (Yin et al. 2021). Because of their unique physical and chemical traits, AuNPs and a sizable number of metal nanoparticles have found use in numerous industries, such as biomedicine, sensors, catalysis, and biosensors. Shape, length, geometry, floor-to-quantity ratio, and aggregation states, all alter the distinct electrochemical and optical traits of those nanoparticles, which might result from their free electron resonance following excitation (enhancement of FRET (Förster resonance energy transfer) within the presence of plasmonic metal as localized surface plasmon resonance) (Gholami et al., 2020).

13.11.9 Graphene-Based Materials

Because of its different physical structure, chemical makeup, and electrical traits, graphene is ideal for application in sensor technology. Because of their excessive signal output and promising increase in commercial worth, totally graphene-based materials like graphene oxide (GO) are getting more and more famous to be used in biosensing applications. Its advanced chemical, mechanical, and electrical traits make it a standard material with a wide range of uses, which include sensors, structural composites, electronics, nanocomposites, and supercapacitors. Because of their stronger signal output, graphene and graphene-based nanomaterials are among the most widely used materials in nano-sensor programs (Banerjee et al. 2015). DNA, bacteria, lipids, peptides, viruses, antibodies, and ions are just a few of the chemical and biological materials that graphene is very effective at detecting. In terms of pathogen detection, graphene-based nanomaterials on antibody biosensors provide a wide range of adaptability. A number of clinically useful graphene-antibody biosensors have recently been created for the early detection of illnesses. Early identification of these diseases with such sensors would help with the diagnosis, prevention, and management of the disease in "high-risk" persons, which would improve patient care and survival (Peña-Bahamonde et al. 2018).

13.11.10 Polymer-Based Nanomaterials

Polymer-based biosensing nanoparticles provide excellent sensitivity and selectivity, as well as long-term consistency and hysteresis. Nowadays, conducting polymers such as polypyrrole (PPy) and polyaniline (PANI) are copolymerized with nanoparticles such as graphene, metal nanoparticles, and carbon nanotubes (CNTs) by a variety of methods. One type of nanomaterial with excellent biodegradability and biological safety is polymer-based NP (Harshita et al., 2023). Polymers are an essential component of our lives and everything around us. They range from the fundamental building blocks of life, such as proteins, nucleic acids, and polysaccharides, to the commercial goods derived from the automotive, construction, and transportation industries, plastic toys and tools, reading glasses, and other everyday items, polymers are essential components of our lives and everything around us. The majority of these materials are polymer composites, which are made up of a combination of one or more materials. Because of their special qualities and environmental friendliness, biopolymers, which come from renewable resources, offer a promising path forward for sensor technology (Nambiar and Yeow 2011; Abady et al. 2024; Pawar et al. 2024).

13.12 Integration of Nanobiosensors with Digital Technologies

Precision agriculture techniques are improved by real-time, high-resolution data collection, made possible by the combination of digital technology and nanobiosensors, which has completely changed agricultural monitoring. Continuous and effective monitoring of crop health, environmental conditions, and soil quality is made possible by the combination of nanotechnology with cutting-edge digital tools, including the Internet of Things (IoT), wireless communication systems, and data-driven decision-making platforms (Yadav et al. 2023). Farmers and agronomists are therefore better able to make proper, knowledgeable choices to increase crop output, decrease resource waste, and reduce the negative outcomes of stressors on plant suitability (Ouaissa et al. 2024).

13.12.1 Internet of Things (IoT) in Agriculture

The Internet of Things (IoT) refers to a network of connected devices that exchange and share data online. The important role that the Internet of Things (IoT) plays is to combine the sensor (including the nanobiosensor) into a single system that can continuously analyze different environmental and agricultural records to make smart farming within the agricultural sector (Singh et al. 2021). Using nanobiosensors incorporated into Internet of Things networks, soil changes in pH, moisture content, pathogen presence, and nutrient levels can be detected, which are sent to centralized platforms to look into. Let us take the example of the IoT-linked soil sensors and nanobiosensors, which could also monitor soil health metrics and inform the farmers of significant discrepancies in real time, thereby activating action. In addition, IoT-based precision agriculture technology enables the automation of pest control, fertilization, and irrigation, thereby ensuring the excellent, feasible use of the sources and the productivity of the crop (Maksimović et al. 2019).

13.12.2 Wireless Communication and Data Transmission

Wireless communication protocols like Wi-Fi, Zigbee, and LoRa (Long Range) are important for allowing smooth data transfer from far off farm regions to centralized databases or cloud platforms according to the functioning of IoT-based networks in agriculture. Wireless nanobiosensors established in the field collect real-time data about crop health metrics from a variety of crops that are sent to processing centers. This makes it possible to monitor crops continuously and non-invasively without requiring physical presence or manual intervention (Aldhaheri et al. 2024; Pratama and Rahayu 2024). Wireless communique networks permit real-time data transfer, which helps in making useful choices when anomalies are spotted and guiding the farmers to react instantly. As shown by Hassan et al. (2023), a nanobiosensor can immediately signal a Wi-Fi conversation community with an alert if it detects the presence of a pathogen in an agricultural field, permitting for targeted remedy.

13.12.3 Real-Time Monitoring Platforms

Real-time tracking platforms are among the noteworthy fulfillments in precision agriculture, where environmental elements and crop health can be routinely evaluated. These platforms combine records from a range of resources, which include satellite photos, weather stations, nanobiosensors, and Internet of Things devices, in order to provide farmers with a full picture of agricultural situations. These dashboards allow farmers to see data on present-day sources of environmental stressors, pest activity, nutrient levels, and soil moisture and establish these data on their mobiles (Bharti et al. 2024; Monteiro et al. 2021). But through reading the collected data and predicting patterns and referencing preventative actions prior to any crop damage, these systems make predictive analytics possible. In addition, cellular packages' integration with Additionally, farmers can receive real-time updates and notifications on their phones through integration with mobile applications, which improves their ability to make choices. This technique also facilitates spatial evaluation and GIS mapping for accurate source distribution in the development of the crop sector (Sennuga et al. 2023), which includes fertilizers or water.

13.12.4 Data-Driven Decision-Making in Precision Agriculture

Data-driven decision-making (choosing on the basis of fact) in real time by means of the use of virtual structures and nanobiosensors is a fundamental guiding principle of precision agriculture. By combining facts from sensors, IoT networks, and great analytics, farmers can make evidence-based selections for crop control, irrigation, fertilization, and pest management. Pratama and Rahayu (2024) state that fact-pushed strategy allows for optimizing agricultural productivity, lowering waste, and using inputs to the maximum. As an example, an algorithm that determines how much fertilizer to apply can be given data from nanobiosensors that measure the amount of nutrients in the soil. It hence minimizes the effect on the environment (Das et al. 2024) while ensuring correct utility. In addition, prediction methods fueled by means of historical and modern measurements make preventive actions possible. For example, farmers would possibly take preventive actions, like focused pesticide utility or irrigation changes, to keep away from crop losses if the sensors discover early indicators of nutrient shortfalls or insect invasions (John et al. 2023). The mixture of data analytics with nanobiosensors ensures extra-worthwhile, efficient, and sustainable farming methods. Precautions are required for nano-dimensional research, and ethical pointers should be developed and followed to prevent any damage to the surroundings or humans. Nanomaterials could absolutely be a game changer for enhancing the operational efficacy of biosensors by using a rigorous, however, watchful method (Mishra 2024).

13.13 Future Prospects

A subfield of nanotechnology called nanobiotechnology (NBT) uses nanoscale dimensions to develop tools, methods, and functional structures that lead to novel uses in health and agriculture in emerging nations, where agriculture is the main economic sector. Recent developments in biosensor technology for agriculture would increase real-time livestock quality and food safety, as well as the sustainability and productivity of crops and natural agricultural resources (Sharma et al. 2022). With their multiplex and real-time sensing capabilities, electrochemical nanosensors, optical nanosensors, electronic tongues and noses, nanobarcode technology, and wireless nanosensors have completely transformed sensing technologies in the food and agricultural industries (Singh 2021). Research on nanobiosensors is primarily focused on developing new technologies to advance the fields of therapeutic compound identification and study, biocatalyst development, nano- and biomaterial characterization, and marker detection of human and animal diseases (Sharma et al. 2022).

13.14 Conclusion

The traditional agricultural sectors have rapidly developed to increase the quality and shelf life of agricultural items due to the growing population and human safety concerns. Therefore, one of the main concerns is the prevention and detection of pollutants and infections. Likewise, maintaining soil quality through the identification of several soil parameters is a difficult undertaking for crops production.

Therefore, the creation or design of nanobiosensors and their possible uses in agricultural industries are highlighted in this chapter. The application of nanobiosensors in agriculture has been transformed by nanotechnology (NT), an interdisciplinary scientific approach at the nanoscale. Therefore, the creation or design of nanobiosensors and their possible uses in agriculture industries are highlighted in this chapter. The application of nano biosensors in agriculture has been transformed by nanotechnology (NT), an interdisciplinary scientific approach at the nanoscale. Agriculture is a broad industry that contributes to the selfsufficiency of any nation's economic growth for the benefit of all people. The various nanobiosensors technologies, including optical nanosensors, electronic technology, wireless nanosensors, and rapid detection systems have been studied. This chapter also elaborated the role of nanobiosensors in agriculture and their applications in digital technology. The various kinds of nanoparticles like silver nanoparticles, gold nanoparticles, quantum dots, carbon nanotubes, etc. have been utilized because of their exclusive chemical, thermal, optical, and mechanical applications. Agriculture is a broad industry that contributes to the self-sufficiency of any nation's economic growth for the benefit of all people. Food processing, food safety, food quality or quality assurance, food security, disaster risk management, diagnosis, and prevention at the local and global levels, all require nanotechnical interventions in agriculture. Farmers who incorporate digital nanotechnology benefit from the utilization of more advanced and precise methods for controlling these challenges, resulting in increased crop outputs and the preservation of resources.

References

Abady MM, Shawky AM, Sakr FA, Mohammed DM, Goda ES (2024) Recent advancements in biosensors using biopolymers. In: Bio-based polymers: farm to industry. Volume 2: current trends and applications. American Chemical Society, pp 81–112

Ahmed F, Dwivedi S, Shaalan NM, Kumar S, Arshi N, Alshoaibi A, Husain FM (2020) Development of selenium nanoparticle based agriculture sensor for heavy metal toxicity detection. Agriculture 10(12):610

Aldhaheri L, Alshehhi N, Manzil IIJ, Khalil RA, Javaid S, Saeed N, Alouini MS (2024) LoRa communication for agriculture 4.0: opportunities, challenges, and future directions. IEEE Internet Things J

Ali Q, Ahmar S, Sohail MA, Kamran M, Ali M, Saleem MH et al (2021) Research advances and applications of biosensing technology for the diagnosis of pathogens in sustainable agriculture. Environ Sci Pollut Res 28:9002–9019

Ansari MH, Lavhale S, Kalunke RM, Srivastava PL, Pandit V, Gade S et al (2020) Recent advances in plant nanobionics and nanobiosensors for toxicology applications. Curr Nanosci 16(1):27–41

Aragay G, Pons J, Merkoçi A (2011) Recent trends in macro-, micro-, and nanomaterial-based tools and strategies for heavy-metal detection. Chem Rev 111:3433–3458. https://doi.org/10.1021/cr100383

Arora K (2018) Advances in nano based biosensors for food and agriculture. In: Nanotechnology, food security and water treatment. Springer, pp 1–52

Banerjee S, Lee JH, Kuila T, Kim NH (2015) Synthesis of graphene-based polymeric nanocomposites. In: Fillers and reinforcements for advanced nanocomposites. Woodhead Publishing, pp 133–155

Berezhetskyy AL, Sosovska OF, Durrieu C, Chovelon JM, Dzyadevych SV, Tran-Minh C (2008) Alkaline phosphatase conductometric biosensor for heavy-metal ions determination. ITBM-RBM 29:136–140. https://doi.org/10.1016/j.rbmret.2007.12.007
Berners-Lee M, Kennelly C, Watson R, Hewitt CN (2018) Current global food production is sufficient to meet human nutritional needs in 2050 provided there is radical societal adaptation. Elem Sci Anth 6:52
Bhalla P, Singh N (2016) Generalized Drude scattering rate from the memory function formalism: an independent verification of the Sharapov-Carbotte result. Eur Phys J B 89:1–8
Bhandari G, Dhasmana A, Chaudhary P, Gupta S, Gangola S, Gupta A et al (2023) A perspective review on green nanotechnology in agro-ecosystems: opportunities for sustainable agricultural practices & environmental remediation. Agriculture 13(3):668
Bharti A, Jain U, Chauhan N (2024) From lab to field: nano-biosensors for real-time plant nutrient tracking. Plant Nano Biol:100079
Bobrinetskiy I, Radovic M, Rizzotto F, Vizzini P, Jaric S, Pavlovic Z et al (2021) Advances in nanomaterials-based electrochemical biosensors for foodborne pathogen detection. Nano 11(10):2700
Brkić S (2016) Optical properties of quantum dots. Eur Int J Sci Technol 5(9):98–107
Cebula Z, Żołędowska S, Dziąbowska K, Skwarecka M, Malinowska N, Białobrzeska W et al (2019) Detection of the plant pathogen pseudomonas syringae pv. Lachrymans on antibody-modified gold electrodes by electrochemical impedance spectroscopy. Sensors 19(24):5411
Dar FA, Qazi G, Pirzadah TB (2020) Nano-biosensors: NextGen diagnostic tools in agriculture. In: Nanobiotechnology in agriculture: an approach towards sustainability, pp 129–144
Das H, Singh SK, Priyadarshini E, Maurya CL, Tiwari AK, Gulaiya S et al (2024) Nanotechnology facilitated real time soil monitoring for optimized crop production. Asian J Soil Sci Plant Nutr 10(2):582–607
DeRosa MC, Monreal C, Schnitzer M, Walsh R, Sultan Y (2010) Nanotechnology in fertilizers. Nat Nanotechnol 5(2):91–91
FAO I (2019) The state of food and agriculture 2019. Moving forward on food loss and waste reduction. FAO, Rome, pp 2–13
Fu Y, Liu T, Wang H, Wang Z, Hou L, Jiang J, Xu T (2024) Applications of nanomaterial technology in biosensing. J Sci Adv Mater Devices:100694
Ghaffar N, Farrukh MA, Naz S (2020) Applications of nanobiosensors in agriculture. Nano:179–196
Gholami M, Salmasi MA, Sohouli E, Torabi B, Sohrabi MR, Rahimi-Nasrabadi M (2020) A new nano biosensor for maitotoxin with high sensitivity and selectivity based fluorescence resonance energy transfer between carbon quantum dots and gold nanoparticles. Journal of Photochemistry and Photobiology A: Chemistry 398:112523
Giardi MT, Piletska EV (2006) Biotechnological applications of photosynthetic proteins: biochips, biosensors and biodevices. Springer
Guisbiers G, Mejía Rosales S, Deepak FL (2012) Nanomaterial properties: size and shape dependencies. J Nanomater 2012:1–2
Hassan M, Park JH, Han MH (2023) Enhancing livestock management with IoT-based wireless sensor networks: a comprehensive approach for health monitoring, location tracking, behavior analysis, and environmental optimization. J Sustain Urban Futures 13(6):34–46
He KC, Wang HR, Yang H, Sun LJ, Liu W, Bao N (2020) Digitalized pencil trace modified electrodes for real time evaluation of salicylic acid in detached Arabidopsis thaliana leaves during regeneration. Anal Chim Acta 1120:59–66
Jin B, Yang Y, He R, Park YI, Lee A, Bai D et al (2018) Lateral flow aptamer assay integrated smartphone-based portable device for simultaneous detection of multiple targets using upconversion nanoparticles. Sensors Actuators B Chem 276:48–56
John MA, Bankole I, Ajayi-Moses O, Ijila T, Jeje O, Lalit P (2023) Relevance of advanced plant disease detection techniques in disease and pest management for ensuring food security and their implication: a review. Am J Plant Sci 14(11):1260–1295

Johnson MS, Sajeev S, Nair RS (2021) Role of nanosensors in agriculture. In: 2021 international conference on computational intelligence and knowledge economy (ICCIKE). IEEE, pp 58–63
Joseph T, Morrison M (2006) Nanotechnology in agriculture and food: a nanoforum report. Nanoforum org
Khaledian S, Nikkhah M, Shams-bakhsh M, Hoseinzadeh S (2017) A sensitive biosensor based on gold nanoparticles to detect Ralstonia solanacearum in soil. J Gen Plant Pathol 83:231–239
Kilic NM, Singh S, Keles G, Cinti S, Kurbanoglu S, Odaci D (2023) Novel approaches to enzyme-based electrochemical nanobiosensors. Biosensors 13(6):622
Koul K, Jawanda IK, Soni T, Singh P, Sharma D, Kumari S (2024) Quantum dots: a next generation approach for pathogenic microbial biofilm inhibition: mechanistic insights, existing challenges, and future potential. Arch Microbiol 206(4):158
Krishnan V, Pandey GR, Pandiyaraj K (2024) Chapter 1: Fundamentals of a biosensor system. In: Veerapandian M, Joseph J, Marimuthu BT (eds) Health and environmental applications of biosensing technologies. Elsevier, pp 1–25
Kulshreshtha NM, Shrivastava D, Bisen PS (2017) Contaminant sensors: nanotechnology-based contaminant sensors. In: Nanobiosensors. Academic, pp 573–628
Kumar P, Chugh P, Ali SS, Chawla W, Sushmita S, Kumar R et al (2024) Trends of nanobiosensors in modern agriculture systems. Appl Biochem Biotechnol:1–24
Lauterwasser C (2005) Small sizes that matter: opportunities and risks of nanotechnologies. Report in co-operation with the OECD International Futures Programme
Lawal AT (2016) Synthesis and utilization of carbon nanotubes for fabrication of electrochemical biosensors. Mater Res Bull 73:308–350
Maheshwaran S, Chen WH, Lin SL, Ghorbani M, Hoang AT (2024) Metal oxide-based electrochemical sensors for pesticide detection in water and food samples: a review. Environ Sci Adv 3(2):154–176
Maksimović M, Omanović-Mikličanin E, Badnjević A (2019) Nanofood and internet of nano things. Springer
Malhotra BD, Ali MA (2018) Nanomaterials in biosensors: fundamentals and applications. In: Nanomaterials for biosensors, vol 1
Malik S, Singh J, Goyat R, Saharan Y, Chaudhry V, Umar A et al (2023) Nanomaterials-based biosensor and their applications: a review. Heliyon 9(9):e19929
Mancini V, Romanazzi G (2014) Seed treatments to control seedborne fungal pathogens of vegetable crops. Pest Manag Sci 70(6):860–868
Mandal N, Adhikary S, Rakshit R (2020) Nanobiosensors: recent developments in soil health assessment. In: Soil analysis: recent trends and applications. Springer, pp 285–304
Mathivanan S (2021) Perspectives of nano-materials and nanobiosensors in food safety and agriculture. Novel Nanomater 197
Mishra H (2024) Nanobiostimulants and precision agriculture: a data-driven approach to farming and market dynamics. In: Nanobiostimulants: emerging strategies for agricultural sustainability. Springer, Cham, pp 365–398
Mishra M, Dashora K, Srivastava A, Fasake VD, Nag RH (2019) Prospects, challenges and need for regulation of nanotechnology with special reference to India. Ecotoxicol Environ Saf 171:677–682
Mishra A, Singh PK, Chauhan N, Roy S, Tiwari A, Gupta S et al (2024) Emergence of integrated biosensing-enabled digital healthcare devices. Sens Diagn 3(5):718–744
Momin JK, Jayakumar C, Prajapati JB (2013) Potential of nanotechnology in functional foods. Emirates J Food Agric 25(1):10–19
Mondal R, Dam P, Chakraborty J, Paret ML, Katı A, Altuntas S et al (2022) Potential of nanobiosensor in sustainable agriculture: the state-of-art. Heliyon 8(12)
Monteiro A, Santos S, Gonçalves P (2021) Precision agriculture for crop and livestock farming—brief review. Animals 11(8):2345
Muenchen DK, Martinazzo J, de Cezaro AM, Rigo AA, Brezolin AN, Manzoli A et al (2016) Pesticide detection in soil using biosensors and nanobiosensors. Biointerface Res Appl Chem 6(6)

Nambiar S, Yeow JT (2011) Conductive polymer-based sensors for biomedical applications. Biosens Bioelectron 26(5):1825–1832

Neethirajan S, Jayas DS (2011) Nanotechnology for the food and bioprocessing industries. Food Bioprocess Technol 4:39–47

Nsibande SA, Forbes PBC (2016) Fluorescence detection of pesticides using quantum dot materials–a review. Anal Chim Acta 945:9–22

Nugaeva N, Gfeller KY, Backmann N, Lang HP, Düggelin M, Hegner M (2005) Micromechanical cantilever array sensors for selective fungal immobilization and fast growth detection. Biosens Bioelectron 21:849–856

Odobašić A, Šestan I, Begić S (2019) Biosensors for determination of heavy metals in waters. In: Biosensors for environmental monitoring. IntechOpen, p 139

Ouaissa M, Ouaissa M, Himer SE, Sathyamoorthy M, Rajasekar V, Dhanaraj RK (2024) Computational intelligence and internet of things in the agriculture sector: an introduction. In: Computational intelligence in internet of agricultural things. Springer, Cham, pp 1–17

Parameswari P, Belagalla N, Singh BV, Abhishek GJ, Rajesh GM, Katiyar D et al (2024) Nanotechnology-based sensors for real-time monitoring and assessment of soil health and quality: a review. Asian J Soil Sci Plant Nutr 10(2):157–173

Park S, Vosguerichian M, Bao Z (2013) A review of fabrication and applications of carbon nanotube film-based flexible electronics. Nanoscale 5(5):1727–1752

Pawar SK, Wadodkar NA, Salunke RS, Patil AM, Shirale DJ (2024) Eco-friendly electrochemical sensor for accurate soil nitrate detection using ZnOx/PANI nanocomposite on nickel foam electrode. Science 8:9

Peña-Bahamonde J, Nguyen HN, Fanourakis SK, Rodrigues DF (2018) Recent advances in graphene-based biosensor technology with applications in life sciences. J Nanobiotechnol 16(1):75

Perdomo SA, Marmolejo-Tejada JM, Jaramillo-Botero A (2021) Bio-nanosensors: fundamentals and recent applications. J Electrochem Soc 168(10):107506

Peterson JJ, Krauss TD (2006) Fluorescence spectroscopy of single lead sulfide quantum dots. Nano Lett 6(3):510–514

Pingarrón JM, Yanez-Sedeno P, González-Cortés A (2008) Gold nanoparticle-based electrochemical biosensors. Electrochim Acta 53(19):5848–5866

Polat EO, Cetin MM, Tabak AF, Bilget Güven E, Uysal BÖ, Arsan T et al (2022) Transducer technologies for biosensors and their wearable applications. Biosensors 12(6):385

Porto LS, Ferreira LF, Dos Santos WTP, Pereira AC (2022) Determination of organophosphorus compounds in water and food samples using a non-enzymatic electrochemical sensor based on silver nanoparticles and carbon nanotubes nanocomposite coupled with batch injection analysis. Talanta 246:123477

Pratama A, Rahayu S (2024) Biosensors enabled by nanotechnology: advancing precision agriculture through IoT monitoring. J Intell Connect Emerg Technol 9(1):25–39

Rahman MM, Ahammad AJ, Jin JH, Ahn SJ, Lee JJ (2010) A comprehensive review of glucose biosensors based on nanostructured metal-oxides. Sensors 10(5):4855–4886

Rai V, Acharya S, Dey N (2012) Implications of nanobiosensors in agriculture. J Biomater Nanobiotechnol 03(02):315–324. https://doi.org/10.4236/jbnb.2012.322039

Rigo AA, Cezaro AMD, Muenchen DK, Martinazzo J, Brezolin AN, Hoehne L et al (2020) Cantilever nanobiosensor based on the enzyme urease for detection of heavy metals. Braz J Chem Eng 36(4):1429–1437

Schrlau MG (2011) Carbon nanotube-based sensors: overview. Elsevier

Sennuga SO, Ujoyi SA, Bamidele J, Onjewu SS, Lai-Solarin WI, Omole AA (2023) Exploring the role of smartphone apps for livestock farmers data management extension and informed decision making in Nigeria. Int J Probiotics Dietetics 3(2):46–53

Sharma P, Pandey V, Sharma MMM, Patra A, Singh B, Mehta S, Husen A (2021) A review on biosensors and nanosensors application in agroecosystems. Nanoscale Res Lett 16:1–24

Sharma D, Teli G, Gupta K, Bansal G, Gupta GD, Chawla PA (2022) Nano-biosensors from agriculture to Nextgen diagnostic tools. Curr Nanomater 7(2):110–138

Shawon ZBZ, Hoque ME, Chowdhury SR (2020) Nanosensors and nanobiosensors: agricultural and food technology aspects. In: Nanofabrication for smart nanosensor applications. Elsevier, pp 135–161

Shi A, Wang J, Han X, Fang X, Zhang Y (2014) A sensitive electrochemical DNA biosensor based on gold nanomaterial and graphene amplified signal. Sensors Actuators B Chem 200:206–212

Singh RP (2021) Recent trends, prospects, and challenges of nanobiosensors in agriculture. In: Biosensors in agriculture: Recent trends and future perspectives. Springer, Cham, pp 3–13

Singh RK, Berkvens R, Weyn M (2021) AgriFusion: an architecture for IoT and emerging technologies based on a precision agriculture survey. IEEE Access 9:136253–136283

Solanki PR, Kaushik A, Agrawal VV, Malhotra BD (2011) Nanostructured metal oxide-based biosensors. NPG Asia Mater 3(1):17–24

Songa EA, Okonkwo JO (2016) Recent approaches to improving selectivity and sensitivity of enzyme-based biosensors for organophosphorus pesticides: a review. Talanta 155:289–304

Srivastava AK, Dev A, Karmakar S (2018) Nanosensors and nanobiosensors in food and agriculture. Environ Chem Lett 16:161–182

Talari FF, Bozorg A, Faridbod F, Vossoughi M (2021) A novel sensitive aptamer-based nanosensor using rGQDs and MWCNTs for rapid detection of diazinon pesticide. J Environ Chem Eng 9(1):104878

Vamvakaki V, Chaniotakis NA (2007) Pesticide detection with a liposome-based nano-biosensor. Biosens Bioelectron 22(12):2848–2853

Velasco-Garcia MN (2009) Optical biosensors for probing at the cellular level: a review of recent progress and future prospects. In: Seminars in cell & developmental biology, vol 20 (1). Academic, pp 27–33

Velasco-Garcia MN, Mottram T (2003) Biosensor technology addressing agricultural problems. Biosyst Eng 84(1):1–12

Verma ML, Rani V (2021) Biosensors for toxic metals, polychlorinated biphenyls, biological oxygen demand, endocrine disruptors, hormones, dioxin, phenolic and organophosphorus compounds: a review. Environ Chem Lett 19(2):1657–1666

Wang X, Cheng M, Yang Q, Wei H, Xia A, Wang L et al (2019) A living plant cell-based biosensor for real-time monitoring invisible damage of plant cells under heavy metal stress. Sci Total Environ 697:134097

Wang X, Luo Y, Huang K, Cheng N (2022) Biosensor for agriculture and food safety: recent advances and future perspectives. Adv Agrochem 1(1):3–6

Yadav A, Yadav K, Ahmad R, Abd-Elsalam KA (2023) Emerging frontiers in nanotechnology for precision agriculture: advancements, hurdles and prospects. Agrochemicals 2(2):220–256

Yin H, Cao Y, Marelli B, Zeng X, Mason AJ, Cao C (2021) Soil sensors and plant wearables for smart and precision agriculture. Adv Mater 33(20):2007764

Zhan F, Wang T, Iradukunda L, Zhan J (2018) A gold nanoparticle-based lateral flow biosensor for sensitive visual detection of the potato late blight pathogen, Phytophthora infestans. Anal Chim Acta 1036:153–161

Zhao F, He J, Li X, Bai Y, Ying Y, Ping J (2020) Smart plant-wearable biosensor for in-situ pesticide analysis. Biosens Bioelectron 170:112636

Zhu J, Zou Z, Shen Y, Li J, Shi S, Han S, Zhan X (2019) Increased ZnO nanoparticle toxicity to wheat upon co-exposure to phenanthrene. Environ Pollut 247:108–117

Chapter 14
Regulations and Ethical Considerations for Nanotechnology in Agri-food Applications

Maida Mobeen, Muhammad Majeed, Samreen Fatima, Akim Oladokoun, Ayesha Akbar, Aftab Umar, Jameel M. Al-Khayri, Waseem Akhtar Quresh, Ambreen Kalsoom, Javed Iqbal, and Banzeer Ahsan Abbasi

Abstract The rapid progress of nanotechnology in agriculture and food production offers significant advantages, such as increased crop yields, more effective pest and disease control, and enhanced food quality and safety. However, the incorporation of nanomaterials into agri-food systems also raises important regulatory and ethical issues that demand careful consideration. This chapter examines the current regulatory environment surrounding the use of nanotechnology in agriculture and food, focusing on international policies, national frameworks, and the complexities of risk assessment. The potential hazards associated with nanomaterials highlight the necessity for rigorous safety evaluations and transparent regulatory standards. Ethical concerns, such as consumer rights, informed consent, environmental justice,

M. Mobeen · M. Majeed (✉)
Department of Botany, University of Gujrat, Gujrat, Punjab, Pakistan
e-mail: m.majeed@uog.edu.pk

S. Fatima
Department of Biological Sciences, University of Baltistan, Gilgit, Pakistan

A. Oladokoun
Department of Geomatics and Ecosystem Modeling, Laboratory of Botany and Plant Ecology (LBEV), University of Lome (UL), Lome, Togo

A. Akbar
Department of Chemistry, Government College University Faisalabad, Faisalabad, Punjab, Pakistan

A. Umar
Department of Botany, University of Agriculture, Faisalabad, Punjab, Pakistan

J. M. Al-Khayri (✉)
Department of Agricultural Biotechnology, College of Agriculture and Food Sciences, King Faisal University, Al-Ahsa, Saudi Arabia
e-mail: jkhayri@kfu.edu.sa

J.-T. Chen (ed.), *Nanobiosensors for Crop Monitoring and Precision Agriculture*, Sustainability Sciences in Asia and Africa,
https://doi.org/10.1007/978-981-96-8335-2_14

and the socioeconomic implications for small-scale farmers, are also explored. The chapter underscores the critical need for interdisciplinary collaboration among scientists, policymakers, and stakeholders to create robust regulatory frameworks that promote the responsible use of nanotechnology in the agri-food sector. Looking ahead, it calls for the development of globally harmonized regulations, greater public involvement, and the establishment of advanced risk assessment methods. By addressing these regulatory and ethical challenges, the safe and sustainable integration of nanotechnology in agriculture can be realized, maximizing its potential benefits while safeguarding human health and the environment.

Keywords Nanotechnology · Agri-food regulation · Risk assessment · Ethical considerations · Consumer safety · Environmental impact · Sustainable agriculture

14.1 Introduction

Nanotechnology has emerged as a transformative force within agriculture and food production, offering innovative solutions to enhance crop yields, improve food quality, and optimize the efficiency of agricultural inputs. The integration of nanomaterials into fertilizers, pesticides, and food packaging has introduced unprecedented possibilities for sustainable farming and food safety. Yet, alongside these advancements, concerns surrounding the regulatory and ethical implications of nanotechnology in the agri-food sector have grown increasingly pronounced. As nanotechnology continues to expand its influence across agricultural and food systems, the establishment of comprehensive regulatory frameworks and ethical guidelines becomes imperative. Regulatory oversight in this domain necessitates a rigorous evaluation of nanomaterials to ensure their safety before market

W. A. Quresh
Cholistan Institute of Desert, The Islamia University of Bahawalpur, Bahawalpur, Punjab, Pakistan
e-mail: waseem.akhtar@iub.edu.pk

A. Kalsoom
Department of Physics, The Government Sadiq College Women University, Bahawalpur, Punjab, Pakistan
e-mail: kalsoom.ambreen@gscwu.edu.pk

J. Iqbal
Department of Botany, Bacha Khan University, Charsadda, Khyber Pakhtunkhwa, Pakistan

B. A. Abbasi
Department of Botany, Rawalpindi Women University, Rawalpindi, Pakistan

introduction. Both national governments and international bodies must develop clear, standardized guidelines to mitigate any potential risks to human health and the environment. Unlike conventional agricultural chemicals, nanoparticles (NPs) possess unique properties that can lead to unpredictable interactions within biological systems. Consequently, exhaustive risk assessments are essential, encompassing toxicity studies, environmental impact evaluations, and long-term exposure analyses. However, regulatory agencies worldwide face challenges in defining nanomaterials, standardizing testing protocols, and establishing universally accepted safety thresholds. Many existing agricultural and food safety regulations were not designed to address the complexities associated with nanoscale materials, resulting in regulatory gaps that must be addressed. Beyond regulatory concerns, ethical considerations are pivotal in shaping the responsible application of nanotechnology in agriculture. Key ethical issues range from consumer rights and informed consent to environmental justice and the socioeconomic ramifications of adopting nanotechnology-based solutions. Consumers have a fundamental right to be informed about the presence of nanomaterials in their food and agricultural products, yet labeling requirements vary significantly across jurisdictions. While some countries mandate the disclosure of nanotechnology use, others lack transparency regulations, raising concerns regarding consumer autonomy and trust (Arif et al. 2021; Rasool et al. 2025). Moreover, the environmental consequences of nanotechnology, including nanoparticle accumulation in soil and water systems, warrant careful scrutiny. Potential ecological disruptions underscore the necessity of sustainable practices throughout nanotechnology's development and deployment.

Another critical ethical consideration pertains to the socioeconomic impact of nanotechnology on small-scale farmers and rural communities. While nanotechnology holds the potential to revolutionize agriculture by increasing efficiency and reducing input costs, it may also exacerbate existing inequalities. Large agribusinesses with greater access to advanced technologies could gain a disproportionate competitive advantage, widening the gap between industrialized and smallholder farming systems. This raises essential questions concerning accessibility, affordability, and the equitable distribution of nanotechnology's benefits. Policymakers must ensure that advancements in agricultural nanotechnology are inclusive and do not disproportionately benefit corporate interests at the expense of marginalized farming communities.

Public perception and acceptance of nanotechnology further shape its regulatory and ethical landscape. Many consumers remain skeptical of nanotechnology, often associating it with synthetic chemicals and genetic modification. Transparent communication and public engagement are crucial to building trust and addressing societal concerns. Collaborative efforts among scientists, policymakers, and industry stakeholders can facilitate the dissemination of evidence-based information regarding nanotechnology's benefits and risks, empowering consumers to make informed choices. The need for a globally harmonized regulatory approach is paramount. Disparities in regulatory stringency across countries create inconsistencies in safety standards and trade barriers. For instance, the European Union (EU) adopts a precautionary approach, requiring extensive safety evaluations and labeling mandates for nanotechnology in food products. Conversely, some countries have yet to establish specific regulations governing nanomaterials in agriculture and food production. This lack of international coordination complicates compliance for multinational agribusinesses and food producers, impeding trade and market access (Arshad et al. 2022; Aziz et al. 2023a; Muhammad et al. 2024b). Establishing globally recognized guidelines for nanotechnology safety assessments and risk management would promote fair and transparent trade while ensuring consumer protection. Advancing risk assessment methodologies is essential to address the uncertainties surrounding nanotechnology in agriculture. Traditional toxicity testing methods may not fully capture the distinct behavior of nanoparticles, necessitating the development of novel analytical techniques. Emerging approaches can enhance the accuracy and efficiency of nanomaterial risk assessments. Additionally, research into the long-term effects of nanoparticle exposure on human health and ecosystems remains a priority. Collaboration between regulatory agencies, academic institutions, and industry players is crucial to advancing scientific knowledge and refining safety evaluation frameworks.

Ethical governance and regulatory oversight should also extend to the research and development phase of nanotechnology applications. Implementing ethical review processes can ensure that scientific advancements align with societal values and environmental sustainability objectives. Funding agencies and research institutions should prioritize projects that address global challenges such as food security, environmental resilience, and social equity, rather than focusing solely on commercial profitability. Encouraging interdisciplinary collaboration among nanoscientists, ethicists, policymakers, and agricultural experts can foster responsible innovation while mitigating potential risks before nanotechnology products reach the market. While the potential benefits of nanotechnology in agriculture are vast, its development must be accompanied by a steadfast commitment to safety, transparency, and ethical responsibility. Regulatory frameworks must evolve in tandem with technological advancements, ensuring that nanomaterials undergo comprehensive safety assessments prior to commercialization (Aziz et al. 2023b; Naveed et al. 2024c). Ethical considerations should remain at the forefront of policy discussions. By fostering a regulatory environment that balances innovation with public welfare, nanotechnology can contribute to the creation of a more sustainable and resilient agri-food system, capable of addressing the global challenges of the twenty-first century.

14.1.1 Background

Nanotechnology has garnered significant attention in the agricultural and food sectors for its potential to transform food production, enhance crop protection, and improve food safety. The integration of nanotechnology into agriculture involves the use of engineered nanoparticles (NPs) in various applications, such as fertilizers, pesticides, food packaging, and disease diagnostics. However, the increasing presence of nanomaterials in the agri-food system has raised concerns about their safety, environmental impact, and ethical implications. These concerns have prompted regulatory bodies worldwide to assess the risks and benefits associated with nanotechnology in agriculture and food production. The regulatory landscape for nanotechnology in agriculture is complex and varies widely across countries. For instance, the United States adopts a case-by-case approach, evaluating nanotechnology applications under existing food and agricultural laws. In contrast, the European Union has implemented stricter regulations, mandating comprehensive safety assessments and labeling for nanomaterials in food (Bashir et al. 2023; Bhattacharya et al. 2023; Naveed et al. 2024a, b). Many other nations, however, lack clear regulatory frameworks, leading to inconsistencies in safety standards, market access, and consumer transparency. A major challenge in regulating nanotechnology is the absence of a universally accepted definition of nanomaterials. Given that nanoparticles exhibit unique physicochemical properties compared to their bulk counterparts, traditional toxicity assessments may not adequately capture their risks.

Ethical concerns further complicate the adoption of nanotechnology in agriculture. A key issue is consumer rights and transparency, as many consumers remain unaware of the presence of nanomaterials in their food and agricultural products due to inconsistent global labeling regulations. This raises questions about informed consent, as individuals should have the right to make informed decisions about their food choices. Moreover, the long-term health effects of nanoparticle consumption are still poorly understood, necessitating further research into their bioaccumulation and potential toxicity. Environmental considerations are another critical aspect of nanotechnology's background in agriculture. The widespread use of nanomaterials in fertilizers and pesticides could lead to their accumulation in soil and water systems. Research indicates that certain nanoparticles can disrupt microbial communities, potentially impairing essential soil functions like nutrient cycling. Additionally, nanoparticles can enter the food chain through plant uptake, raising concerns about their persistence in ecosystems and potential bio magnification (Haq et al. 2022a; Barwant et al. 2024). The lack of standardized environmental risk assessment protocols complicates efforts to predict the long-term ecological impacts of nanotechnology on biodiversity and ecosystem stability.

The socioeconomic implications of nanotechnology in agriculture also warrant attention. While nanotechnology holds promise for boosting crop productivity and reducing chemical inputs, it may exacerbate economic disparities within the agricultural sector. Large agribusinesses with access to nanotechnology-based solutions could gain a competitive edge, potentially marginalizing smallholder farmers who lack the resources to adopt such advanced technologies. This raises ethical questions about fairness, inclusivity, and the equitable distribution of nanotechnology's benefits. Policymakers must ensure that regulations foster responsible innovation while addressing these socioeconomic challenges.

As nanotechnology continues to evolve, international collaboration is essential to develop harmonized regulatory frameworks. Discrepancies in nanotechnology regulations across countries create barriers to global trade, complicating compliance for businesses. Collaborative efforts among governments, scientific institutions, and industry stakeholders can help establish standardized testing protocols, safety thresholds, and transparent labeling systems. Furthermore, public engagement and education are crucial to building awareness and trust in the use of nanotechnology in agriculture and food production (Haq et al. 2022b). By addressing these regulatory, ethical, and socioeconomic challenges, the responsible and sustainable integration of nanotechnology into agriculture can be achieved, maximizing its benefits while minimizing risks to human health and the environment.

14.1.2 Significance

The regulation and ethical considerations surrounding nanotechnology in agriculture and food production are of paramount importance, given their profound implications for public health, environmental sustainability, consumer rights, and global trade. Robust regulatory oversight ensures that nanomaterials employed in agriculture are safe for human consumption and do not inadvertently harm ecosystems. In the absence of clearly defined regulations, the unchecked expansion of nanotechnology within the agri-food sector could lead to unforeseen consequences, thereby eroding public confidence in food safety and agricultural practices. A principal reason for stringent nanotechnology regulation lies in the protection of human health. Nanoparticles, due to their enhanced bioavailability and reactivity, may exhibit toxicity levels that surpass those of their conventional counterparts. Exposure through dietary intake or environmental contact could engender long-term effects, potentially disrupting human metabolism, cellular function, and immune responses (Hussain et al. 2023; Hassan et al. 2022a; Majeed et al. 2024b). Therefore, effective regulatory frameworks must delineate rigorous guidelines for toxicity testing, establish exposure limits, and institute comprehensive risk assessment protocols to ensure that only safe nanotechnology applications are sanctioned.

Environmental protection represents another critical facet of nanotechnology regulation. The deliberate introduction of engineered nanoparticles into soil, water, and air poses the risk of disturbing ecological balance. For example, nanoparticles designed to boost nutrient uptake in crops might inadvertently alter microbial communities or interact unpredictably with other soil components. Regulatory measures are essential to define best practices for the responsible utilization and disposal of nanomaterials, thereby preventing environmental contamination and reducing their ecological footprint. Without such oversight, nanotechnology could contribute to pollution, biodiversity loss, and soil degradation—undermining the very sustainability objectives it seeks to promote. Furthermore, consumer rights and transparency are integral to the discourse on nanotechnology regulation. Many consumers remain unaware that their food may contain nanoparticles, as labeling standards differ widely among jurisdictions. This opacity raises ethical concerns, as individuals are deprived of the information necessary to make informed dietary choices. Mandatory labeling and public education initiatives are crucial to bridge this information gap and empower consumers to exercise their rights effectively. Additionally, ethical considerations in the realm of nanotechnology research and development must ensure that innovations align with societal values and do not disproportionately advantage certain groups over others. From an economic perspective, nanotechnology regulation is instrumental in shaping the future landscape of agricultural markets and international trade. Nations with well-defined nanotechnology policies tend to cultivate environments that not only foster innovation but also uphold safety and accountability (Hassan et al. 2022b; Jamil et al. 2022; Rasool et al. 2024; Majeed et al. 2024a). Conversely, discrepancies in regulatory standards between countries can lead to trade barriers and complicate the efforts of multinational corporations to market nanotechnology-based agricultural products on a global scale. The establishment of harmonized international standards for nanotechnology risk assessment and labeling is, therefore, imperative for facilitating smoother trade relations while safeguarding consumer safety.

In summary, while the promise of nanotechnology in agriculture is immense, its successful integration depends on a delicate balance between innovation and precaution. Comprehensive regulatory frameworks, underpinned by rigorous risk assessments and ethical guidelines, are essential to harness the benefits of nanotechnology without compromising public health, environmental integrity, or consumer trust.

14.1.3 Limitations

Despite the potential benefits of nanotechnology in agriculture, several limitations hinder its regulation and ethical oversight. One of the foremost challenges is the absence of a universally accepted definition for nanomaterials. Various regulatory agencies employ differing size thresholds to define nanoparticles, thereby complicating efforts toward standardization. Without a clear, shared definition, it becomes exceedingly difficult to establish uniform safety assessments, labeling requirements, and risk management strategies across different countries and industries. Another significant limitation is the insufficient data on the long-term health and environmental effects of nanomaterials. Although some studies have investigated nanoparticle toxicity, our understanding of how prolonged exposure impacts human health and ecological systems remains limited. Regulatory bodies frequently face challenges in evaluating these risks due to the scarcity of comprehensive long-term studies. This uncertainty makes it difficult to develop robust guidelines that safeguard consumer health without unduly hindering technological innovation. Furthermore, the high cost associated with nanotechnology regulation and compliance presents another obstacle. Extensive risk assessments, toxicity studies, and environmental impact evaluations require substantial financial resources. Small and medium-sized enterprises (SMEs) may struggle to meet these demands, potentially curtailing their ability to compete with larger corporations that can more readily absorb these costs. This economic imbalance could lead to a monopolization of nanotechnology-driven agricultural solutions, thereby reducing market diversity and stifling innovation. Public perception and a general lack of awareness about nanotechnology also pose regulatory and ethical challenges. Many consumers remain skeptical about nanotechnology, often conflating it with genetic modification or synthetic chemicals. This fear of unknown risks can lead to resistance against its adoption in agriculture (Khoja et al. 2022; Khan et al. 2022; Rani et al. 2024). Without effective public engagement and education, pervasive misconceptions about nanotechnology may impede its acceptance, slowing regulatory progress and curtailing innovation.

Finally, regulatory frameworks often struggle to keep pace with the rapid advancements in nanotechnology. The dynamic nature of nanoscience means that new nanomaterials and applications are constantly emerging, frequently outstripping the capacity of regulatory bodies to evaluate them comprehensively (Elabiad 2025). This regulatory lag creates uncertainty for industry stakeholders and raises concerns about potential gaps in safety oversight.

14.2 Nanotechnology in Agri-Food Systems

Nanotechnology has emerged as a groundbreaking innovation with transformative potential across various industries, including healthcare, electronics, and environmental science. In recent years, its application in the agriculture and food sectors has gained considerable traction. By manipulating materials at the nanoscale, nanotechnology enhances food safety, boosts crop yields, and optimizes resource efficiency. With the global population projected to surpass 9 billion by 2050, food security has become a pressing concern, driving the need for innovative solutions like nanotechnology to improve agricultural productivity and food preservation. In agri-food systems, nanotechnology is primarily applied in three key areas: precision agriculture, food processing, and food packaging. In precision agriculture, nanomaterials are integrated into smart sensors, fertilizers, and pesticides to enable precise nutrient delivery and pest control while minimizing environmental harm. These advancements support sustainable farming practices by reducing the excessive use of chemical inputs, which often contribute to soil degradation and water pollution (Majeed et al. 2020; Hussain et al. 2024; Shafqat et al. 2024). Additionally, nanobiosensors can detect plant diseases and soil nutrient deficiencies in real time, allowing farmers to take proactive measures before significant crop losses occur.

In food processing, nanoparticles such as nanoemulsions, nanocapsules, and nanoliposomes are employed to enhance nutrient bioavailability, improve food texture, and extend shelf life. Encapsulation techniques using nanotechnology enable the controlled release of vitamins, minerals, and antioxidants, ensuring better nutrient retention in processed foods. Furthermore, nanotechnology facilitates the development of antimicrobial coatings that prevent bacterial contamination; thereby enhancing food safety. Food packaging has also seen significant advancements through the incorporation of nanomaterials. Nanocomposites and nanocoatings provide superior barrier properties against oxygen, moisture, and UV light, preserving food quality and reducing spoilage. Some advanced nanotechnology-based packaging materials even feature antimicrobial properties that actively inhibit the growth of harmful microorganisms, raising food safety standards. Intelligent packaging, equipped with nanosensors, can monitor temperature, humidity, and spoilage levels in real time, offering consumers greater transparency and quality assurance. Despite its many benefits, the use of nanotechnology in the agri-food sector raises concerns about safety, ethical implications, and regulatory challenges. The potential toxicity of nanoparticles and their long-term effects on human health and the environment remain incompletely understood. Additionally, consumer skepticism and limited awareness about nanotechnology applications in food underscore the importance of regulatory transparency and public engagement (Majeed et al. 2021a; Nasar et al. 2024b). Governments and scientific institutions worldwide are working to establish standardized guidelines for risk assessment and ethical considerations, ensuring the responsible development of nanotechnology in agriculture and food production. The integration of nanotechnology into agri-food systems represents a paradigm shift in how food is grown, processed, and preserved. While it offers solutions to critical challenges such as food security and sustainability, a careful evaluation of its risks and benefits is essential. As research advances and regulatory frameworks evolve, nanotechnology holds the potential to revolutionize the global food industry, paving the way for safer, more efficient, and sustainable agricultural practices.

14.2.1 The Role of Nanotechnology in Modern Agriculture

Nanotechnology has revolutionized multiple sectors, and its integration into modern agriculture is proving to be a game-changer (Fig. 14.1). As the global population continues to rise, increasing food demand puts immense pressure on agricultural systems to enhance productivity, improve efficiency, and ensure sustainability. Conventional farming methods, though effective, often contribute to resource wastage, soil degradation, and environmental pollution due to excessive use of fertilizers and pesticides. Nanotechnology offers innovative solutions by enabling precision farming, improving crop protection, and enhancing soil and plant health, ultimately addressing key challenges in the agricultural sector. One of the primary applications of nanotechnology in modern agriculture is nano-based fertilizers and pesticides. Traditional fertilizers are often applied in excess, leading to nutrient leaching, water contamination, and poor soil health. Nano-fertilizers, formulated with nanoparticles, allow for controlled and targeted release of nutrients, ensuring that crops receive the required nutrients at the right time. This minimizes nutrient loss, enhances plant absorption, and increases yield efficiency. Similarly, nano-pesticides provide more effective pest control while reducing chemical runoff, thus mitigating environmental hazards. Some nanomaterials have antimicrobial and insecticidal

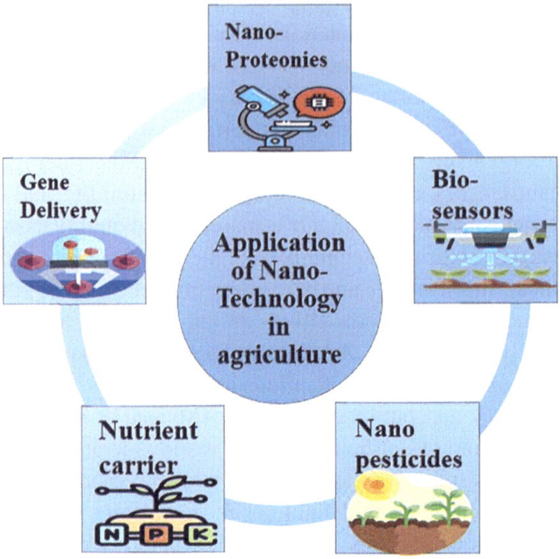

Fig. 14.1 Applications of nanotechnology in agriculture

properties, reducing dependency on synthetic chemicals. Another breakthrough is the development of nanosensors for precision farming. These sensors can detect soil moisture levels, nutrient deficiencies, and plant diseases in real time. By providing farmers with accurate data, nanosensors facilitate timely intervention, reducing crop losses and optimizing resource use. For example, carbon nanotube-based sensors can identify pathogens or stress conditions in plants at an early stage, allowing for prompt treatment.

Additionally, nano-enabled smart delivery systems are being developed to transport nutrients, herbicides, and pesticides directly to plant roots, ensuring maximum efficiency with minimal wastage. Water management is another critical area where nanotechnology is making a significant impact. Nano-enhanced water filtration systems improve irrigation efficiency by removing contaminants, desalinating water, and ensuring optimal water quality for crops. Nanoparticles such as titanium dioxide and silver nanoparticles are used in filtration membranes to eliminate harmful bacteria and chemicals from irrigation water. Moreover, nanotechnology is being used to develop hydrogel-based soil conditioners that retain moisture and release it gradually to crops, ensuring better water conservation, especially in drought-prone regions. Post-harvest losses remain a major concern in agriculture, leading to food wastage and economic losses. Nanotechnology offers solutions such as nano-based food preservatives and packaging materials that extend the shelf life of agricultural produce. Nanocoatings with antimicrobial properties protect fruits and vegetables from microbial decay, reducing spoilage during transportation and storage. Additionally, intelligent packaging systems embedded with nanosensors can detect contamination, spoilage, or gas emissions, providing real-time information about food quality. This innovation enhances food safety and minimizes waste along the supply chain. Despite the numerous benefits, the widespread adoption of nanotechnology in agriculture is not without challenges. Concerns regarding the toxicity of nanoparticles, their impact on soil microbiota, and potential risks to human health remain areas of active research. Regulatory bodies are working to develop guidelines and safety protocols for the use of nanomaterials in agriculture. Transparency in research and public awareness initiatives will play a crucial role in fostering trust and acceptance among farmers and consumers (Majeed et al. 2021b; Khan et al. 2024; Nasar et al. 2024a). Nanotechnology has the potential to redefine modern agriculture by making farming more efficient, sustainable, and environmentally friendly. From precision nutrient delivery to advanced pest control, water management, and post-harvest preservation, the applications of nanotechnology are vast. As scientific advancements continue and regulations evolve, nanotechnology is set to become a cornerstone of future agricultural practices, ensuring global food security while mitigating environmental impact.

14.2.2 Potential Benefits and Risks of Nanotechnology in Food and Farming

Nanotechnology has the potential to revolutionize the agriculture and food industry by offering solutions that enhance efficiency, sustainability, and productivity. The incorporation of nanomaterials in farming practices and food production can increase crop yields, improve food quality, and reduce environmental pollution. However, despite its promising applications, there are concerns regarding the safety of nanoparticles, their potential toxicity, and long-term environmental impacts. Understanding both the benefits and risks of nanotechnology in food and farming is crucial for its responsible development and acceptance. One of the most significant advantages of nanotechnology in agriculture is the development of nano-fertilizers and nano-pesticides (Engelmann 2024). Traditional fertilizers and pesticides often lead to overuse, causing nutrient runoff, soil degradation, and water contamination. In contrast, nano-based fertilizers release nutrients in a controlled and targeted manner, ensuring better plant absorption and minimizing waste. Similarly, nano-pesticides provide efficient pest control with lower chemical doses, reducing harmful environmental residues. These innovations contribute to sustainable agricultural practices by optimizing input use and minimizing ecological damage. Another major benefit is real-time soil and crop monitoring through nanosensors. Traditional farming methods often rely on manual soil testing, which can be time-consuming and inaccurate. Nanosensors embedded in the soil can detect nutrient deficiencies, moisture levels, and plant diseases instantly, enabling precision farming. By providing farmers with real-time data, these technologies allow for targeted interventions, reducing the need for excessive fertilizers, pesticides, and water usage. Nanotechnology is also transforming food preservation and packaging. The food industry faces significant losses due to spoilage and contamination. Nanocoatings and nanocomposites used in food packaging improve barrier properties, extending the shelf life of perishable products. Some nanomaterials, such as silver and zinc oxide nanoparticles, have antimicrobial properties that prevent bacterial growth on food surfaces, enhancing food safety. Additionally, intelligent packaging equipped with nanosensors can detect food spoilage, microbial contamination, or gas emissions, providing consumers with real-time quality assessments. Water scarcity is a major global concern, and nanotechnology offers innovative solutions for water purification and management in agriculture. Nano-filtration membranes can remove contaminants, salts, and heavy metals from water, making it suitable for irrigation. Some nanomaterials also have self-cleaning properties, reducing the need for chemical treatments and maintenance. Moreover, nano-enhanced hydrogels can improve water retention in the soil, ensuring better crop hydration in drought-prone areas. Despite its numerous benefits, the use of nanotechnology in agriculture and food systems is associated with potential risks that require careful evaluation (Majeed et al. 2021c; Waheed et al. 2023). One of the primary concerns is the toxicity of nanoparticles and their potential impact on human health. When nanoparticles enter the food chain through nano-fertilizers, nano-pesticides, or nano-packaging materials, they may accumulate in the body and cause unintended health effects. Research on the long-term consequences of nanoparticle consumption is still in its early stages, and there is uncertainty regarding safe exposure levels. Another major risk is the impact of nanomaterials on soil microbiota and ecosystems. Soil microorganisms play a vital role in nutrient cycling, plant growth, and maintaining soil health. The introduction of engineered nanoparticles into the soil may alter microbial communities, disrupt ecological balance, and potentially lead to unforeseen consequences. Additionally, the persistence of certain nanomaterials in the environment raises concerns about their bioaccumulation in plants, animals, and aquatic systems. Regulatory challenges also pose a significant barrier to the widespread adoption of nanotechnology in food and farming. There is currently no universal regulatory framework governing the use of nanomaterials in agriculture and food products. Different countries have varying levels of regulations, and the lack of standardized testing methods makes it difficult to assess the safety of nano-enabled agricultural products. Without clear guidelines, there is a risk of consumer skepticism and market resistance to nanotechnology-based food products. Public perception and ethical considerations also play a crucial role in the acceptance of nanotechnology in food production. Many consumers are hesitant to accept nano-engineered foods due to concerns about safety, transparency, and potential health risks. The lack of labeling requirements in some regions further fuels mistrust, as consumers are often unaware of the presence of nanomaterials in their food. Ethical concerns also arise regarding the monopolization of nanotechnology by large corporations, which could marginalize small-scale farmers and create disparities in agricultural advancements. Nanotechnology holds immense promise for transforming agriculture and food production by improving efficiency, sustainability, and food security. From precision nutrient delivery to real-time monitoring and advanced packaging solutions, the benefits of nanotechnology are substantial. However, the potential risks associated with nanoparticle toxicity, environmental impact, and regulatory challenges cannot be overlooked (Waheed et al. 2022). To ensure the responsible integration of nanotechnology into food and farming, comprehensive risk assessments, standardized regulations, and transparent communication with consumers are essential. As research progresses, it will be crucial to strike a balance between innovation and safety, ensuring that nanotechnology contributes positively to the future of agriculture and food security.

14.2.3 Regulatory and Ethical Challenges in Agri-Food Nanotechnology

The rapid advancements in nanotechnology have revolutionized the agricultural and food industries, offering innovative solutions for precision farming, food safety, and sustainable agriculture. However, as the use of nanomaterials increases, regulatory and ethical concerns arise regarding their potential risks to human health, environmental sustainability, and consumer rights. The absence of a well-defined global regulatory framework for nanotechnology in agri-food systems has led to inconsistencies in safety assessments, market approvals, and public acceptance. Ethical issues, including transparency, informed consumer choices, and equitable access to nanotechnology, further complicate its widespread adoption. This section explores the regulatory and ethical challenges associated with the integration of nanotechnology into food and agricultural systems. One of the biggest challenges in nanotechnology regulation is the lack of universal standards for defining and assessing nanomaterials. Different countries and regulatory bodies classify and evaluate nanotechnology-based products differently, creating disparities in safety assessments and approval processes. For instance, while the European Union (EU) has implemented strict regulations requiring nano-based food products to undergo rigorous risk assessments, other countries, such as the United States and China, have more flexible approaches. The absence of a harmonized global framework makes it difficult for manufacturers, researchers, and policymakers to develop uniform safety protocols for nano-enabled agricultural products. Another challenge is the difficulty in detecting and monitoring nanoparticles in food and soil systems. Traditional food safety testing methods are not always effective in identifying nanoparticles, especially when they interact with other substances. The small size and high reactivity of nanomaterials make it challenging to determine their toxicity, persistence, and long-term effects on human health and the environment. This lack of precise testing methods raises concerns about the adequacy of current safety evaluations and the potential for unintentional exposure to harmful nanoparticles. The slow adaptation of existing food and agricultural regulations to nanotechnology innovations also poses a significant challenge. Many regulatory frameworks were developed long before the emergence of engineered nanomaterials, meaning they do not fully account for the unique properties of nanoparticles. For example, traditional pesticide regulations are based on chemical composition and toxicity levels, but nanomaterials behave differently in biological and environmental systems. The absence of specific guidelines for nano-based fertilizers, pesticides, and food additives increases uncertainty among farmers, food producers, and consumers regarding the safety and efficacy of these products. Intellectual property (IP) and patenting issues are also a major regulatory concern. Large corporations often secure patents for nano-enabled agricultural technologies, creating barriers for small-scale farmers and startups who may lack the resources to compete in the market. This raises concerns about corporate monopolization of agricultural nanotechnology, potentially leading to higher costs, restricted access, and unequal distribution of benefits (Majeed et al. 2022a; Tassadduq et al. 2022; Ullah et al. 2022). Policymakers must address these disparities to ensure that nanotechnology serves the interests of all stakeholders, including small farmers and consumers in developing countries. Beyond regulatory issues, ethical concerns surrounding the use of nanotechnology in agriculture and food production are becoming increasingly significant. One of the primary concerns is consumer transparency and informed choice. Many consumers are unaware of the presence of nanomaterials in their food and farming systems due to the lack of mandatory labeling in several countries. Without clear labeling requirements, consumers cannot make informed decisions about whether they want to consume nano-enabled food products. This lack of transparency contributes to public skepticism and mistrust toward nanotechnology in food production. Another major ethical dilemma is the potential health risks of consuming nanoparticles. While nanotechnology has the potential to enhance food safety and shelf life, the long-term effects of ingesting engineered nanomaterials remain uncertain. Some studies suggest that certain nanoparticles, such as silver or titanium dioxide, may accumulate in human tissues and cause oxidative stress or inflammation. Without long-term safety studies, the precautionary principle should be applied to prevent unintended health consequences. Ethical considerations demand that regulatory bodies prioritize human health and safety over commercial interests when approving nano-based food products. The impact of nanotechnology on traditional farming practices also raises ethical concerns. As nano-enabled precision agriculture becomes more widespread, small-scale and organic farmers may struggle to compete with industrialized farms that have access to advanced nanotechnologies. This could widen the gap between large agribusinesses and smallholder farmers, exacerbating socioeconomic inequalities in the agricultural sector. Additionally, the use of engineered nanoparticles in soil and water systems may disrupt natural ecosystems, potentially harming beneficial soil microbes and biodiversity. The role of public perception and misinformation also plays a critical part in the ethical discourse surrounding nanotechnology in food and agriculture. Public opposition to genetically modified organisms (GMOs) has demonstrated how scientific advancements can face resistance due to a lack of clear communication and public engagement. If consumers associate nanotechnology with uncontrolled risks and corporate manipulation, they may reject its applications, regardless of its potential benefits (Majeed et al. 2022b; Tariq et al. 2023). Ethical responsibility requires that scientists, policymakers, and industry leaders engage in transparent discussions about the risks, benefits, and real-world implications of nanotechnology in food and farming. Developing a global regulatory framework—International organizations such as the World Health Organization (WHO), Food and Agriculture Organization (FAO), and Codex Alimentarius Commission should collaborate to create harmonized guidelines for assessing the safety of nanomaterials in agriculture and food production. Improving nanoparticle detection and risk assessment techniques—Governments and research institutions should invest in advanced toxicological studies and risk assessment models to evaluate the long-term impact of nanoparticles on human health and the environment. Mandatory labeling of nano-based food products—Regulatory agencies should enforce transparent labeling requirements to empower consumers with knowledge about the presence of nanomaterials in food. Encouraging ethical business practices—Policies should be introduced to prevent monopolization of nanotechnology by large corporations and ensure equitable access for small-scale farmers and agribusinesses (Zamri 2023). Enhancing public engagement and education awareness campaigns and open discussions between scientists, policymakers, and consumers can address misinformation and improve trust in nanotechnology applications.

14.3 Global Regulatory Frameworks for Nanotechnology in Agriculture

The integration of nanotechnology in agriculture and food systems has revolutionized modern farming, improving productivity, precision, and sustainability. However, as nanomaterials are increasingly utilized in fertilizers, pesticides, food packaging, and food additives, regulatory concerns have emerged regarding their safety, environmental impact, and long-term effects on human health. Countries and international organizations have adopted different regulatory approaches to manage nanotechnology in the agri-food sector, but a lack of universal standards has created inconsistencies and challenges (Majeed et al. 2021c; Sinha et al. 2023). This section explores the global regulatory landscape for agricultural nanotechnology, the challenges in establishing universal guidelines, and case studies of successful and failed regulatory approaches.

14.3.1 Regulatory Approaches in Different Countries (United States, EU, China, Etc.)

Different countries have developed distinct regulatory frameworks for nanotechnology applications in agriculture and food systems. While some regions enforce strict risk assessment procedures, others have more flexible regulatory mechanisms

(Majeed et al. 2023a; Muhammad et al. 2024a). Below is an overview of key regulatory approaches in major global markets.

14.3.1.1 United States

The United States Food and Drug Administration (FDA), Environmental Protection Agency (EPA), and United States Department of Agriculture (USDA) regulate nanotechnology applications in food and agriculture. The FDA classifies nanomaterials under the Generally Recognized as Safe (GRAS) framework, but does not have specific labeling requirements for nano-based food products. The EPA regulates nano-enabled pesticides under the Federal Insecticide, Fungicide, and Rodenticide Act (FIFRA) but faces challenges in assessing long-term environmental risks (Majeed et al. 2023b; Muhammad et al. 2023). The USDA oversees the use of nanotechnology in fertilizers, soil enhancers, and crop protection materials, focusing on environmental safety and sustainability.

14.3.1.2 European Union (EU)

The European Food Safety Authority (EFSA) has one of the world's most stringent regulatory frameworks for nanotechnology in food and agriculture. The EU Novel Food Regulation mandates that all food products containing engineered nanomaterials undergo pre-market risk assessments. The EU Registration, Evaluation, Authorization, and Restriction of Chemicals (REACH) regulation also requires detailed safety assessments for nanomaterials used in agriculture. Additionally, mandatory labeling regulations ensure that consumers are informed about the presence of nanomaterials in food and agricultural products (Majeed et al. 2023c). However, the complexity of EU regulations has slowed down the approval process for nano-based agricultural innovations.

14.3.1.3 China

China has become a global leader in agricultural nanotechnology research and development, with the Ministry of Agriculture and Rural Affairs overseeing regulations. China's regulatory system is less stringent than the EU's, allowing for faster commercialization of nano-enabled fertilizers, pesticides, and food products. However, the lack of comprehensive risk assessment policies has raised concerns about potential health and environmental impacts (Jamshaid et al. 2024). The Chinese government has invested in nanotechnology safety research, but regulatory gaps still exist.

14.3.1.4 India

India has made significant progress in nanotechnology applications in agriculture, particularly in precision farming and nano-fertilizers. The Food Safety and Standards Authority of India (FSSAI) regulates nano-based food products, while the Indian Council of Agricultural Research (ICAR) evaluates the safety and efficacy of nano-enabled agrochemicals (Kumari et al. 2023). However, India lacks specific guidelines for nanoparticle exposure limits, risk assessments, and environmental monitoring, which could pose future regulatory challenges.

14.3.1.5 Other Countries and International Organizations

Countries such as Australia, Brazil, and Japan have varying degrees of regulations for nanotechnology in agriculture. The World Health Organization (WHO), Food and Agriculture Organization (FAO), and Codex Alimentarius Commission have proposed global guidelines for nanotechnology risk assessments (Feitshans and Vassara 2023), but their implementation remains inconsistent across different regions.\.

14.3.2 Challenges in Developing Universal Standards for Nanomaterials

Developing consistent global standards for nanotechnology in agriculture remains a challenge, despite regional regulations having advanced for various elements. There is a fluctuation in hazard perception approaches, and therefore, different regulatory approaches result from some countries disagreeing on the possible hazards of nanomaterials. The United States and China adopt/utilize a more flexible/adjustable approach to nano-based agricultural products, while the EU follows the preventive principle, Deficiency of Standard Investigation Methods. There are no universally recognized testing rules for evaluating/assessing the persistency, toxicity, and bioaccumulation of nanoparticles in water, soil, and food chains. Complexity/intricacy of nanomaterials: It is challenging to create a one-size-fits-all regulatory framework that is entirely applicable due to the various ranges of sizes, shapes, and compositions of nanomaterials. Regulatory overlaps and gaps: Numerous regulatory agencies (FDA, EPA, EFSA, FAO, WHO) supervise various features of agricultural nanotechnology, leading to inefficiency, delays, and overlaps. Resistance from industry stakeholders (Kaur et al. 2024): Due to concerns about enhanced costs, longer approval timelines, and market limitations, some agricultural corporations fight stricter regulations.

14.3.3 Case Studies of Regulatory Success and Failures

One of the most efficient worldwide models is considered to be the EU's rigorous regulatory framework for nanotechnology in food and agriculture. The EU has assured greater clarity and consumer security by requiring a comprehensive/complete hazard evaluation and compulsory labeling. Nevertheless, professional debate that prolonged approval processes slow down innovation in the European agricultural sector. The unregulated use of nano-fertilizers, pesticides, and food additives is allowed in numerous developing nations due to a deficiency of specific regulations for nanotechnology in agriculture. This has led to concerns about unexplored health hazards, water pollution, and soil contamination. Farmers and consumers in this part may be exposed to adverse nanoparticles if proper/appropriate management is not applied. Faster exploitation of nano-enabled agricultural goods allowed by nanotechnology is made possible by the US regulatory scheme, which strikes a balance between innovation and safety. Even so, the deficiency of compulsory labeling and uneven testing protocols has led to consumer mistrust/distrust and doubt about the safety of nano-based food goods. While the global regulatory viewpoint on nanotechnology in agriculture is altering, important challenges remain in harmonized standards, rising safety assessments, and ensuring public reliance. While a few countries like China and the United States prioritize plasticity and creativity, others like the EU adopt rigorous regulations. Governments, research organizations, industry stakeholders, and international organizations will need cooperative efforts or will struggle to develop a universal regulatory framework (Singh and Kumar 2023). Agricultural nanotechnology must be safe, clear, and sustainable in order to maximize its goodness and minimize hazards.

14.4 Ethical Considerations in Agri-Food Nanotechnology

The fast development of nanotechnology in food systems and agriculture raises significant ethical concerns related to food security, social justice, consumer rights, and environmental issues. While agricultural productiveness and sustainability may be better by nano-enabled fertilizers, pesticides, and food additives, it is difficult to oversee ethical implications or associations encompassing safety, clarity, public perception, and equitable rights. Policies, regulations, and public attitude toward agricultural advancements based on nanotechnology are severely influenced by ethical considerations (Chowdhury et al. 2023). This section investigates the ethical challenges related to the biotechnology quandary, consumer clarity, and the consequences of public conceptuality on nanotechnology acceptance.

14.4.1 Consumer Rights and Transparency in Nanotech-Based Food Products

One of the essential ethical concerns in agricultural nanotechnology is the lack of clarity in labeling and consumer consciousness. Many consumers are unconscious/unaware that nanomaterials are being used in food production, packaging, and crop protection. The lack of clear labeling regulations in various countries raises ethical queries about familiar consumer decision-making and expected health hazards. Consumers have a basic right to cognize whether the food they eat comprises engineered nanomaterials (ENMs). Numerous countries, including the European Union, mandate labeling necessitate for nano-enabled food goods. Nevertheless, in areas like the United States and China, nano-based food additives and agricultural chemicals do not need special labeling. This lack of clarity prevents consumers from making informed decisions about their dietary consumption regarding the possible impact of nanomaterials. Another ethical concern is the insecurity encompassing long-term health concerns of utilizing food products that have been nano-engineered (Abd-Elsalam 2024). While few nanomaterials have been considered safe, others have shown expected toxicity in animal studies. Ethical concern arises when companies introduce nano-based agricultural products without comprehensive long-term safety studies. Regulatory bodies must check that nanotechnology is utilized responsibly and that its goodness does not exceed the potential health hazard. Large agribusiness corporations play an important role in the development and exploitation of food products based on nanotechnology. When corporate/total interests prioritize net income over consumer health and environmental sustainability, ethical concerns appear. Some companies have been defendant of deduction research data regarding the hazards related to the influence of the nanoparticle. Consumer safety, clarity in research, and collective responsibility should all be controlled by ethical frameworks.

14.4.2 Ethical Dilemmas in Agricultural Biotechnology and Nanotechnology

Genetic modification, environmental hazard, and the possibility for collective monopolization are only an insufficient part of the ethical quandary increased by the integration of biotechnology and nanotechnology in agriculture. By allowing for much finer gene editing in crops and livestock, nanotechnology is being utilized to intensify the genetic engineering method in agriculture. Nevertheless, this raises ethical queries about ecological balance, biodiversity, and unplanned genetic consequences (Isibor 2024). Some bio-ethicists debate that the creation of genetically modified organisms (GMOs) via nanotechnology might disturb both natural ecosystems and conventional farming activity. Pesticides and fertilizers utilizing nanotechnology may be more effective, but pose/bearing possible hazard to water ecosystems and soil. The ethical query is raised: Should agricultural productiveness be prioritized over environmental sustainability? Nanomaterials could have an effect on soil microorganisms, aquatic life, and plant biodiversity if they are put into the environment. Environmental responsibility and invention should be in equilibrium in ethical agricultural activity. Another ethical concern is that big multinational corporations are mainly responsible for the development of nanotechnology in agriculture, which prevents small-scale and local farmers from utilizing it. The economic differences between large agribusinesses and smallholder farmers may broaden as an outcome of the great costs related to nano-enabled fertilizers, precision farming instruments, and smart sensors (Merck et al. 2022a, b). A clean/fair approach to nanotechnology innovations should be checked by ethical regulations, and corporate monopolization that limits affordability/ for farmers in developing areas should be prevented.

14.4.3 The Role of Public Perception and Misinformation in Nanotech Adoption

The achiever or failure of nanotechnology adoption/approval in agriculture and food production is mainly reliant on public perception. Public resistivity against nano-based innovations is triggered by misinformation, concern, or a deficiency of awareness, and ethical concerns originate. Numerous consumers are either unaware of nanotechnology applications in food or grip misconceptions/misinterpretations about its hazards. Media sensationalism frequently represents nanotechnology as possibly adverse without supplying a well-balanced aspect of its goodness and regulatory precaution (Ghosh et al. 2024). When fear-based substance overshadows scientific evidence, Ethical concerns originate. To educate consumers about the actual hazard and goodness of nanotechnology in agriculture, a public awareness effort should be planned. Several governments and regulatory agencies adopt/follow the preventive principle, which means they pause nanotechnology adoption until its safety is entirely established. While this approach prioritizes consumer safety, it can also prevent technological advancement and hinder resolution for global food security. There are ethical debates about the required for further scientific certainty/evidence before nanotechnology can be widely acceptable/utilized in food and agriculture. In order to intensify public reliance on nano-enabled agriculture, researchers, policymakers, and companies must choose an ethical communication scheme. Having/engaging discussions with consumers and farmers about ethical implications and regulation of nanotechnology. Apart/ away from scientific advancements and regulatory frameworks, the ethical considerations of nanotechnology in agriculture and food systems are wider. They cover public cognition/awareness, collective responsibility, consumer rights, and environmental sustainability (Singh et al. 2023). To ensure responsible and sustainable nanotechnology applications in the agri-food sector, secure clarity in labeling, equitable approach to nano-based innovations, and ethical decision-making in agricultural biotechnology will be necessary.

14.5 Discussion

Important progress in environmental sustainability, precision farming, and food safety has been made from the utilization of nanotechnology in the agri-food sector. Scientists and agricultural experts have formulated nano-enabled fertilizers, pesticides, and food packaging solutions that intensify productivity and decrease resource loss by employing engineered nanomaterials. These innovations offer favorable goodness, such as better nutrient assimilation in crops, targeted pest control, and prolonged shelf life of biodegradable foods. Even so, concerns regarding the long-term effects of nanoparticles on human health, soil ecosystems, and biodiversity remain mostly unsolved. Nanomaterials' capacity to store/persist in the environment and food chains, leading to unexpected toxicological hazards, increases ethical and regulatory challenges. Though the scientific community is actively investigating ways to measure and decrease these hazards, it has been challenging to set up standardized safety directions for nanotechnology in agriculture due to inconsistency in global regulatory frameworks (Deka and Kalita 2024). Worldwide regulatory agencies have effort to develop coherent policies to manage the utilization of nanomaterials in farming/agriculture and food. Various countries have adopted multifaceted approaches, with few focusing on rigorous hazard assessments and compulsory labeling, while others have taken a much easier posture, allowing nano-enabled products to go into the market without broad examination. These inequalities in regulation build challenges in public reliance and ensuring consumer. Concerns about ethics are also important, especially regarding the clarity of nano-based food labeling and the availability of these technologies to small-scale farmers. The economic space between developing agricultural sectors and industrialized countries could broaden if large agri-businesses monopolized nanotechnology (Shah et al. 2024). Governments, industry stakeholders, and researchers need a combined effort to address these disparities by making comprehensive policies that equilibrium innovation with consumer protection and environmental sustainability. Given that nanoparticles can modify soil chemistry, interrupt microbial communities, and pose a hazard to aquatic ecosystems through water runoff, the environmental implications of nanotechnology also need urgent attention. The long-term aggregation of engineered nanomaterials in nature could have unexpected consequences/effects on biodiversity and plant growth. Despite these hazards, nanotechnology holds huge potential to add to sustainable agriculture by reducing chemical overexploitation, intensifying crop resilience, and minimizing food waste. The development of eco-friendly, biodegradable nanomaterials that can bring the goodness of nanotechnology without long-term environmental harm must be a priority for researchers moving forward (Merck et al. 2022a, b). In addition, global cooperation in ethical policymaking and regulatory oversight will be necessary to ensure the responsible adoption of nanotechnology in agriculture.

14.6 Future Recommendations

Various captious paths must be taken at scientific, regulatory, moral, and environmental levels to ensure the trustworthy and sustainable integration of nanotechnology in the agri-food sector. The formation of harmonized global regulations that supply a clear-cut explanation and standardized safety assessment protocols for nanomaterials is one of the most urgent needs. Before allowing their universal utilization, regulatory agencies must utilize strict testing mechanisms to assess the effect of nanoparticles on soil ecosystems, water bodies, and human health. To secure clarity for consumers, compulsory labeling of nano-enabled food products should be implemented. Governments can avoid variance in safety standards, decrease public doubt, and boost ethical commercialization of nanotechnology in agriculture by establishing an international regulatory framework. In order to fully realize the long-term effects of nanoparticles, scientific research must go beyond regulation and advance hazard assessment methodologies (Rouhi et al. 2022). This includes conducting broad studies on their toxicity, bioaccumulation, and possible interactions with biological systems. Alternative nanomaterials that are environmentally safe, biocompatible, and biodegradable should be examined by researchers in order to reduction concerns over continual nanoparticle pollution. To develop safer, more sustainable nano-based agricultural products, cooperation between environmental scientists, nanotechnology experts, scientific agriculture, and toxicologists will be central (More et al. 2021). Moreover, investment in next-generation nano biosensors could intensify real-time monitoring of food stuff, pesticide residual, and soil nutrients, enabling farmers to make data-driven decisions while minimizing ecological consequences. To secure impartial access to nanotechnology innovations, ethical considerations must also be addressed. Compared to industrialized agribusinesses, Small-scale farmers in developing countries often deficiency the resources and knowledge to adopt nano-agriculture technologies, placing them at a competitive disadvantage. Policies should be planned to make nanotechnology inexpensive and approachable to all farming communities, irrespective of economic condition, in order to bridge/link this space. Campaigns to increase consumer consideration about the goodness and hazards of nanotechnology should also be set up in order to increase decision-making. In addition, open-access enquiry and impartial market competition policies should prevent corporate monopolization of nano-based agricultural products. To secure that nanotechnology contributes to food safety and environmental sustainability without intensifying economic inequality, scientific innovation must be balanced with ethical responsibility.

14.7 Conclusion

The use of nanotechnology in the agri-food sector has the potential to advance agriculture and offer a resolution for imperative challenges such as food security, environmental sustainability, and efficient resource usage. By rising crop production, reducing waste product, and intensifying soil health, the development of nano-enabled fertilizers, pesticides, and biosensors has already begun to optimize farming practices. In addition, food packaging solutions based on nanotechnology have better food safety, prolonged shelf life, and decreased decay. Still, despite these advancements, the general application of nanotechnology in food production and agriculture raises important concerns regarding environmental safety, regulatory governance, human health,, and ethical considerations. To ensure the responsible and sustainable adoption of nanotechnology in the agricultural sector, these challenges must be addressed/determined. The possible health hazard posed by nanoparticles is one of the major concerns encompassing nanotechnology. Nanoparticles' small size and high reactivity permit them to move into biological systems easily, possibly be stored in human tissues, and cause unexpected health effects. The continuity/persistence of nanoparticles in soil and water systems can interrupt microbial communities and change ecological balances in agricultural environments. This increase in concerns about long-term toxicity and bioaccumulation requires extensive research into the interactions of nanomaterials and environmental occurrence. To decrease harmful ecological impacts, scientists must prioritize the development of biodegradable and non-toxic nanomaterials. In addition, real-time observation through advanced nanobiosensors can aid in tracing the existence of nanoparticles in water systems and soil, allowing farmers to adopt preventative measures and decrease hazards effectively. Normally, smallholder farmers lack the financial resources and technical knowledge to support the application of nanotechnology-driven agricultural strategies, especially in developing countries. This may lead to a combative unbalance that technologically modern farming operations benefit disproportionately, departure conventional farmers struggle to compete. Policymakers must create strategies to ensure that nano-agricultural technologies are treated fairly through fair-market policies, subsidies/grants, and educational enterprise. Furthermore, compulsory labeling of food products with nano-enabled ingredients can raise clarity and responsibility in the food industry while empowering consumers to make informed decisions. Adoption of nanotechnology depends critically on environmental sustainability. Although fine nutrient transfer and targeted pest control are provided by nano-enabled fertilizers and pesticides, excessive or unstructured use of nanomaterials can have unintentional ecological consequences. When discharged into soil and water systems, nanoparticles have the potential to store over time, altering soil chemistry, affecting plant growth, and possibly contaminating food chains. The safety, efficiency, and environmental sustainability of rising nanotechnologies depend on continuing investment in research and development. Regulatory frameworks must be developed to keep pace with technological advancements, and ensure that new agricultural products based on nanotechnology undergo strict safety assessment before market approval. In order to avoid socioeconomic inequality and boost responsible commercialization, ethical considerations, such as consumer rights and an equitable approach, must stay at the front of policymaking. When utilized/applied wisely, nanotechnology has the potency to revolutionize the agricultural landscape, making farming much cost-efficient, sustainable, and resilient in the aspect of global challenges such as food safety and climate change. Nanotechnology can pave the way for a more sustainable and technologically modern agricultural sector that benefits farmers, consumers, and the environment, similarly, by addressing regulatory space, boosting comprehensive development, and decreasing health hazards.

Acknowledgments The authors extend their appreciation for the support of the Deanship of Scientific Research, Vice Presidency for Graduate Studies and Scientific Research, King Faisal University, Saudi Arabia [Grant No. KFU000000].

References

Abd-Elsalam KA (2024) Special issue: agricultural nanotechnology. Plants 13(4):489. https://doi.org/10.3390/plants13040489

Arif U, Bhatti KH, Ajaib M, Wagay NA, Majeed M, Zeb J, Hameed A, Kiani J (2021) Ethnobotanical indigenous knowledge of tehsil Charhoi, district Kotli, Azad Jammu and Kashmir, Pakistan. Ethnobot Res Appl 22:10.32859/ERA.22.50.1-24

Arshad F, Waheed M, Harun N, Fatima K, Khan BA, Fatima K, Abbas Z, Jabeen S, Majeed M (2022) Indigenous farmer's perception about fodder and foraging species of semi-arid lowlands of Pakistan: a case study of district Kasur, Pakistan. Taiwania 67(4):510–523. https://doi.org/10.6165/tai.2022.67.510

Aziz A, Anwar MM, Majeed M, Albanai JA, Almohamad H, Abdo HG (2023a) Quantifying the impacts of urbanization on urban green, evidences from Maga City, Lahore Pakistan. Discov Sustain 4(1):1–11. https://doi.org/10.1007/s43621-023-00169-z

Aziz A, Anwar MM, Majeed M, Fatima S, Mehdi SS, Mangrio WM, Elbouzidi A, Abdullah M, Shaukat S, Zahid N, Mahmoud EA, Casini R, Yessoufou K, Elansary HO (2023b) Quantifying landscape and social amenities as ecosystem Services in Rapidly Changing Peri-Urban Landscape. Land 12(2). https://doi.org/10.3390/land12020477

Barwant MM, Rasool A, Mannam VK, Hussain A, Abdullah M, Nisa Memon Z et al (2024) Nanotechnology applications in livestock farming and animal health. In: Harnessing nanoomics and nanozymes for sustainable agriculture. IGI Global, pp 413–442. https://doi.org/10.4018/979-8-3693-1890-4.ch020

Bashir SM, Altaf M, Hussain T, Umair M, Majeed M, Mangrio WM, Khan AM, Gulshan AB, Hamed MH, Ashraf S, Amjad MS, Bussmann RW, Abbasi AM, Casini R, Alataway A, Dewidar AZ, Al-Yafrsi M, Amin MH, Elansary HO (2023) Vernacular taxonomy, cultural and ethnopharmacological applications of avian and mammalian species in the vicinity of Ayubia National Park, Himalayan Region. Biology 12(4). https://doi.org/10.3390/biology12040609

Bhattacharya R, Bhattacharya R, Majeed M, Bhandari S, Aziz R, Sinha D et al (2023) Environmental rehabilitation of industrial waste dumping site. In: Biohydrometallurgical processes. CRC Press, pp 44–89. https://doi.org/10.1201/9781003451457-4

Chowdhury AR, Sarkar DJ, Singh BK, Sarkar B (2023) Nanotoxicological issues in agriculture and related regulatory framework, pp 217–231. https://doi.org/10.1007/978-981-99-3629-8_13

Deka RC, Kalita MC (2024) Nanotechnology in agriculture, pp 105–115. https://doi.org/10.58532/v3becs21p6ch3

Elabiad S (2025) Navigating the ethical and sustainable. Front Nanotechnol. https://doi.org/10.2139/ssrn.5079782

Engelmann W (2024) Regulation of nanomaterials in agriculture, food, and environment. Elsevier BV, pp 57–61. https://doi.org/10.1016/b978-0-443-15570-3.00010-7

Feitshans IL, Vassara C (2023) Law and ethics of nanotechnology safety and health in food nanotexnology. In: 2023 conference workshop, University of Aristotle, Greece July 6 2023. https://doi.org/10.22541/au.169777839.91696287/v1

Ghosh R, Bakshi S, Bhattacharya P, Kanthal S (2024) Use of nanotechnology in modern agriculture. Int J Agric Ext Soc Dev. https://doi.org/10.33545/26180723.2024.v7.i4sb.532

Haq SM, Tariq A, Li Q, Yaqoob U, Majeed M, Hassan M, Fatima S, Kumar M, Bussmann RW, Moazzam MFU, Aslam M (2022a) Influence of edaphic properties in determining Forest Community patterns of the Zabarwan Mountain range in the Kashmir Himalayas. Forests 13(8). https://doi.org/10.3390/f13081214

Haq SM, Yaqoob U, Majeed M, Amjad MS, Hassan M, Ahmad R et al (2022b) Quantitative ethnoveterinary study on plant resource utilization by indigenous communities in high-altitude regions. Front Vet Sci 9:944046

Hassan M, Haq SM, Ahmad R, Majeed M, Sahito HA, Shirani M, Mubeen I, Aziz MA, Pieroni A, Bussmann RW, Alataway A, Dewidar AZ, Al-Yafrsi M, Elansary HO, Yessoufou K (2022a) Traditional use of wild and domestic Fauna among different ethnic groups in the Western Himalayas—a cross cultural analysis. Animals 12(17). https://doi.org/10.3390/ani12172276

Hassan M, Haq SM, Majeed M, Umair M, Sahito HA, Shirani M, Waheed M, Aziz R, Ahmad R, Bussmann RW, Alataway A, Dewidar AZ, El-Abedin TKZ, Al-Yafrsi M, Elansary HO, Yessoufou K (2022b) Traditional food and medicine: ethno-traditional usage of fish Fauna across the valley of Kashmir: a Western Himalayan region. Diversity 14(6). https://doi.org/10.3390/d14060455

Hussain S, Raza A, Abdo HG, Mubeen M, Tariq A, Nasim W, Majeed M, Almohamad H, Al Dughairi AA (2023) Relation of land surface temperature with different vegetation indices using multi-temporal remote sensing data in Sahiwal region, Pakistan. Geosci Lett 10(1):33. https://doi.org/10.1186/s40562-023-00287-6

Hussain F, Salam IU, Memon ZN, Abdullah M, Abbas G, Akbar M et al (2024) Occurrence of fungal microbial contamination in drinking water of megacity of Karachi (Pakistan) and their physico-chemical control. Heliyon 10(7)

Isibor PO (2024) Regulations and policy considerations for nanoparticle safety, pp 295–316. https://doi.org/10.1007/978-3-031-54154-4_14

Jamil MD, Waheed M, Akhtar S, Bangash N, Chaudhari SK, Majeed M, Hussain M, Ali K, Jones DA (2022) Invasive plants diversity, ecological status, and distribution pattern in relation to edaphic factors in different habitat types of district Mandi Bahauddin, Punjab, Pakistan. Sustainability (Switzerland) 14(20). https://doi.org/10.3390/su142013312

Jamshaid A, Basim M, Atta A, Haseeb MA, Ullah S, Saleem MB, Walait M (2024) Nanotechnology in food: processing, packaging, and preservation. Diet Fact (J Nutr Food Sci). https://doi.org/10.54393/df.v5i03.122

Kaur H, Thakur A, Thakur RC, Kumar A (2024) Ethical and safety considerations in nanomaterial use. In: Advances in chemical and materials engineering book series. IGI Global, pp 326–356. https://doi.org/10.4018/979-8-3693-7640-9.ch013

Khan AM, Li Q, Saqib Z, Khan N, Habib T, Khalid N, Majeed M, Tariq A (2022) MaxEnt modelling and impact of climate change on habitat suitability variations of economically important Chilgoza pine (Pinus gerardiana wall.) in South Asia. Forests 13(5). https://doi.org/10.3390/f13050715

Khan AM, Altaf M, Hussain T, Hamed MH, Safdar U, Ayub A et al (2024) Ethnopharmacological uses of fauna among the people of Central Punjab, Pakistan. Front Vet Sci 11:1351693. https://doi.org/10.3389/fvets.2024.1351693

Khoja AA, Haq SM, Majeed M, Hassan M, Waheed M, Yaqoob U, Bussmann RW, Alataway A, Dewidar AZ, Al-Yafrsi M, Elansary HO, Yessoufou K, Zaman W (2022) Diversity, ecological and traditional knowledge of Pteridophytes in the Western Himalayas. Diversity 14(8). https://doi.org/10.3390/d14080628

Kumari R, Suman K, Karmakar S, Mishra VK, Lakra SG, Saurav GK, Mahto BK (2023) Regulation and safety measures for nanotechnology-based Agri-products. Front Genome Edit 5. https://doi.org/10.3389/fgeed.2023.1200987

Majeed M, Bhatti KH, Amjad MS, Abbasi M, Id RWB, Nawaz F, Rashid A, Mehmood A, Id MM, Khan WM, Id SA (2020) Ethno-veterinary uses of Poaceae in Punjab, Pakistan. PLoS One 15:1–28. https://doi.org/10.1371/journal.pone.0241705

Majeed M, Bhatti KH, Amjad MS (2021a) Impact of climatic variations on the flowering phenology of plant species in jhelum district, Punjab, Pakistan. Appl Ecol Environ Res 19(5):3343–3376. https://doi.org/10.15666/aeer/1905_33433376

Majeed M, Bhatti KH, Pieroni A, Sõukand R, Bussmann RW, Khan AM, Chaudhari SK, Aziz MA, Amjad MS (2021b) Gathered wild food plants among diverse religious groups in Jhelum District, Punjab, Pakistan. Food Secur 10(3). https://doi.org/10.3390/foods10030594

Majeed M, Tariq A, Anwar MM, Khan AM, Arshad F, Mumtaz F, Farhan M, Zhang L, Zafar A, Aziz M, Abbasi S, Rahman G, Hussain S, Waheed M, Fatima K, Shaukat S (2021c) Monitoring of land use–land cover change and potential causal factors of climate change in Jhelum district, Punjab, Pakistan, through GIS and multi-temporal satellite data. Land 10(10). https://doi.org/10.3390/land10101026

Majeed M, Khan AM, Habib T, Anwar MM, Sahito HA, Khan N, Ali K (2022a) Vegetation analysis and environmental indicators of an arid tropical forest ecosystem of Pakistan. Ecol Indic 142. https://doi.org/10.1016/j.ecolind.2022.109291

Majeed M, Lu L, Haq SM, Waheed M, Sahito HA, Fatima S, Aziz R, Bussmann RW, Tariq A, Ullah I, Aslam M (2022b) Spatiotemporal distribution patterns of climbers along an abiotic gradient in Jhelum District, Punjab, Pakistan. Forests 13(8). https://doi.org/10.3390/f13081244

Majeed M, Lu L, Anwar MM, Tariq A, Qin S, El-Hefnawy ME, El-Sharnouby M, Li Q, Alasmari A (2023a) Prediction of flash flood susceptibility using integrating analytic hierarchy process (AHP) and frequency ratio (FR) algorithms. Front Environ Sci 10. https://doi.org/10.3389/fenvs.2022.1037547

Majeed M, Muhammad M, Hussain T, Ali M, Naz T, Nawaz S et al (2023b) Nanopriming-based management of biotic stresses for sustainable agriculture. In: Nanopriming approach to sustainable agriculture. IGI Global, pp 263–289. https://doi.org/10.4018/978-1-6684-7232-3.ch012

Majeed M, Muhammad M, Nawaz S, Naz T, Iqbal MM, Zahid N et al (2023c) Nanopriming for crop management for sustainable agriculture: an overview. In: Nanopriming approach to sustainable agriculture. IGI Global, pp 110–141. https://doi.org/10.4018/978-1-6684-7232-3.ch005

Majeed M, Gulshan AB, Muhammad M, Hussain M, Aziz R, Abdullah M et al (2024a) An investigation into the reactions of biochar in heavy metal polluted soils. In: Biochar-assisted remediation of contaminated soils under changing climate. Elsevier, pp 247–276. https://doi.org/10.1016/B978-0-443-21562-9.00010-4

Majeed M, Noor W, Rasool A, Saleem M, Bashir I, Fatima R et al (2024b) Nanotechnology for soil remediation and land restoration. In: Harnessing nanoomics and nanozymes for sustainable agriculture. IGI Global, pp 60–77. https://doi.org/10.4018/979-8-3693-1890-4.ch004

Merck A, Grieger K, Kuzma J (2022a) How can we promote the responsible innovation of nano-agrifood research? Environ Sci Pol 137:185–190. https://doi.org/10.1016/j.envsci.2022.08.027

Merck A, Grieger K, Cuchiara MR, Kuzma J (2022b) What role does regulation play in responsible innovation of nanotechnology in food and agriculture? Insights and framings from U.S. Stakeholders 42(3):85–103. https://doi.org/10.1177/02704676221102066

More SJ, Bampidis V, Benford D, Bragard C, Halldorsson TI, Hernandez-Jerez AF, Bennekou SH, Koutsoumanis K, Lambré C, Machera K, Naegeli H, Nielsen SS, Schlatter JR, Schrenk D, Silano V, Turck D, Younes M, Castenmiller J, Chaudhry Q et al (2021) Guidance on technical

requirements for regulated food and feed product applications to establish the presence of small particles including nanoparticles. EFSA J 19(8). https://doi.org/10.2903/J.EFSA.2021.6769
Muhammad M, Waheed A, Wahab A, Majeed M, Nazim M, Liu YH et al (2023) Soil salinity and drought tolerance: an evaluation of plant growth, productivity, microbial diversity, and amelioration strategies. Plant Stress 100319. https://doi.org/10.1016/j.stress.2023.100319
Muhammad M, Basit A, Arooj A, Dixit G, Majeed M, Sinha D et al (2024a) Roles of gibberellins in plant defense against biotic and abiotic stress. In: Plant growth regulators to manage biotic and abiotic stress in agroecosystems. CRC Press, pp 183–210
Muhammad M, Basit A, Majeed M, Shah AA, Ullah I, Mohamed HI et al (2024b) Bacterial pigments and their applications. In: Bacterial secondary metabolites. Elsevier, pp 277–298. https://doi.org/10.1016/B978-0-323-95251-4.00019-3
Nasar S, Muhammad M, Siddiqi EH, Majeed M, Ameen N, Ullah S et al (2024a) OMICS-based knowledge for achieving food and nutritional security. In: OMICs-based techniques for global food security. Wiley, pp 67–90. https://doi.org/10.1002/9781394209156.ch4
Nasar S, Rasool A, Ikram W, Shujat S, Majeed M, Muhammad M, Saleem S (2024b) Nano-osmics: a new approach to plant breeding. In: Sustainable agriculture: nanotechnology and biotechnology for crop production and protection, p 129. https://doi.org/10.1515/9783111234694-008
Naveed M, Majeed M, Jabeen K, Hanif N, Naveed R, Saleem S, Khan N (2024a) Plant disease diagnosis with artificial intelligence (AI). In: Microbial data intelligence and computational techniques for sustainable computing. Springer, Singapore, pp 217–234. https://doi.org/10.1007/978-981-99-9621-6_15
Naveed M, Memon ZN, Abdullah M, Makhdoom SI, Azeem A, Mehmood S et al (2024b) Use cases and future aspects of intelligent techniques in microbial data analysis. In: Microbial data intelligence and computational techniques for sustainable computing. Springer, Singapore, pp 259–280. https://doi.org/10.1007/978-981-99-9621-6_17
Naveed M, Zahid N, Fatima I, Saleem A, Majeed M, Abid A et al (2024c) Sustainable AI-driven applications for plant care and treatment. In: Microbial data intelligence and computational techniques for sustainable computing. Springer, Singapore, pp 235–258. https://doi.org/10.1007/978-981-99-9621-6_16
Rani J, Rasool A, Sher H, Fatima R, Hussain J, Majeed M et al (2024) Precision farming with nanoscale sensors: monitoring soil, water, and crop health. In: Harnessing nanoomics and nanozymes for sustainable agriculture. IGI Global, pp 78–99. https://doi.org/10.4018/979-8-3693-1890-4.ch005
Rasool A, Tariq M, Asif M, Shah GM, Fatima R, Majeed M et al (2024) Nanotechnology for improved crop resilience in challenging environments. In: Harnessing nanoomics and nanozymes for sustainable agriculture. IGI Global, pp 149–175. https://doi.org/10.4018/979-8-3693-1890-4.ch008
Rasool A, Feroz E, Qayyum Khan Z, Qayyum Khan M, Majeed M, Kousar S (2025) Chapter 19: Sustainable agriculture: a nanoosmic-based management. In: Rajput V, Singh A, Ghazaryan K, Minkina T, Al-Tawaha A (eds) Sustainable agriculture: nanotechnology, biotechnology, management and food security. De Gruyter, Berlin/Boston, pp 411–430. https://doi.org/10.1515/9783111369754-019
Rouhi M, Nateghi L, Aghamirzaei M, Yousefi S, Garavand F, Feyzi S, Nooshkam M, Jafari SM (2022) Safety, standards and regulations of nanotechnology in food. In: Encapsulation in food processing and fermentation. CRC Press eBooks, pp 335–358. https://doi.org/10.1201/9780429324918-13
Shafqat A, Rasool A, Fatima R, Majeed M, Rahman KU, Narejo AH et al (2024) Nanomaterials for sustainable pest and disease management in agriculture. In: Harnessing nanoomics and nanozymes for sustainable agriculture, pp 100–125. https://doi.org/10.4018/979-8-3693-1890-4.ch006
Shah MA, Shahnaz T, Masoodi JH, Nazir S, Qurashi A, Ahmed GH (2024) Application of nanotechnology in the agricultural and food processing industries: a review. Sustain Mater Technol 39:e00809. https://doi.org/10.1016/j.susmat.2023.e00809

Singh R, Kumar S (2023) Regulatory and safety concerns regarding the use of active nanomaterials in food industry. In: Nanotechnology advancement in agro-food industry. Springer, Singapore, pp 269–306. https://doi.org/10.1007/978-981-99-5045-4_8

Singh U, Pramanik A, Sinha A, Pal AK, Rajput P, Chaubey KK, Yadav S, Shrivastava C, Hariharan S, Bachheti RK, Bachheti A (2023) Safety regulation of Nanoemulsion in the food, agriculture and biomedical sector, pp 299–311. https://doi.org/10.1007/978-981-99-5398-1_16

Sinha D, Maurya AK, Abdi G, Majeed M, Agarwal R, Mukherjee R, Ganguly S, Aziz R, Bhatia M, Majgaonkar A, Seal S, Das M, Banerjee S, Chowdhury S, Adeyemi SB, Chen JT (2023) Integrated genomic selection for accelerating breeding programs of climate-smart cereals. Genes 14(7) NLM (Medline). https://doi.org/10.3390/genes14071484

Tariq A, Mumtaz F, Majeed M, Zeng X (2023) Spatio-temporal assessment of land use land cover based on trajectories and cellular automata Markov modelling and its impact on land surface temperature of Lahore district Pakistan. Environ Monit Assess 195(1). https://doi.org/10.1007/s10661-022-10738-w

Tassadduq SS, Akhtar S, Waheed M, Bangash N, Nayab DE, Majeed M, Abbasi S, Muhammad M, Alataway A, Dewidar AZ, Elansary HO, Yessoufou K (2022) Ecological distribution patterns of wild grasses and abiotic factors. Sustainability (Switzerland) 14(18). https://doi.org/10.3390/su141811117

Ullah I, Aslam B, Shah SHIA, Tariq A, Qin S, Majeed M, Havenith HB (2022) An integrated approach of machine learning, remote sensing, and GIS data for the landslide susceptibility mapping. Land 11(8). https://doi.org/10.3390/land11081265

Waheed M, Arshad F, Majeed M, Fatima S, Mukhtar N, Aziz R, Mangrio WM, Almohamad H, Al Dughairi AA, Al-Mutiry M, Abdo HG (2022) Community structure and distribution pattern of Woody vegetation in response to soil properties in semi-arid Lowland District Kasur Punjab, Pakistan. Land 11(12). https://doi.org/10.3390/land11122145

Waheed M, Arshad F, Majeed M, Haq SM, Aziz R, Bussmann RW, Ali K, Subhan F, Jones DA, Zaitouny A (2023) Potential distribution of a noxious weed (Solanum viarum Du-nal), current status, and future invasion risk based on MaxEnt modeling. Geol Ecol Landsc. https://doi.org/10.1080/24749508.2023.2179752

Zamri NB (2023) Nanotechnology in agriculture: a review of innovative utilization. Malaysian Nano 3(2):50–64. https://doi.org/10.22452/mnij.vol3no2.4

The manufacturer's authorised representative in the EU is Springer Nature Customer Service Centre GmbH, Europaplatz 3, 69115 Heidelberg, Germany. If you have any concerns regarding our products, please contact ProductSafety@springernature.com

Printed and bound by CPI Group (UK) Ltd, Croydon, CR0 4YY
01/10/2026
02225922-0012